GERMANY'S HIGH SEA FLEET IN THE WORLD WAR

ADMIRAL SCHEER

Germany's High Sea Fleet in the World War

By
ADMIRAL SCHEER

With a Portrait and Twenty-eight Plans

CASSELL AND COMPANY, LTD
London, New York, Toronto and Melbourne
1920

PREFACE

THE victor has the privilege of writing the story of the war; for one mistrusts the vanquished, because he will try to palliate and excuse his defeats. But we are victors and vanquished at one and the same time, and in depicting our success the difficult problem confronts us of not forgetting that our strength did not last out to the end.

Exceptionally tragic is the fate of our Fleet. It embodied the sense of power resulting from the unification of the Empire, a sense which was conscious of its responsibility to provide for the suitable security of our immensely flourishing political and economical expansion. By creating a fleet we strengthened our claim to sea-power, without which the Empire must wither away, we remained a thorn in the side of the British, and their ill-will was the constant accompaniment of our growth. The freedom of the seas, which we strove for in line with our evolution, England was never willing to grant, even if it had to come to a world-war on the point.

In the four years' struggle which Germany waged against the desire of its enemies to destroy it, the Fleet was able, beyond all foreign expectations, to hold its own, and what is more, it was our conduct of the naval war that succeeded in forcing the stubborn enemy to the brink of destruction. But, nevertheless, we have lost the war, and with the surrender of the German Fleet the expectations of an independent shaping of our destiny have vanished for long enough.

To the history of the naval war, as it presented itself to me and was for some years carried on under my guidance, this book will add a contribution. I should like, however, along with the description of my war experiences, to give the assurance to the German people that the German Fleet, which ventured to boast of

Preface

being a favourite creation of the nation, strove to do its duty, and entered into the war inspired only by the thought of justifying the confidence reposed in it and of standing on an equal footing with the warriors on land. The remembrance of the famous deeds which were accomplished on the sea will henceforth preserve over the grave of the German Fleet the hope that our race will succeed in creating for itself a position among the nations worthy of the German people.

SCHEER.

Weimar, September, 1919.

CONTENTS

* " The Hoofden " is a German term applied to the area of the North Sea that lies below the latitude of Terschelling.

Contents

PART III
The U-boat Campaign

Folding Plans

INTRODUCTION

THE origin of the world-war lies in the opposition between the Anglo-Saxon and the German conceptions of the world. On the former side is the claim to the position of unrestricted primacy in sea-power, to the dominion of the seas, to the prerogative of ocean-trade and to a levy on the treasures of all the earth. "We are the first nation of the world" is the dogma of every Englishman, and he cannot conceive how others can doubt it.

English history supplies the proof of the application—just as energetic as inconsiderate—of this conception. Even one of the greatest eulogists of the English methods in naval warfare—which best reflect English history—the American, Captain Mahan, made famous through his book, "The Influence of Sea Power upon History," characterises it in his observations on the North American War of Independence, which ended in 1783: "To quote again the [French] summary before given, their [the Allies—America, France and Spain] object was 'to avenge their respective injuries, and to put an end to that tyrannical Empire which England claims to maintain upon the ocean.' The revenge they had obtained was barren of benefit to themselves. They had, so that generation thought, injured England by liberating America; but they had not righted their wrongs in Gibraltar and Jamaica. The English fleet had not received any such treatment as would lessen its haughty self-reliance, the armed neutrality of the Northern Powers had been allowed to pass fruitlessly away, and the English Empire over the seas soon became as tyrannical and more absolute than ever."

Still, England has in process of time understood how to create an almost universal recognition of its claim. Its whole policy, based on the authority of its Fleet and the favourable situation of the British Isles, has always been adapted to the principle that all that may contribute "*ad majorem gloriam Britanniae*" is of advantage also to the progress of mankind.

The principal feature of the English character is markedly materialistic and reveals itself in a striving for power and profit. The commercial spirit, which animates the individual Englishman, colours the political and military dealings of the whole people. Their

Introduction

claims, to themselves a matter of course, went so far always that they never granted advantages to another, even if their utilisation was not possible to themselves at the time, but might perhaps be so later. That has manifested itself most clearly in the Colonial sphere.

The edifice of English world-importance and might has rested for a hundred years on the fame of Trafalgar, and they have carefully avoided hazarding it. They have besides, with skill and success, left untried no means of accentuating the impression of power and using it. What we should consider boastful was to the British only the expression of their full conviction and an obvious means to their end. In support of this we may mention such expressions as: "We have the ships, we have the men, we have the money, too," as well as ships' names, such as *Irresistible, Invincible, Indomitable, Formidable,* and many others.

This method, fundamentally, is really as the Poles asunder from ours, but still it does not fail to leave an impression on many Germans owing to its pomposity and the customary embroidery of commonplaces about promoting the happiness of mankind.

On the opposite side Prussia—Germany! Its whole history filled with struggle and distress, because the wars of Europe were carried on by preference on its territory. It was the nation of the Categorical Imperative, ever ready for privations and sacrifices, always raising itself again, till it seemed at last to have succeeded through the unification of the Empire in being able to reap the fruits of its hard-won position of power. The victory over the hard times it had to pass through was due to its idealism and to its tried loyalty to the Fatherland under the oppressions of foreign rule. The strength of our defensive power rested above all things on our conscientiousness and thoroughness acquired by strict discipline.

In contrast to the inaccessibility of the English island-position was our Continental situation in the heart of Europe, in many respects without natural defence on the frontiers. Instead of having wealth pouring in from all quarters of the globe, we had to toil in the sweat of our brows to support our people on the scanty native soil; and yet we succeeded, in defiance of all difficulties, in elevating and advancing in undreamt-of fashion the economic status of the people while at the same time effecting their political unification.

In such a situation and after such experiences, schemes of conquest were utterly absent from the minds of the German people. They sought to find satisfaction for their need of expansion in peaceful fashion, so as not to hazard lightly their hard-won position of

Introduction

power. That we should be regarded as an unwelcome intruder in the circle of nations who felt themselves called upon to settle the fate of Europe and the world was due, apart from the deeply-wounded vanity of the French, to the mistrust of the British, to whose way of thinking our harmlessness appeared incredible.

In order to retain our position and to ensure the maintenance of our increasing wealth we had no other choice than to secure the ability to defend ourselves according to the old-established principle of the Wars of Liberation : by efficiency to compensate for what was lacking in numbers. How could we establish armies superior to those of our neighbours otherwise than by efficiency?

With the same fundamental motive we turned to the building-up of our sea power, as, owing to the increased dependence of our Administration on foreign countries and to the investment of vaster sums in German property on and oversea, our development unquestionably required protection.

The intention imputed to us of wishing to usurp British world-power never existed; our aims were much more simply explained by the provisions of the Navy Bills of a limited number of ships, which nowhere approached the English total. Nevertheless England considered herself threatened and saw in us a rival who must at any cost be destroyed. That this sentiment prevailed over there lay indeed less in the fact of the appearance of a sea Power of the *second rank* in a corner of the North Sea far removed from the world-oceans, than in the estimate of its worth. They foresaw the exercise in it of a spirit of progress which characterised the German nature, by which England felt herself hampered and prejudiced.

It is not disputed that through our fleet-construction a sharper note was introduced into our relations with England than would have resulted from peaceful competition alone, but it is not a just judgment, nor one going to the foundation of the German-English relations, if the disaster of the world-war or even of the unsuccessful result of the war is attributed simply to the building of the German Fleet. To that end it is necessary to consider the justification of our fleet-building and the reasons why the war was lost and what prospects existed for us of winning it. In that way we shall recognise the decisive rôle which fell to naval power after this struggle of nations grew into a world-war through England's accession to the side of Russia and France.

The mere apprehension of falling out with England could not and dare not form ground for refusing to such an important part

Introduction

of our national wealth as had accumulated in the undertakings bound up with our sea interests the necessary protection through a fleet, which the townsman, dependent on inland activities, enjoyed in the shape of our army and accepted as quite a matter of course.

The Empire was under an obligation to support and protect in their projects the shippers and merchants who undertook to dispose of the surplusage of our industrial energy in foreign lands and there establish new enterprises bringing profit to Germany. This connection with overseas was securing us universal benefit in so far as, by its means, the Homeland was enabled to employ and to feed all its inhabitants, so that, in spite of the great increase of population, emigration was no longer required as a safety-valve for the surplus man-strength. What the maintenance of the man-strength of a country means when converted into work, the last ten years and the war-years have shown us quite remarkably.

It is expected of every small State that it should make whatever efforts lie in its power to justify a claim to consideration of its independence. On this is based the guarantee, won in the international life of peoples with the advance of civilisation, that the weaker will not unjustifiably be fallen upon by the stronger.

The conduct of a Great Power which left its sea-interests without protection would have been as unworthy and contemptible as dishonourable cowardice in an individual; but it would have been most highly impolitic also, because it would have made it dependent on States more powerful at sea. The best army we could create would lose in value if Germany remained with the Achilles-heel of an unprotected foreign trade amounting to thousands of millions.

Although the purpose of our competition on a peaceful footing followed from the modesty of our colonial claims, our policy did not succeed in removing England's suspicion; but, considering the diversity of the claims of both peoples, having its roots in their world views, all the art of diplomacy could not have succeeded in so far bridging over the antagonisms that the recourse to arms would have been spared us.

Was there perchance still another method of creating for ourselves the necessary protection against attacks at sea, which did not bear the provocative character that in England was attributed to the building of our High Sea Fleet? Just as the desire for a German Fleet had for a long time been popular, so has the average German had little idea of the meaning of sea power and of its practical application. This is not to be wondered at, in view of the

Introduction

complete absence of national naval war-history. It will hence be necessary, in order to answer the question whether we chose the suitable naval armament for the condition of affairs in which the new Germany saw itself placed, to enter somewhat more closely into the peculiarities of naval warfare.

It has been held as an acknowledged axiom, proved from war-history, that the struggle at sea must be directed to gaining the mastery of the sea, *i.e.* to removing all opposition which stands in the way of its free and unhindered use.

The chief resisting strength lies in the enemy Fleet, and a successful struggle against it first renders possible the utilisation of the mastery of the seas, for thereupon one's own Fleet can go out with the object of attacking the enemy coasts or oversea possessions, of carrying out landings or preparing and covering the same on a larger scale (invasion). Finally, it can further shut off the enemy by means of a blockade from every sort of import from overseas and capture his merchant-ships with their valuable cargoes, until they are driven off the open sea. Contrary to the international usage in land warfare of sparing private property, there exists the principle of prize-right at sea, which is nothing more than a relic of the piracy which was pursued so vigorously in the form of privateering by the freebooters in the great naval war a hundred years ago.

The abrogation of the right of prize has hitherto always been frustrated by the opposition of England, although she herself possesses the most extensive merchant-shipping trade. For she looks for the chief effect of her sea power to the damaging of the enemy's sea-trade. In the course of time England, apparently yielding to the pressure of the majority of the other maritime States, has conceded limitations of the blockade and naval prize-rights—with the mental reservation, however, of disregarding them at pleasure—which suited the predominant Continental interest of these States. It deserves especially to be noticed that England has held inflexibly to this right to damage enemy (and neutral) trade because she was convinced of her superiority at sea. When our trade-war began, unexpectedly, to be injurious to the island-people they set all the machinery possible in motion to cause its condemnation.

It is possible in certain circumstances for the less powerful maritime States, according to position, coast-formation and ocean traffic, to protect themselves at their sensitive and assailable points by measures of coast-defence.

With us this course has found its zealous champions, first on

Introduction

account of cheapness, partly from a desire not to provoke the more powerful States, and finally on the ground of strategical considerations which lay in the same direction as those of the *jeune école* in France. The idea was to check an opponent by means of guerilla warfare and through direct attack on enemy trade, but the only result of the *jeune école* in France has been that the French Navy has sunk into insignificance. A system of guerilla warfare remains a struggle with inadequate means, which does not guarantee any success. England rightly did not at all fear the cruiser-war on her shipping trade, otherwise she would have given way on the question of the naval prize-right. As regards coast defence, we did not consider that policy, as it could not hinder the English from harming us, while it in no way affected them, seeing that our coasts do not impinge on the world-traffic routes, and did not come within the range of operations.

If the damage caused to one's own sea-trade (including that of the Colonies) becomes intolerable, as in our own case, means of coast-defence provide no adequate protection.

If it comes to the point that one must decide antagonisms by arms, the foremost consideration is no longer "how can I defend myself?" but "how can I hit the enemy most severely?" Attack, not defence, leads most quickly to the goal.

The best deterrent from war is, moreover, to impress on the enemy the certainty that he must thereby suffer considerably.

The method adopted by us of creating an efficient battle-fleet, an engagement with which involved a risk for England, offered not only the greatest prospect of preventing war, but also, if war could not be avoided, the best possibility of striking the enemy effectively. Of the issue of a fleet action it could with certainty be stated that the resultant damage to the English supremacy at sea would be great and correspond proportionately with our losses. Whilst we at need could get over such a sacrifice, it must exercise an intolerable effect on England, which relied on its sea power alone. How far these considerations, on which the construction of our Fleet was based, were recognised as correct on the English side, can be judged from the tactics of England's Fleet in the world-war, which through-out the struggle were based on the most anxious efforts to avoid suffering any real injury.

How our Fleet conducted itself in opposition to this, and succeeded in making the war at sea an effective menace to England will be evident from the following account of the war.

PART I

The First Two Years of the War to the
Battle of the Skagerrak

Germany's High Sea Fleet in the World War

CHAPTER I

THE OUTBREAK OF WAR

THE visit of an English squadron for the Kiel Week in June, 1914, seemed to indicate a desire to give visible expression to the fact that the political situation had eased. Although we could not suppress a certain feeling of doubt as to the sincerity of their intentions, everyone on our side displayed the greatest readiness to receive the foreign guests with hospitality and comradeship.

The opportunity of seeing great English fighting-ships and their ships' companies at close quarters had become so rare an event that on this account alone the visit was anticipated with the liveliest interest. All measures were taken to facilitate the entrance of the English into Kiel Harbour and make it easy for them to take up their station and communicate with the shore, and it goes without saying that they were allotted the best places in the line, close to the Imperial yacht. Accustomed as we were from early times to regard the English ships as models, the external appearance of which alone produced the impression of perfection, it was with a feeling of pardonable pride that we now had an opportunity of making comparisons which were not in our disfavour. The English ships comprised a division of four battleships under the command of Vice-Admiral Sir George Warrender, who was flying his flag in the battleship *King George V.*, which was accompanied by *Audacious*, *Ajax*, and *Centurion*, and a squadron of light cruisers, *Southampton*, *Birmingham*, and *Nottingham*, under Commodore Goodenough.

While the time of the senior naval officers was fully taken up with official visits and ceremonies, the juniors largely made use of the facilities afforded them to visit Hamburg and Berlin by rail. Friendly relations were soon established between the men, after the

3

way of seafaring folk, and these were further promoted by games and festivities to their taste.

The feeling of camaraderie which, as my experience went, had marked intercourse between German and English naval officers, as men of similar ways of thought and capacity, up to the year 1895, had now disappeared as a result of the attitude of hostility towards our progress which had been displayed by English statesmen, especially in recent years. Every attempt to sham a relationship to which our inmost feelings did not correspond would have compromised our dignity and lowered us in the eyes of the English. It is also easy to realise that there could be no question of making an impression by a full-dress muster of every possible ship. For this occasion only those of our ships were assembled at Kiel which were based thereon.

As our Fleet increased, it had become necessary to distribute the various squadrons between the two main bases, Kiel and Wilhelmshaven, both with a view to using simultaneously the available docking facilities and also to keeping the ships' companies in touch with their nucleus crews on land. The families, too, resided at the headquarters of these nucleus crews, to which the long-service men, especially the warrant and petty officers, returned on receiving a special order and there awaited fresh employment. The ships spent the unfortunately all too short periods which the annual training permitted, at their bases.

The disturbing element in this gay and peaceful picture, in which the only note of rivalry was sounded by competitions in skill in the realms of sport, was the news of the murder of the Austrian heir, the Archduke Franz Ferdinand. The Kaiser left Kiel the very next day and travelled to Berlin. The English ships departed on June 29, their light cruisers using the Kaiser Wilhelm Canal. They thus had an opportunity of making a close acquaintance with the new waterway which had only been completed a few weeks before. Whether it could be also used by our heavy ships was one of their questions which must be laid to the account of untimely curiosity. The deepening and widening of the Canal and the construction of the new locks at the entrances had been completed only just in time. They had become necessary to permit the passage of the big ships, the building of which had been imposed upon us by the introduction of the "Dreadnought" type. The unsuitability of this highway for battle-cruisers like the *Blücher* and the battleships of the "Nassau" class had been a matter of much concern to our naval

4

The Outbreak of War

High Command since 1909, on account of the injurious effect on the strategic situation. It also involved laying an unnecessary burden on our main base in the North Sea, which could not keep pace with the growing number of ships assigned to it.

About a week later the Kaiser returned to Kiel, and on July 5 started out for his usual cruise to Norway. As the situation could by no means be considered reassuring, exhaustive conferences were held between the Naval authorities in Berlin and the Fleet to discuss the various contingencies of war. As subsequent events showed, the most noteworthy of these was the hypothesis that England would remain neutral in the collision with Russia, and most probably her Ally, France, with which we were threatened. It was on this account that the Fleet was allowed to leave for the summer cruise to Norway at the time provided for in the annual scheme.

This decision, as indeed that of the Kaiser, can only be attributed to carelessness or an intention to show no nervousness. That intention, in turn, can only have been due to a firm conviction of England's neutrality.

In the annual scheme the summer cruise represented the high watermark of the development attained. As a reward for the effort shown in daily work, the individual training of the ships and the handling of separate squadrons as well as the whole Fleet, it ended with a visit to foreign ports instead of a sojourn in our own harbours.

This excursion abroad not only served the purpose of keeping up interest in the work but also helped us to maintain our political prestige by showing the flag, especially when an impression of power was thereby created.

When a single gunboat turned up on a distant shore to show the German flag there, the foreigner at once professed to regard it as obvious that this ship was the emissary of the Imperial Government which, for the matter of that, had at home an imposing Fleet and a great Army to secure our position in Europe. A corresponding display of power on the spot was far more convincing and at the same time revealed the capabilities of our shipbuilding industry and refuted the widespread legend that England alone had the best and largest ships.

In view of the uncertain political situation since the summer of 1909 we had discontinued the practice of sending the whole Fleet, or substantial parts of it, to great distances such as the Mediterranean, to Spanish or Portuguese harbours, Cape Verde and the Azores. Thus for our purpose the principal country for us to visit

Germany's High Sea Fleet

was Norway, in the numerous fjords of whose coast it was possible to distribute the ships to the satisfaction of all concerned and avoid overwhelming the inhabitants with a mass of sailors on leave. The distribution also made a greater variety of excursions available to the men, as each ship had its particular place of call.

There had only been one break—in the summer of 1912—in our annual visit to the Norwegian coast since 1910. In this year, 1914, the general political situation required that the visit of the Kaiser and the Fleet should have its usual objective. A cruise to the coasts of the eastern Baltic, even a hasty call at our harbours in that region, does not appear to have been in keeping with the policy we were pursuing at this critical moment.

With the cruise to Norway we abandoned the chance of sending our Fleet east and thus bringing pressure to bear on Russia to induce her to stop her preparations for war. The use of the floating army, which requires no special mobilisation, is ideal for such a purpose. In that case Danzig Bay would have offered us a first-class base, as the larger units could have deployed from there with extreme ease in contrast to the difficult exits from the estuaries of the North Sea rivers—the Elbe, Weser, Jade and Ems, while the light forces attached to the Fleet would have found a fortified base in the harbour of Neufahrwasser.

How Norway could have been chosen for the goal of our cruise in the situation at that moment seems incredible and gives one the impression that we deliberately intended to shut our eyes to the danger. The chance of appearing with a strong naval force, first as a demonstration and later in dead earnest, in our eastern waters was from the start not given the consideration its importance merited.

On July 14 Squadron II, of which I had assumed command at the beginning of February in the previous year, in succession to Vice-Admiral von Ingenohl, who had been appointed Commander-in-Chief of the Fleet, left Kiel Bay to rendezvous off Skagen with the ships coming from Wilhelmshaven and then carry out extensive fleet exercises which were principally concerned with the solution of tactical problems. Through the addition of a third squadron to the High Sea Fleet these exercises were of particular importance for this cruise, as this newly-formed third squadron had as yet had no chance of taking part in combined exercises.

The practical application of theoretical tactics to the circumstances arising out of battle is inexhaustible and provides fresh material from year to year.

The Outbreak of War

The new squadron required training in that respect. In war games, indeed, very useful preliminary work can be done in this department, but that tactical insight which knows how to exploit a favourable situation is itself first trained on the open sea and in the last resort it is the sum of the impressions received which first enables the commander to come to the right decision in the time available, which is often only a matter of seconds. For such decisions there are no rules, however valuable certain tactical principles may be, which have been sanctified by experience.

In the era of sailing ships it was a simple matter, owing to the slow deployment for battle and the small range of the guns. But to-day it is altogether different, in view of the great speed of the ships and the huge range of the guns. The first shells usually arrive the moment the enemy is seen, and we have known cases in which the impact of the enemy's projectiles is the first notification of his being in the vicinity, and he has not become visible until some time afterwards.

With regard to England, we were faced with a particularly difficult, indeed almost insoluble problem. We had to deal with our enemy in such a way as to give greater effect to our smaller calibre guns at short range and be able to use a torpedo wherever possible. From the English we had to expect that in view of the greater speed possessed by their ships of every type and their heavier artillery, they would select the range that suited them and fight a "holding-off" action. That, indeed, is exactly what happened in the war. The necessity of practical training in this department illustrates the importance of the addition of a third squadron.

Further, Squadron III, comprising our latest battleships, was not at full strength, but just formed a division consisting of the *Prinz Regent Luitpold* (flagship), and the battleships *Kaiser, Kaiserin*, and *König Albert*. In the course of the winter, beginning at the end of December, the *Kaiser* and *König Albert* had been away on a longish cruise in foreign waters. The ships had paid a visit to our colonies—the Cameroons and German South-West Africa—visited the harbours of Brazil and the Argentine, and then passed through the Straits of Magellan to the west coast of South America and Chile. The ships had behaved very well on the distant cruise, which was particularly arduous on account of the long sojourn in the tropics. In particular, the engine-room personnel had had an opportunity of becoming thoroughly familiar with the internal arrangements. On the other hand battle-practice could not be carried out to the extent that it was at home, where no diversions were involved.

At the same time as we were starting on our Scandinavian cruise the English Fleet had assembled for a great test-mobilisation at Spithead. It was thus ready and thereafter continued so.

On our way north two French destroyers which we passed on July 16 so close that we could make out their names—*Stilette* and *Trombeau*—reminded us that the President of the French Republic, Poincaré, was on his way from Dunkirk to St. Petersburg in the battleship *France*, accompanied by the cruiser *Jean Bart*, and might pass us at any time. We did not like the prospect of having to show him the usual courtesies on the high seas—a salute—prescribed by international usage, so we drew ahead in order to avoid any chance of a meeting.

Our battle-practice was continued until July 24, on which day the high cliffs of the Norwegian coast were for the most part visible, thanks to the clear, fine weather. On July 22 we had crossed the 60th degree of latitude, which forms the boundary of home waters, but not for long. We stayed quite a short time in Norwegian waters, in fact just long enough to allow coaling from colliers sent to meet us at certain anchorages. My flagship *Preussen* and the battleship *Schlesien*, which together formed one division, were looked after by the Dutch steamer *Willi*. The First Division was in the Nordfjord by Olde, the Second, comprising *Hessen* and *Lothringen*, was also in the Nordfjord, by Sandene, while the other half of the squadron, the Fourth Division, had called at Molde. In the same way the battle-cruisers and light cruisers of the Fleet, as well as the battleships of Squadrons I and II, were distributed among other inlets, notably the Sogne and Hardanger Fjords. The very day we left, Saturday, July 25, the news reached us of the Austrian ultimatum to Serbia. In view of that we were not at all surprised to get an order to hold ourselves ready to put to sea immediately. In the afternoon of the next day, Sunday, we left for the rendezvous appointed for the whole Fleet, about 250 nautical miles from the entrance to the Nordfjord.

After the Fleet had assembled the Flag Officers of the squadrons had a conference on the Fleet-Flagship, at which Admiral von Ingenohl explained the political situation and the necessity for our being prepared for the immediate outbreak of war. He also told us that England would probably remain neutral. On this subject we had received a report that King George of England had

The Outbreak of War

expressed himself in that sense to Prince Henry of Prussia. Notwithstanding this, every possible warlike precaution was taken for the rest of our homeward journey. But the Fleet was divided in such a way that Squadron I, under the command of Vice-Admiral von Lans, and comprising the four ships of the "Ostfriesland" class and the four of the "Nassau" class, with the battle-cruisers, steamed to Wilhelmshaven through the North Sea, while Squadrons II and III with the Fleet-Flagship returned to Kiel through the Kattegat. This distribution of the Fleet is manifest proof of our confidence that no attack threatened us from the side of England. It was only in the East that danger was visible, and accordingly it seemed inadvisable to remove all our big ships from the Baltic.

On July 29 the ships lay in Kiel Harbour and were engaged in effecting the pre-arranged measures which as a rule precede a regular mobilisation, measures which were ordered on account of the increasing tension of the political situation.

All our preparations were inspired by the impression that what we had to face was a war with Russia and France. Fuelling and taking in supplies took up the whole of July 29. We had not yet recalled the men on leave, as all hope of the maintenance of peace had not by any means yet been abandoned. It was only on the following day that the news became more menacing and England's attitude more hostile. Squadron III accordingly made preparations to go through the Canal into the North Sea, while the final steps were now taken to make the ships ready for the change to battle conditions, which might at any time become necessary.

On July 31 the Commander-in-Chief in the *Friedrich der Grosse* passed into the Canal on his way to the North Sea. It was obvious from this step that for us the centre of gravity of the war at sea now lay in the west. Shortly before his departure I had an interview with Admiral von Ingenohl in which he told me that in case of war my task with Squadron II would be to deal with Russia.

It is easy to understand that this commission, which put me in a position to lead and execute the first naval enterprises independently, had a great attraction for me. The appointment of a new Commander-in-Chief for the Baltic in the person of Prince Henry of Prussia had no material effect on my freedom of action at sea, once we had set out for enemy waters; and, besides, Prince Henry's professional knowledge, his whole mode of thought and conception of responsibility offered a guarantee that his appointment could only serve a useful purpose. It may here be said at once that the

9

Germany's High Sea Fleet

royal Commander-in-Chief grasped and carried out in the most typical fashion the difficult and thankless task of our defensive operations in the Baltic, for which we disposed of very limited resources, both as regards numbers and efficiency, after England had appeared on the scene as the principal enemy. A Russian invasion like that of East Prussia, which might easily have been followed by another from the sea, and would have meant the total destruction of numerous important and beautiful places on the Baltic coast, was spared us.

But our hopes of an independent Baltic operation were destroyed the very same day by the order to Squadron II to follow the others immediately to the North Sea. The High Sea Fleet was accordingly concentrated in the Jade on August 1 and at 8 o'clock in the evening the mobilisation order arrived, which was greeted by the crews of the ships with loud cheers.

Meanwhile, opinion had veered round completely as to the probable attitude of England, and it was accepted as certain in the Fleet that she would join the two opponents with whom we had alone been concerned at the outset. This view corresponded to the temper prevailing in the Fleet. We were fully aware of the seriousness of the situation, and that we should now be faced with a contest in which an honourable defeat might well be our only prospect. But nowhere was there the slightest sign of despondency over the enemy's overwhelming superiority, but rather a burning enthusiasm and lust of battle, worked up by the feeling of indignation at the oppression which that superiority had meant, and the conviction that our duty was now to put in our last ounce of strength lest we leave the Fatherland in the lurch. The crews needed no special exhortation to give of their best, for the joy of battle shone in their eyes. The leaders, calmly weighing up the prospects of battle, could only feel that the men's confidence in victory encouraged them to dare to the uttermost. The whole service was carried away by the feeling that we were under a duty to fulfil the expectations to which expression had many a time been given in peace.

During its history of barely more than fifty years, the Prussian and German Fleet had not been permitted an opportunity of matching itself in a serious campaign with European opponents of equal standing, apart from individual affairs which justified the brightest hopes. Our ships had shown what they could do mainly in co-operating in the acquisition of our colonial possessions or maintaining respect for and upholding the prestige of the German flag against the encroachments of half-civilised or savage races. We had no personal experience of commanding and handling in battle the big ships which had recently come into existence. Nor, for the matter of that, had our most important opponent at sea, England.

The English Fleet had the advantage of looking back on a hundred years of proud tradition which must have given every man a sense of superiority based on the great deeds of the past. This could only be strengthened by the sight of their huge fleet, each unit of which in every class was supposed to represent the last word in the art of marine construction. The feeling was also supported by the British sailor's perfect familiarity with the sea and with conditions of life on board ship, a familiarity which took for granted all the hardships inseparable from his rough calling.

In our Fleet reigned a passionate determination not to fall behind our comrades of the Army, and a burning desire to lay the foundation-stone of a glorious tradition. Our advantage was that we had to establish our reputation with the nation, while the enemy had to defend his. We were urged on by the impulse to dare all, while he had to be careful that he did not prejudice his ancient fame.

There was only one opinion among us, from the Commander-in-Chief down to the latest recruit, about the attitude of the English Fleet. We were convinced that it would seek out and attack our Fleet the minute it showed itself and wherever it was. This could be accepted as certain from all the lessons of English naval history, and the view was reinforced by the statement, so often made on the English side, that the boundaries of the operations of their fleet lay on the enemy's coasts. It was also confirmed by an earlier remark of the Civil Lord, Lee: "If it ever comes to war with Germany the nation will wake up one morning and find that it has possessed a fleet." All this pointed to the intention of making a quick and thorough job of it.

Right up to the last moment in which there was the remotest possibility of keeping England out of the war everything was avoided which could have provided a superficial excuse for the existence of a crisis. The Heligoland Bight was left open to traffic so far as it was not commanded by the guns on the Island; elsewhere there were none which had a sufficient range to stop traffic. We had never regarded it as possible that the English Fleet would be held back from battle and, as a "fleet in being," be restricted solely to blockading us from a distance, thereby itself running no risks.

Germany's High Sea Fleet

The test mobilisation to which I have already referred and the advanced stage of preparation thus involved also seemed to indicate that offensive operations were to be expected immediately. This mobilisation at the same time afforded a proof of the resolve of the English Government not to be afraid of increasing the existing tension, and to add the weight of their Fleet, fully prepared for war, to the concentration of the Russian armies.

CHAPTER II

RELATIVE STRENGTHS AND THE STRATEGIC SITUATION

OUR High Sea Fleet was concentrated in the North Sea. Since February, 1913, it had been under the command of Admiral von Ingenohl, who was flying his flag in the battleship *Friedrich der Grosse*. The High Sea Fleet was composed of three squadrons, cruisers and destroyers:

SQUADRON I

Vice-Admiral von Lans (*In Command*).
Rear-Admiral Gaedecke (*Second in Command*).

BATTLESHIPS

Ostfriesland.　　*Thüringen.*　　*Helgoland.*　　*Oldenburg.*
Posen.　　*Rheinland.*　　*Nassau.*　　*Westfalen.*

SQUADRON II

Vice-Admiral Scheer (*In Command*).
Commodore Mauve (*Second in Command*).

BATTLESHIPS

Preussen.　*Schlesien.*　*Hessen.*　*Lothringen.*　*Hannover.*
Schleswig-Holstein.　*Pommern.*　*Deutschland.*

SQUADRON III

Rear-Admiral Funke (*In Command*).

BATTLESHIPS

Kaiser.　　*Kaiserin.*　　*König Albert.*　　*Prinz Regent Luitpold.*

CRUISERS

Rear-Admiral Hipper (*In Command*).
Rear-Admiral Maass (*Second in Command*).
Rear-Admiral Tapken.

Germany's High Sea Fleet

BATTLE-CRUISERS

Seydlitz.   *Moltke.*   *Von der Tann.*

LIGHT CRUISERS.

Köln.   *Mainz.*   *Stralsund.*   *Kolberg.*   *Rostock.*
Strassburg.

SEVEN DESTROYER FLOTILLAS

(In peace these were only occasionally under the orders of the High
Sea Fleet.)

TENDERS

Hela (small cruiser of no fighting value).   *Pfeil.*   *Blitz.*

At this point I must say something about the organisation of
the Fleet in order to present a picture of its fighting value. As is
well known, our Navy Bills had provided for a total of 41 battleships, 20 battle-cruisers, 40 light cruisers, 12 destroyer flotillas and
4 submarine flotillas. This fleet was divided into the Home Fleet and
the Foreign Fleet. The nucleus of the Home Fleet was the High
Sea Fleet which was principally concerned with preparing itself for
battle in case of war. In order to devote ourselves wholly to that
purpose and be in a condition to be sent wherever required—that
is, be permanently mobile—it was relieved of all other tasks and
these were assigned to special ships (Training, Gunnery, and
Specialist). The result of this was that a continuously high standard
of preparedness in battle-practice was not to be attained under our
system because every year a portion of each crew went to the Reserve
and had to be replaced by recruits who for the most part came to
the sea service as utter novices. The most varied efforts to tide
over the period of weakness that was thus involved every autumn
had hitherto led to no conclusive results. From our point of view
the fact that this war broke out in summer was thus peculiarly
unfavourable. The training, gunnery and specialist ships were
used for the education of the rising generation of officers and
embryo officers (cadets and midshipmen) and the training of gunnery, torpedo, and mine specialists, as well as coast survey and
fishery protection. As a rule, these duties were assigned to older
ships which were no longer fit to take their place in the first battle
line. For example, the old armoured cruisers *Herta, Hansa, Freya,*

Relative Strengths and the Strategic Situation

Vineta, and *Viktoria Luise* were employed as training ships. It had not been found possible to avoid calling on the modern ships for the special purpose of gunnery and torpedo practice, although the Commander-in-Chief of the Fleet was very reluctant to part with them because the training of these ships for war purposes was limited to a very short period of the year. Our weakness in cruisers with the High Sea Fleet—for the requirements of foreign stations had to be satisfied as well—was particularly deplorable. We had abroad a cruiser squadron in Eastern Asia and two cruisers (*Goeben* and *Breslau*) in the Mediterranean, in addition to a few old gunboats stationed permanently at various places. The cruiser squadron under Count Spee consisted of the two battle-cruisers *Scharnhorst* and *Gneisenau*, and the light cruisers *Nürnberg*, *Emden*, *Dresden*, and *Leipzig*. In this connection importance was attached to sending the best that we had in the way of light cruisers to foreign seas. As regards battle-cruisers, *Scharnhorst* and *Gneisenau*, which were a match for any pre-Dreadnought cruiser, had to suffice, while we had only three battle-cruisers in home waters, as *Goeben* was in the Mediterranean, and *Derfflinger* and *Lützow* were not yet ready. Another battle-cruiser, *Blücher*, was being employed in gunnery practice. With her twelve 21-cm. guns and the speed of 25 knots, she was considerably inferior in fighting value to the first of the English battle-cruisers of the "Invincible" class, which dated from two years later and carried eight 30.5-cm. guns.

Besides the ships commissioned for training and experimental purposes there were a certain number of other ships in home waters which, as provided by the Navy Bills, were to form the Reserve Fleet. As the provisions of the Navy Bills had not yet been worked out, of these formations only a nucleus in the shape of the battleship *Wittelsbach* could be kept permanently in commission. Another ship of the same class, the *Wettin*, was used as a gunnery training school, while the rest were docked and received only as much attention as was required to keep their engines, structure, and armaments in proper condition.

On mobilisation, all training and experimental ships stopped their work and passed under the command of the High Sea Fleet. Out of the ships in reserve in dock, Squadrons IV, V and VI were formed. The battleships of the "Wittelsbach" class formed Squadron IV under the former Inspector of Gunnery, Vice-Admiral Ehrhard Schmidt; the ships of the older "Kaiser" class made up Squadron V (Vice-Admiral Grepow); while the old coast defence cruisers of the "Siegfried" class formed Squadron VI (Rear-Admiral Eckermann).

Thanks to careful preparation, the ships were put on a war footing without the slightest hitch. Of course, it took some further time before the ships' companies of Squadrons IV, V, and VI were so advanced in training, either as individual units or in combination, that they could be used for war purposes. With a view to increasing the peace establishment, the crews of the High Sea Fleet received on mobilisation an extra quota of men, who joined the ships in the first days and were a very welcome reinforcement.

While steaming at full speed was seldom permitted in peace time, in order to economise coal and save the engines, in war a ship must be in a condition, as soon as she gets to sea, to develop the utmost capacity of her engines, and so all the boilers must be used continuously. With a crew of about a thousand men, which is normal for battleships and battle-cruisers, it is essential to make allowance for a certain percentage of sick and other casualties. Such deficiencies were made good by the mobilisation "supplement," which amounted to about 10 per cent. of the peace establishment. As the war proceeded, the system proved its usefulness by enabling us to let the men go on leave without lowering the standard of the ships' readiness for battle to a disadvantageous degree. The reinforcement was particularly important to the battle-cruisers, which, in view of their enormous consumption of coal in order to attain the very highest speed, were not in a position, with the engine-room complement allowed by establishment, to bring the coal from the more distant bunkers to the stokehold, so that help had to be requisitioned from the sailors. As far as possible, the bunkers in the immediate vicinity of the stokehold were left untouched, in readiness for action, when not a man on board could be spared from his action station.

The system of command is a question of special importance to the organisation of a navy. The bulk of the ships in home waters were under the command of a single authority, the Commander-in-Chief of the High Sea Fleet. Of course, the ships at distant stations abroad could not be under his command, and certain ships in home waters, operating in a theatre which had no absolutely direct connection with the operations in the main theatre, had a Commander-in-Chief of their own. The number of ships combined under one command must not be so large that their commander cannot control and lead them in action, for one of the most material differences between fighting on land and at sea is that in the latter case the commander himself goes into the firing line. But command goes hand in hand with responsibility for the execution of all plans, and it was therefore a doubtful policy to establish an authority above the Commander-in-Chief of the Fleet who had the most important forces under his command. In view of the peculiarities of naval warfare, the higher authority cannot be in a position to settle beforehand the details of time and method of any particular enterprise which has been decided upon, in the same way as this is both possible and essential for the command of operations on land.

However, the demands of the various theatres in which fighting took place in this war made some central authority necessary which could distribute the number of ships required for all purposes, and which could also have strong influence on the conduct of operations in the individual theatres. The authority for this purpose was the Naval Staff, in which the preliminary work on the plan of operations had already been done. The Chief of the Naval Staff had the duty of laying the proposed orders for the operations before the Supreme War Lord, to whom the Constitution gave the supreme command over all our forces on land and sea. After these orders had received the Imperial approval, the Chief of the Naval Staff had to transmit them to the Fleet.

The functions of the Naval Staff assumed particular importance in this war, in which the closest co-operation of the Fleet and Army for the common end was of quite special importance. The development of the Navy, which had grown to the status of a great war machine in the last decades, had not, however, admitted of the simultaneous satisfaction of the requirements in personnel which made themselves felt in all quarters. The working of the Naval Staff had suffered from this cause in peace time and it produced its effect in war. In peace the influence of the State Secretary of the Imperial Naval Administration was paramount, especially when that office was held by a personality like Grand Admiral von Tirpitz, who by his outstanding abilities had gained an influence which no naval officer had ever before exercised in the history of our Navy. In war, on the other hand, he had no direct influence on the conduct of operations.

The development of our Navy had not taken place without numerous differences of opinion about the best method of its construction. At the front and in the Naval Staff the principal requirement was considered to be that the existing Fleet should be so complete in all its details, and therefore so ready for war, that all differences would be made good. The Secretary of State, on the other hand, who had a great programme in mind and steadily pursued its realisation, attached more importance to having all the essential elements ready, and as regards secondary matters, trusting more or less to improvisation if war came before the final development of the Fleet had been realised. He accordingly promoted the construction of battleships and destroyers primarily, bearing in mind the root principle from which our Navy Bills had sprung, that with the Fleet we should create a weapon which should be strong enough to fight against a superior hostile fleet. The course of the war has proved the soundness of that principle.

Only in one material point were our strategical views based on an assumption which proved unfounded, the assumption that the English Fleet, which had kept ahead of ours in its construction at every stage, would seek battle in the German Bight in the North Sea, or would force its way to wherever it hoped to find the German Fleet. On that account we had attached particular importance to the greatest defensive and offensive powers, and considered we might regard speed and radius of action as secondary matters. The difference between our type of ships and that of the English shows that in both Fleets strategic ideas governed the method of construction. The English were content with less armour, but attached importance to higher speed and the largest possible calibre of gun so that they could impose on their opponent their own choice of battle area.

Side by side with the Commander-in-Chief of the High Sea Fleet a special command was introduced for the Baltic forces. The commanders of ships in foreign waters were of course independent and received their orders through the Chief of the Naval Staff, whose co-operation in the business of procuring coal and supplies for the conduct of cruiser warfare could not be dispensed with.

Thus for the first time in German history sea power also was to play a mighty part in the great fight for existence with which our nation was faced. As regards the handling of our Fleet, we had not only to consider how we could bring about the most favourable opportunity of winning the victory, but also what tasks, within the framework of the combined operations, fell to our share. The strategical plans of the Army had a decisive influence on the functions of the Fleet. The Navy had the duty of supporting the Army in its uphill task of fighting a superior enemy on two fronts in such a way that its rear was unconditionally secured against any danger threatening from the north. So long as it was only a question of fighting the Dual Alliance the Army was relieved of all anxiety from that direction, as the Fleet was quite equal to its task. The Army had made its plans in such a way that victory could be expected from an offensive, and the full weight of that offensive would at first be directed to one spot. It followed from this that at the outset a defensive attitude would be adopted on the other front, and all preparations for defence would have to be made in that quarter.

The third front, the sea front, acquired a special importance when England joined the ranks of our opponents. But so far as can be seen from the course of the war no material change was made in the fundamental principles underlying our strategic operations on land. As I was then only holding the position of commander of a squadron, I did not know whether, in view of the increasing hostility of England, the idea was considered of adopting a fresh joint plan of operations for the Army and the Fleet, which would be based on the notion of improving our defensive prospects against England. This could have been obtained by the speediest possible acquisition of the sector of the French coast which commanded the Dover-Calais line. In this way the English cross-Channel transport service, as well as the trade routes to the Thames, would have been seriously threatened. If only we had realised from the start that the influence of England's sea power on the course of the war would be as great as it turned out to be later, to our disadvantage, a higher importance would have been attached to this question at the outset. As it happened, the course of the campaign in France forced us into a position in which we were nothing but the flank protection of the right wing of our Army which stretched to the sea and therefore brought us the Flemish coast as our starting point, though nothing like so valuable, for attacks against England. The Navy had to spring into the breach and take up the defence against English sea power. It appeared obvious that the entry of England into the ranks of our enemies would not divert the Army from its task. The Army considered it much more obvious that the Navy should support it by hindering the passage of transports across the Channel.

The protection of these transports was one of the principal functions of the English Fleet. We could only interfere with it at the price of a decisive battle with the English Fleet, and even if the encounter took a favourable course there was no guarantee that we should attain our end of permanently and effectively interrupting supplies from overseas. We shall have to go into the feasibility of such plans at a later stage.

Even without the inauguration of a comprehensive and detailed plan of operations for the Army and Navy the military situation required that the movements of the Navy should be adapted to the progress of the Army's operations lest the failure of some naval undertaking should put the Army in the dilemma of having to relax its own offensive or perhaps break it off altogether.

The enemy, too, cannot have failed to realise the importance of the German Fleet for a favourable development of the war on land. If the enemy ever succeeded in securing the command of the Baltic and landing Russian troops on the coast of Pomerania our Eastern front must have collapsed altogether and brought to naught our plan of campaign, which consisted of a defensive attitude in the East and the rapid overthrow of the French Army. The command of the Baltic rested on the power of the German Fleet. If we had destroyed the Russian Fleet our danger from the Baltic would by no means have been eliminated, as a landing could have been carried out just as easily under the protection of English naval forces if the German Fleet no longer existed to hinder it. For such a purpose the English Fleet had no need to venture into the Baltic itself. They had it in their power to compel us to meet them in the North Sea immediately they made an attack upon our coast. In view of such an eventuality we must not weaken ourselves permanently, as we could not help doing if we attempted to eliminate the danger which the Russian Fleet represented for us in the Baltic.

It was all the more probable that the English Fleet would attack because the combined enemy fleets would then have a free hand against our coasts. It was improbable that England would seek battle with the German Fleet—which she was bound to regard as her primary naval objective—in the Baltic where all the advantages were on our side.

For this reason the concentration area of our Fleet was the North Sea. It was from there that we could threaten the east coast of England and therefore tie up the English Fleet in the North Sea. We could always deal in time (by using the Kaiser Wilhelm Canal) with any attempt of the English to penetrate into the Baltic. At the outset somewhat weak observation forces had to suffice against the Russians, and these forces had to try to intimidate the Russians

Relative Strengths and the Strategic Situation

into the same course of action by adopting offensive methods
wherever possible. Mines could do us good service in that respect.
This method of intimidation, however, could only be effective so
long as we could still employ a superior force against the Russians,
and we should abandon that superiority out of hand if we attempted
to seek battle with the English Fleet under unfavourable circumstances, because, to say the least of it, the result was doubtful. In
view of the high state of preparedness and the superiority of the
English Fleet probabilities pointed to a failure for us which would
have a fateful effect on the final result of the war.

Apart from the fact that these considerations urged caution,
at the beginning of the war we were without any certain data as
to the whereabouts of the English Fleet, and could only acquire
some by observation of the movements of the enemy. We had to
expect an attack in the greatest possible strength because our unfavourable strategic situation, which was due to the geographical
formation of the North Sea theatre, put us at a disadvantage at the
outset. Our position in the North Sea suffered from the fact that
for any enterprise we had only one point of exit : in that far corner
which faces the mouths of the Elbe and the Weser. From it alone
could the Fleet emerge for an attack, and to it must return again
to seek the shelter of our bases in the estuaries of the Jade and Elbe.
The route round Skagen and the Belt was closed to us, as the Danes
had laid minefields in these waters. The sides of the "Wet
Triangle," the apex of which can be imagined at Heligoland,
ended at Sylt in the north and the mouth of the Ems in the west.
The left bank of the Ems is in Dutch, and therefore neutral,
territory. All movements of ships there could accordingly be
observed and the observation brought to the knowledge of the enemy
in the shortest time. The channel at Sylt is navigable solely for
destroyers and light cruisers, and then only in favourable conditions
of wind and tide.

On the other hand, the east coast of England offered a whole
series of safe anchorages for large ships, indeed for the whole Fleet.
As appears from the map, the English coast takes a westerly
direction the farther north it gets, so that on our attacks against
the northern bases our distance from home is increased, to the great
advantage of the enemy.

While we could be taken in flank from the south if we attacked
the English Fleet, thinking it to be in the north, and taken in the
flank from the north if we made our attack in the south, the English

were in the favourable position that as they approached our coast they need expect danger from only one quarter immediately ahead, the German Bight. They could send out submarines against the one base from which we should have to emerge, to do us all the damage they could on our way out and home, and need only keep that one point under observation. That relieved them of the obligation of detaching special observation forces.

THE BRITISH BATTLE FLEET *

FLAGSHIP
Iron Duke.

FIRST BATTLE SQUADRON
BATTLESHIPS

Marlborough.	*St. Vincent.*	*Colossus.*	*Hercules.*
Neptune.	*Vanguard.*	*Collingwood.*	*Superb.*

SECOND BATTLE SQUADRON
BATTLESHIPS

King George V. *Orion.* *Ajax.* *Audacious.* *Centurion.*
Conqueror. *Monarch.* *Thunderer.*

THIRD BATTLE SQUADRON
BATTLESHIPS

King Edward VII. *Hibernia.* *Commonwealth.* *Zealandia.*
Dominion. *Africa.* *Britannia.* *Hindustan.*

FOURTH BATTLE SQUADRON
BATTLESHIPS

Dreadnought. *Temeraire.* *Bellerophon.* *Agincourt.* *Erin.*
Queen Elizabeth. *Warspite.* *Valiant.* *Barham.*

---

* The Third Battle Squadron consisted of ships of the pre-Dreadnought period, the First, Second and Fourth Battle Squadrons of " Dreadnoughts." In the Fourth Battle Squadron the ships from the *Agincourt* onwards were not ready for sea at the outbreak of war.

Relative Strengths and the Strategic Situation

FIRST BATTLE-CRUISER SQUADRON

Battle-Cruisers.

Lion.   Princess Royal.   Queen Mary.   New Zealand.
Invincible.   Inflexible.   Indomitable.   Indefatigable.

SECOND CRUISER SQUADRON.

Shannon.   Achilles.   Cochrane.   Natal.

THIRD CRUISER SQUADRON.

Antrim.   Argyll.   Devonshire.   Roxburgh.

FIRST LIGHT CRUISER SQUADRON.

Southampton.   Birmingham.   Nottingham.   Lowestoft.
Destroyer Flotillas (number and composition unknown).

The above ships formed The Grand Fleet under the command of Admiral Sir John Jellicoe.

THE SECOND BRITISH FLEET.

Flagship.

Lord Nelson.

FIFTH BATTLE SQUADRON.

Battleships.

Prince of Wales.   Agamemnon.   Bulwark.   Formidable.
Implacable.   Irresistible.   London.   Queen.
Venerable.

SIXTH BATTLE SQUADRON.

Battleships.

Russell.   Cornwallis.   Albemarle.   Duncan.   Exmouth.
Vengeance.

FIFTH CRUISER SQUADRON.

Light Cruisers.

Carnarvon.   Falmouth.   Liverpool.

Germany's High Sea Fleet

SIXTH CRUISER SQUADRON.

Drake. Good Hope. King Alfred. Leviathan.

THE THIRD BRITISH FLEET.

SEVENTH BATTLE SQUADRON.

EIGHTH BATTLE SQUADRON.

Eight ships of the "Majestic" class.

Six ships of the "Canopus" class.

SEVENTH CRUISER SQUADRON.

NINTH CRUISER SQUADRON.

TENTH CRUISER SQUADRON.

ELEVENTH CRUISER SQUADRON.

TWELFTH CRUISER SQUADRON.

They comprised older cruisers, such as :

Cressy. Aboukir. Hogue. Hawke. Theseus.
Crescent. Edgar. Endymion. Gibraltar. Grafton.
Royal Arthur.

The Second and Third British Fleets were combined into the Channel Fleet under a special Commander-in-Chief.

With these enormous forces England was certainly in a position to make us feel the weight of her sea power. The most effective method of doing so would be the destruction of our Fleet. This was also the view of the English Commander-in-Chief at that time, who put it in these words :

"The above objects are achieved in the quickest and surest manner by destroying the enemy's armed naval forces, and this is therefore the first objective of our Fleet. The Fleet exists to achieve victory."*

The English Fleet did not live up to these proud words, in spite of its strength and the geographical position. Yet our belief that

---

* In the book published in January, 1919, " The Grand Fleet, 1914-16," by Admiral Viscount Jellicoe of Scapa, the tasks of the British Fleet were set out as follows : (1) To ensure for British ships the unimpeded use of the seas, this being vital to the existence of an island nation, particularly one which is not self-supporting in regard to food. (2) In the event of war to bring steady economic pressure to bear on our adversary by denying to him the use of the sea, thus compelling him to accept peace. (3) Similarly, in the event of war, to cover the passage and assist any army sent overseas and to protect its communications and supplies. (4) To prevent invasion of this country and its overseas dominions by enemy forces.

it would act thus was thoroughly justified, and we had to decide our attitude accordingly.

In the War Orders which were issued to the Commander-in-Chief of the High Sea Fleet the task before him was framed as follows: The objective of the operations must be to damage the English Fleet by offensive raids against the naval forces engaged in watching and blockading the German Bight, as well as by mine-laying on the British coast and submarine attack, whenever possible. After an equality of strength had been realised as a result of these operations, and all our forces had been got ready and concentrated, an attempt was to be made with our Fleet to seek battle under circumstances unfavourable to the enemy. Of course, if a favourable occasion for battle presented itself before, it must be exploited. Further, operations against enemy merchant ships were to be conducted in accordance with Prize Court regulations, and the ships appointed to carry out such operations in foreign waters were to be sent out as soon as possible.

The order underlying this plan of campaign was this: The Fleet must strike when the circumstances are favourable; it must therefore seek battle with the English Fleet only when a state of equality has been achieved by the methods of guerilla warfare.

It thus left the Commander-in-Chief of the High Sea Fleet freedom of action to exploit any favourable opportunity and put no obstacles in his way, but it required of him that he should not risk the whole Fleet in battle until there was a probability of victory. Moreover, it started from the assumption that opportunities would arise of doing the enemy damage when, as was to be expected, he initiated a blockade of the German Bight which was in accordance with the rules of International Law. It is also to be emphasised that a submarine offensive was only required "whenever possible." The achievements of our U-boats absolutely exceeded all expectations, thanks to the energy with which the command faced the most difficult problem and the resolution of the commanders and crews, on their own initiative, to do more than was required of them.

As regards operations in the Baltic, the War Orders to the Commander-in-Chief of the High Sea Fleet contained no instructions, as a special Commander-in-Chief had been appointed for this area. If the English Fleet tried to carry the war into the Baltic, the condition precedent (a favourable opportunity for attack) laid down in the War Orders to the High Sea Fleet would materialise in the simplest fashion.

CHAPTER III

AWAITING THE ENEMY'S OFFENSIVE

ON August 2 the Commander-in-Chief had summoned all the commanders of the three battleship squadrons, cruisers, destroyers and submarines to the flagship, and there explained to them the task set in the War Orders and his intentions with regard to it. Instructions had just been received from the Naval Staff that on the express wish of the Foreign Office no hostile action should be taken against English warships and merchant ships, as all hope of England's neutrality had not yet been abandoned. In his desire to keep England out of the war the Imperial Chancellor had gone so far as to enter into an obligation, through our ambassador in London, not to conduct operations in the Channel or against the north coast of France if England remained neutral. In this way she would be released from her own obligation to protect the north coast of France with the English Fleet. The same day, however, we received subsequent instructions that the English cable communications with the Continent had been broken off, and that we had to anticipate hostile action on the part of England.

How universal was the conviction that the English Navy would immediately take the offensive is illustrated by the fact that after the conference with the Commander-in-Chief the Commander of Squadron I advised me very strongly to take Squadron II to the Elbe during the night, instead of waiting till next morning as had been arranged, as otherwise we might arrive too late. However, we adhered to the original decision to take up the anchorage appointed for us in the Elbe the next day (August 3), and, as the necessary precautions were taken by sending a mine-sweeping division ahead, the movement was carried out without any mishap.

To take up its anchorage in the Altenbruch Roads, between Cuxhaven and Brunsbuttel, Squadron II had to pass through the minefield which had meanwhile been laid at the mouth of the Elbe. In this river there was a dangerous congestion of vessels which were trying to get out as fast as they could. Among them were some English steamers which would not pay attention to the warnings of the pilot ship, so that there was a dangerous crush in this difficult and narrow channel. The English steamer *Wilfred* paid for its temerity by running on the mines and was sunk by two explosions following closely on each other. We thus had an opportunity of observing a practical demonstration of the effect of mines. After this occurrence the Commandant of the fortress of Cuxhaven, who was responsible for the security of the estuary, gave orders that all ships were to be sent back to Hamburg so that their knowledge of the position of the minefield should not be turned to the advantage of the enemy.

The next day brought us the English declaration of war. A few hours later the first English submarine was reported in the German Bight. The security of the Heligoland Bight required prompt information of the enemy's intentions so that we could meet him in strength with our naval forces without ourselves suffering from the enemy's counter-measures on our way out. This object could be attained by submarines or mine-layers of which the latter could slip out under cover of darkness and sow the exits from the estuaries with mines. We had also to expect that floating mines would be sown in the mouths of the rivers with a view to their drifting up stream with the tide and endangering our ships lying at anchor. We knew of one type of English mine which drifted with the rising tide only, sank to the bottom when the tide ebbed and then rose again and floated farther up stream. Mines of this kind would have been able to get much farther—in fact to the anchorage of our ships —instead of drifting backwards and forwards in a limited area through the action of ebb and flow, and thereby being stranded in due course.

We had also to anticipate that enemy submarines would penetrate into the rivers. Although the depth of water was not great the passage of submarines, when submerged, was by no means impossible. It was only later, when the depth charge had been evolved, that submarines needed greater depth to escape their effect. Even if the enemy shrank somewhat from such venturesome enterprises as these, it was enough for him to haunt the neighbourhood of the estuaries to operate against our big ships the moment we attempted to gain the open sea.

It is true that we had two types of protection against these dangerous possibilities; first, the initiation of technical defence measures such as mines, nets and so forth, and secondly, the sharpest lookout on the part of the ships engaged in observation duties. If the enemy tried to bring on an action in the neighbour-hood of Heligoland—and we assumed he would—we suffered from the outset under a disadvantage if we had to deploy for it out of the estuaries. The narrow exits from the Elbe and the Jade prescribed the line of deployment and compelled the ships to follow in line ahead, a formation which provides splendid opportunities for lurking submarines. For this reason prompt knowledge of the enemy's approach as well as his strength was of particular importance in enabling us to go out and meet him in the open sea with the necesary forces. In the first days of August we attained such a state of preparedness that all the big ships were kept under steam all day, ready to weigh anchor at any moment. We could not concentrate in the outer roads because the submarine obstructions had not yet been laid.

The time from the receipt of a report about the enemy to the issue of the appropriate orders, and then again from the first execution of those orders to the arrival at the appointed rendezvous at sea, was not inconsiderable. According to the state of readiness of the ships and the choice of anchorage it might take hours, during which the enemy would continue his approach unimpeded. Thus arose the necessity of getting the report as soon as possible. But the greater the distance from Heligoland of the arc which had to be covered by our reconnaissance and observation patrols, the less carefully could it be watched. The greater distance either demanded more ships or involved less reliable information when the line was held too thinly.

The use of wireless telegraphy came in extraordinarily handy for intelligence purposes. Unfortunately a large number of the older destroyers which had now been attached to the mine-sweeping division had not yet been fitted with this highly ingenious piece of equipment. The result was that in certain circumstances very valuable time might be lost.

The establishment of a protective system was entrusted to the Commander of the scouting forces, Vice-Admiral Hipper, and all the destroyer flotillas, U-boats, mine-sweeping divisions, aero-planes and airships were placed under his orders. From these forces a protective zone was formed which by day consisted of several circles at varying distances from the lightship "Elbe I." The outermost line, 35 nautical miles (of 1,852 metres) was held by destroyers. Six nautical miles behind there were submarines, and a further six miles back the inmost line was

28

Awaiting the Enemy's Offensive

patrolled by mine-sweeping divisions. Two to four light cruisers were distributed behind the two wings of this protective zone, east and south of Heligoland. At night the U-boats and the destroyers on the outermost line were withdrawn, and only the inner one was held. The result of this was that we had all the more destroyers at hand for nocturnal enterprises.

This whole system, however, was more useful for protection than for reconnoitring. It did not extend far enough for the latter purpose. Even if the approach of serious enemy forces at a distance of fifteen miles was reported from the outermost line, these ships, by steaming at full speed, could be within range of the fortress of Heligoland in about an hour and a half. In that time only the ships lying in the outer Jade could gain the open sea. The ships lying in the Elbe at Cuxhaven or in the Wilhelmshaven Roads in the Jade needed longer. If we had depended on this system alone we should have found ourselves in the condition either of being surprised by the enemy and having to meet him in insufficient strength, or having to keep the whole Fleet in a perpetual state of readiness. The latter alternative was impossible in the long run. The duties of the destroyers and cruisers in the protective zone and the necessity of relieving them every few days (for the strain of this anxious service on the personnel at sea would otherwise have worn them out) absorbed such a large force of light units that their principal task of seeking out and attacking the enemy far away in the North Sea before he got to close quarters with us was seriously affected.

Our commanders were therefore faced with a many-sided problem which was made more difficult by the limited resources at our disposal: to avoid any chance of surprise, to prevent the safety of the Bight being endangered by mines or submarines in such a way that the Fleet would not have the necessary freedom of movement to get out of harbour, and finally to seek out the enemy himself in the North Sea and do him as much damage as possible by guerilla operations. It was, therefore, a very proper decision to entrust all these tasks to one commander who had to make his dispositions with an eye to wind and weather, breakdowns, injuries and the absences these involve, and question of coaling, as well as the multifarious duties laid upon him. In view of the relatively little bunker capacity of the smaller ships, it was continuously necessary to replenish supplies. Their ships' companies also suffered from heavy weather far more than those of the big ships, and therefore required relief sooner.

Germany's High Sea Fleet

Nor was it a simple matter to regulate the system of transmission of orders and intelligence by wireless in such a way as to be certain of getting messages accurately and promptly, and avoiding confusion through the operations of other stations, especially such as were in a different sphere of command.

In our situation aeroplanes and airships played a particularly important part. Unfortunately, their number was very small at the start. Heligoland was fitted up as an aviation station, but at first disposed of only five aeroplanes. The number was subsequently increased to eight. In the early days we had only the one airship, "L 3," for distance reconnaissance. The most zealous efforts were made to cruise in all kinds of weather, and so praiseworthy was the persistence shown that these cruises often extended to within sight of the Norwegian coast.

Side by side with the organisation of the protective zone, the organisation of the defences of the North Sea islands, the most important of which was Heligoland, was completed under the direction of the Headquarters of the North Sea Naval Stations, Vice-Admiral von Krosigk, at Wilhelmshaven. It was also the duty of this authority to carry out the evacuation of the native population, who did not at all like leaving their island, and arranging their transfer to the mainland. They had been previously prepared for this eventuality, and their transport presented no special difficulties. The establishment of minefields and the substitution of buoys to mark the war channels for those of the peace-time channels was also the business of the Naval Stations Headquarters.

Another of its duties was the removal of landmarks which could be seen far out to sea, and would thus be known to the enemy and might enable him to find his bearings.

One victim of this bitter necessity was the venerable old church tower of Wangeroog, the island adjacent to the Jade channel. From time immemorial it had been an object of affectionate familiarity to seafarers. It had stood so long that the whole island had gradually slid past beneath its feet, in consequence of the movement from west to east which is peculiar to the sands of the North Sea. It was now so close to the west side of the island that its walls were washed by the waves.

Harbour flotillas were formed to watch the minefields and guard the entrances to our own rivers. These flotillas were within the sphere of action of the fortresses, and therefore were likewise under the command of the Naval Stations Headquarters. The release of the

Awaiting the Enemy's Offensive

Fleet from such duties definitely proved a sound idea, and thanks to the understanding and co-operation of all services, all further requirements which cropped up as time went by were generously met.

The organisation of the lightship system was of great importance. As soon as war threatened, all the lights in the lightships were extinguished, and the light-buoys removed, so that the whole coast was in darkness. It was impossible to do without lights at night altogether when cruising by the dangerous North Sea coast and navigating the strong current off the mouths of the Elbe, Weser and Ems. Further, lights that were easily recognisable had to be shown to indicate the position of the minefields and the channels through them. Yet in spite of the difficulties of navigation, darkness had the immense advantage that it enabled us to slip out unnoticed, and therefore without great risk, so that night time was preferred for such operations. Of course, the lights must not be shown a moment longer than was necessary for the purposes of navigation. Further, it must be possible for incoming ships to show their lights and be safe against any tricks on the part of the enemy. The main thing was that the light should be shown exactly at the right moment. The outer lightships at the mouths of the Jade and Elbe, which also served as observation stations and had military personnel, certainly had no easy task in the long and stormy nights of the four and a half years' war. We depended on their reliability just as much as on that of all the other posts which existed to assist the navigation of our Fleet, whether the safety of a single steamer or that of a whole squadron was at stake. Special thanks are due to the officers of the Imperial Pilot Service and its chief, Commander Krause. They were always reliable advisers to the commanders of squadrons and ships.

Our view of the whole situation and the War Orders issued to the Fleet made it imperative to get at the outset data as to the movements of the enemy. While the North Sea islands and the estuaries were being put into a state of defence, the primary requirement was security against surprise. The battleship squadron and battle-cruisers (at their anchorages) used these few days to prepare for action. With Squadron II, in the construction of whose ships less importance had been attached to the use of fireproof material than in the later ships, it was a question of removing everything that was dangerous from that point of view and could at all be dispensed with. This had a very adverse effect on the comfort of the ward-rooms and cabins as well as the men's quarters, in which all the wooden beams were removed from the thin sheet-iron partitions as well as the sides of the ship.

The removal of wooden chairs, tables, curtains, tablecloths, easy chairs and such like, the scraping off of paint which was too thick, the transfer of clothes and supplies of all kinds to the space under the armoured decks where they could not easily be got at, took up a lot of time and produced a good deal of noise and discontent. However, the work of destruction was carried out with as much devotion as if it were the enemy himself who was being destroyed, and in the certain expectation that we should not have to wait long for the actual meeting.

Although in peace time everything possible had been thought of which might prove useful or necessary in the emergencies of action, there were always fresh possibilities of perfecting measures and preparing for all conceivable occurrences with things such as rafts, steel nets, anchor cables, lifebelts, and so on. As the Flag Officers of the squadron passed from ship to ship in order to supervise the work that was in progress and make further suggestions, they noted what seemed to them useful on any particular ship and handed the information on to the others.

In this work were associated the newly-joined seamen ratings, mostly reservists, who had served on the same ships not long before, and among whom I recognised many old acquaintances, for, with the exception of a break of one year, I had been with the Fleet continuously since 1907. My pleasure at meeting them again was mingled with a feeling of pride at the sight of the manly, healthy, and robust figures which had developed out of the former recruits or ordinary seamen. It went to my heart to see with what a straightforward sense of duty these men, whose resolve to stand on their own feet through industry and efficiency was plain to the eyes, had left behind them everything they loved and cherished in order to be present when the day came to meet the foe.

We spent the first days of suspense and expectation in this essential work. The opening days of the war gained a particular interest from the varying reports of home-coming steamers or our patrols, the series of false alarms about aeroplanes and submarines, firing at night, or the showing of lights in improbable directions, the explosion of mines in shallow spots in the Elbe (phenomena which subsequently found a natural explanation, though at first attributed to enemy activity), our isolation from all human intercourse—although we could see the cows grazing peacefully on the banks of the Elbe 300 yards away—and the organisation of the watches. In the further distance there was no visible sign of any change in the wonted scenes of peace, for there was still a lively movement of ships in the Elbe, and every incoming German steamer had a particularly warm welcome for having succeeded in getting safely home. But the wireless messages flashing to and fro might at any moment summon us out to meet the foe.

Preparations for the offensive were not neglected during the days in which England was making up her mind what her attitude was to be, and when at 7.47 P.M. on August 4 we received the message, "Prepare for war with England," we also heard the order to the auxiliary cruiser *Kronprinz Friedrich Wilhelm* to put to sea immediately. At 9.30 P.M. the auxiliary minelayer *Königin Luise* also left the Ems on the way to the Thames estuary. Thus began the first essay in cruiser warfare and the introduction of guerilla operations on the English coast. In the wireless room of the Flagship we listened hopefully for further news of the progress of the first two enterprises against the enemy. Would the great ocean greyhound be forced back, or would she succeed in getting unchallenged into the ocean? She remained dumb, and that could justly be taken as a favourable sign.

The wireless message to the *Königin Luise* had run: "Make for sea in Thames direction at top speed. Lay mines near as possible English coasts, not near neutral coasts, and not farther north than Lat 53°." The task assigned to the *Königin Luise* gave little ground for the hope that she could escape the watchfulness of the English; but, with a supreme contempt of death, the ship, under the command of Commander Biermann, held on her way. The steamer which usually plied in summer to the watering-places of the North Sea islands was engaged about 11 A.M. next morning by enemy cruisers and destroyers, and was sunk by a torpedo.

She had had time to sow her mines, however, with the result that the cruiser *Amphion* (3,500 tons, launched 1911), which was pursuing her, fell a victim to them and followed the *Königin Luise* to the bottom with a loss of 131 men. Thus the first day of the war (August 5) had brought losses to both sides, and the first attack on the English coast had been a success for us.

However, the sacrifice it had involved had not been incurred in vain. It was not merely that it had cost the enemy a new cruiser. Far more important was the impression that this proof of a bold spirit of enterprise must have made on friend and foe alike. The situation at the outset thus appeared in such a light that in view of these aggressive operations the enemy thought that he could best protect himself by withdrawing to northern waters, and did not take the other alternative of closing our sally ports himself. Throughout the whole war not a single mine was sown in our estuaries, notwithstanding the thousands upon thousands which were employed in the open waters of the North Sea.

As the next few days passed without incident, and aeroplanes and airships had made no discoveries, while incoming steamers reported that English battleships were only to be seen at a great distance (by Aberdeen) from the German Bight, our business was now to discover the whereabouts of the enemy and get to close quarters with him if we were to bring about an equalisation of strength. For this purpose we had at our disposal the destroyers and submarines which could be spared from the defensive organisation of the Heligoland Bight.

Commander Bauer, in command of the U-boats, was convinced that the defensive employment of submarines in a narrow circle round Heligoland was useless, as there was only a slight probability that the enemy would approach so close, and even if he did it was doubtful whether the boats would get a chance to shoot. The necessity of perpetually coming in and going out of the harbour of Heligoland, a difficult process in view of the methods employed in the defensive system, led to a useless strain on the material and injury to the boats. He therefore represented to the Commander-in-Chief of the Fleet that only the offensive use of U-boats could bring about a change. The number of boats employed must certainly be larger, but the prospects of success would be greater still.

The justice of this argument was recognised, and a decision was taken which was extremely important for the further course of the war. Nor was there much hesitation in carrying it into execution, for the U-boats received orders to proceed on August 6 against English battleships, the presence of which was suspected in the North Sea. These ships were supposed to be about 200 nautical miles from Heligoland and charged with the duty of intercepting some of our battleships which ought to be on their way from Kiel round Skagen into the North Sea because the passage of the Kaiser Wilhelm Canal presented too great difficulties Ten U-boats were assigned to this enterprise, and six days were allowed for it.

Awaiting the Enemy's Offensive

This cruise was to carry the ships across the entire North Sea and as far north as the Orkneys. The boats were left to their own devices, as the cruisers *Hamburg* and *Stettin*, parent ships of the U-boat flotilla, could, of course, not accompany it the whole way. They were only to cover the first run of the boats, a hundred miles or so, and endeavour to draw off any enemy light craft from the U-boats in the direction of Heligoland. The submarines themselves were not to pay any attention to such ships, as their goal was the enemy battleships. It was only for the return journey that the boats were left a free hand to do the enemy all the damage they could. The weather being thick and rainy, and the visibility poor, was not favourable for the enterprise, and indications pointed to its becoming worse. As the latter eventuality did not materialise, however, the commander gave the order to put to sea.

In so great an area, and taking into account the rapid changes which experience shows may be expected, it is very difficult to forecast the weather in the North Sea. The decision was, therefore, a brilliant tribute to the fiery enthusiasm of the new weapon, which had never been faced with a task of such magnitude in peace. The course was to be taken in such a way that the submarines, in line ahead with seven-mile intervals between them, first negotiated a stretch of 300 nautical miles in a north-westerly direction, then turned and went back to a line directly between Scapa Flow and Stavanger, which they were to reach about seventy-two hours after putting to sea. They were to remain on this line until 6 o'clock in the evening of the next day—in all about thirty-nine hours—and then return to Heligoland. One boat had to return when 225 nautical miles from Heligoland, on account of trouble in her Diesel engines. Two others, commanded by Lieutenant-Commanders Count Schweinitz and Pohle, were lost. All the rest carried out their allotted task and were back by August 11.

Nothing was seen of the enemy, with the exception of a four-funnelled cruiser which emerged out of the mist for a short time. All that was known of the lost boats was that one of them was still in wireless communication early on August 8. On the 9th the region in which the U-boats were lying was shrouded in mist, and the wind was blowing with force 6. It was only on August 15 that we learnt that a large part of the English Fleet had been in the same area and had there destroyed six German herring-boats after taking their crews on board. Fog and the amount of sea that wind of a force 6 means are the most unfavourable conditions conceivable for a submarine, in view of the fact that the conning-tower is so low down in the water. It is to be assumed that the missing boats had been surprised by English cruisers in weather of this kind and rammed before they had time to dive.

It was certainly regrettable that at the very moment of meeting the English Fleet was protected by mist, that two of our boats had fallen victims, and that this first enterprise, so smartly carried out, had not been crowned by the success it deserved. The loss of two boats had no depressing effect whatever on the crews. It rather increased their determination to do even better.

The course of this six-day cruise cleared the way for the further exploitation of the U-boat weapon, the great importance of which lay in its power of endurance and its independence, two characteristics which appeared at their true value for the first time in this cruise under war conditions. In these two respects the U-boats were superior to all surface vessels in the Fleet. The destroyers, in particular, were not to be compared with them for their ability to remain at sea. Their fuel capacity was too small for that purpose, and when going at high speed the consumption of coal increased out of all proportion. Further, as the big ships needed the co-operation of the smaller as submarine-screens and mine-sweepers, these, too, were dependent on their smaller consorts for the length of time they could remain at sea, especially when they were in areas in which regard had to be paid to the submarine danger.

Our naval operations took a decisive turn as a result of this cruise, and though the change was gradually introduced, it dates from this enterprise. For that reason it has been described in rather more detail than would be justified, seeing that a tangible success was not achieved. The first proof of the ability of the submarine to remain at sea for a long time had been given, and progress was made along the lines I have mentioned, thanks to the greatest perseverance, so that the submarine, from being merely a coastal-defence machine, as was originally planned, became the most effective long-range weapon.

The other splendid quality of the submarine is its independence, by which I mean that it is not dependent on the support and co-operation of ships or craft of other types. Whilst a force of surface ships comprises various classes, according to the presumed strength of the enemy, the submarine needs no help to attack, and in defence is not so dependent on speed as the surface ships, as it has a sure protection in its ability to dive. This again increases its radius of action, for whereas a surface ship, meeting a superior enemy, has no other resource but to make use of its speed—and that means a large consumption of fuel—diving means a very great economy in engine-power. In the submarine there is no question of driving the engines too hard in such a situation, as the boat can escape from the enemy by diving. The engines need not therefore be constructed to stand perpetual changes of speed.

It is not surprising that the special importance of these technical advantages was not recognised until the war came, for they first came to light thanks to the energy of the personnel, who seemed to despise all difficulties, although going to sea in these small craft involves incredible personal discomforts of all kinds. The advantages of the submarine service first became of practical value through the fact that human strength of will brought men voluntarily to display such endurance as was shown in our boats. Patriotism was the motive-power of the ships' companies.

The fact that an English offensive did not materialise in the first weeks of the war gave cause for reflection, for with every day's grace the enemy gave us he was abandoning some of the advantage of his earlier mobilisation, while our coast defences were improved. The sweep of light-cruisers and destroyers which, starting out star-wise from Heligoland, had scoured the seas over a circumference of about 100 sea miles had produced nothing. Yet while the U-boats were on that cruise to the north which has already been discussed, four other U-boats went on a patrol about 200 miles west, until they were on a level with the Thames estuary. They discovered several lines of destroyers patrolling on about Lat. 52°, but of larger ships nothing was seen. The impression must have been forced on the Commander-in-Chief, as indeed all of us, that the English Fleet was following a strategic plan other than that with which we were inclined to credit it. It appeared probable that the 2nd and 3rd Fleets were concentrated to protect the transport of troops in the English Channel.

The bulk of the 1st English Fleet must be supposed to be in the northern part of the North Sea, to which our light forces had not yet penetrated. Further, we had not yet heard anything from the ten U-boats sent out in that direction, so apparently they too had seen nothing. Should we now attempt to bring the English 1st Fleet to action? We had at our disposal 13 " Dreadnoughts,"

Germany's High Sea Fleet

8 older battleships, 4 battle-cruisers (counting in *Blücher*), a few light cruisers, and 7 destroyer flotillas. With these the Commander-in-Chief intended to give battle, with full confidence in victory. What held him back was the reflection that the whereabouts of the 1st English Fleet was absolutely unknown, and it was therefore questionable whether it could be found in the time at our disposal—which could not be more than two days and nights on account of the fuel capacity of the destroyers. In the meantime, the German Bight would be without any protection against minelaying and other enterprises, and there would be no flank protection on the west. On the other hand, our ships might suffer losses from the operations of enemy submarines, for which there would be no compensation in the way of victory if the English Fleet were not found. We knew from various sources that we had to reckon with English submarines. Such an attempt was therefore abandoned, and in its place a series of patrolling and minelaying operations were set on foot which carried the war right to the English coast in the following weeks.

With this decision began the trying period of waiting for the battleship squadrons, and a start was made with the operations intended to equalise the opposing forces, operations which, apart from mine successes, rested on the anticipation that our destroyers would find opportunities for attack in their nocturnal raids. The lack of scouts—for the new battle-cruisers *Seydlitz*, *Moltke*, and *Von der Tann* could not be put to such uses if they were to be held ready for battle—made it essential that U-boats should be employed on reconnaissance duties.

As early as August 14 new tasks were assigned to the U-boats which had returned from their cruise to the West on the 11th; and, indeed, the boats under the command of Lieutenant-Commanders Gayer and Hersing were to cross the North Sea from the Norwegian Coast (by Egersund) in the direction of Peterhead, while a third U-boat (Hoppe) observed the English forces patrolling before the Humber with a view to securing data for minelaying. They brought valuable information about the enemy's defensive measures, but they had not seen any large ships. The length of time they had spent under water was remarkable. For instance, Gayer's ship had been compelled by destroyers to remain under water six and a quarter hours on August 16, eleven and three-quarters on the 17th, and eleven and a quarter on the 18th.

Awaiting the Enemy's Offensive

Let us now cast a glance at the chances for attack which presented themselves to the enemy. It could not possibly be unknown to him that the German Fleet was concentrated in the North Sea. The reports of spies from Holland and Denmark could not have left any doubt about that. If the English Fleet made a demonstration against Sylt or the East Frisian Islands it would have compelled our Fleet to come out of the estuaries unless we were prepared to allow them a bombardment without retaliation, and they would thus have an opportunity of using their submarines which were patrolling at the mouths of the Jade and Elbe. A success for their submarines would be satisfaction enough for them if we did not follow them out to sea. They could arrange their approach in such a way that they took up a favourable position in the early morning hours to offer battle to our fleet as it came up, or if they appeared with only part of their forces they could promptly retire before a superior German force and limit themselves to the operations of their submarines. The only danger in such an attack lay in the possibility of a nocturnal meeting with our destroyers. This danger was not to be overestimated, as the English could plan their entrance into the German Bight in such a way that our destroyers, which were dependent on darkness, would be already on their way back to the Bight at the time the enemy was approaching. Further, no very serious danger was to be anticipated from our U-boats, as most of them were away on distant enterprises.

The English High Command, however, must have had a much higher estimate of the damage our destroyers and U-boats could do than was actually the case. It appears also that their confidence in the achievements of their own submarines, which were the foundation for the execution of any such plan, was not very great. At the outset, therefore, considerations prevailed on both sides which led the Commands to hold back their fleets from battle. The overestimate of the submarine danger played a most important rôle.

The German Commander-in-Chief, Admiral von Ingenohl, gave expression to his view of the general situation on August 14 in the following Order of the Day :—

"All the information we have received about the English naval forces points to the fact that the English Battle Fleet avoids the North Sea entirely and keeps far beyond range of our own forces. The sweep of our brave U-boats beyond the Lat. 60° in the north and as far as the entrance to the English

Germany's High Sea Fleet

Channel in the south, as well as the raids of our destroyers and aeroplanes, have confirmed this information. Only between the Norwegian and Scottish coasts and off the entrance to the English Channel are English forces patrolling. Otherwise in the *rest of the North Sea not a single English ship has been found hitherto.*

"This behaviour on the part of our enemy forces us to the conclusion that he himself intends to avoid the losses he fears he may suffer at our hands and to compel us to come with our battleships to his coast and there fall a victim to his mines and submarines.

"*We are not going to oblige our enemy thus. But they must, and will, come to us some day or other.* And then will be the *day of reckoning.* On that day of reckoning we must be there with all our battleships.

"Our immediate task is therefore to cause our enemy losses by all the methods of guerilla warfare and at every point where we can find him, so that we can thus compel him to join battle with us.

"This task will fall primarily to our light forces (U-boats, destroyers, mine-layers and cruisers) whose prospects of success increase the darker and longer the nights become.

"The bold action of our mine-layer *Königin Luise,* which did the enemy material damage before she came to her glorious end, and the audacious cruises of our U-boats have already made a beginning. Further enterprises will follow.

"*The duty of those of us in the battleships of the Fleet is to keep this, our main weapon, sharp and bright for the decisive battle which we shall have to fight.* To that end we must work with unflinching devotion to get our ships perfectly ready in every respect, to think out and practise everything that can be of the slightest help and prepare for the day on which the High Sea Fleet will be permitted to engage a numerically superior enemy in battle for our beloved Emperor who has created this proud Fleet as a shield for our dear Fatherland, in full confidence in the efficiency which we have acquired by unflagging work in time of peace.

"The test of our patience, which the conduct of the enemy imposes upon is, is hard, having regard to the martial spirit which animates all our ships' companies as it animates our army also, a spirit which impels us to instant action.

Awaiting the Enemy's Offensive

"The moment the enemy comes within our range he shall find us waiting for him. Yet we must not let him prescribe the time and place for us *but ourselves choose what is favourable for a complete victory.*

"*It is therefore our duty not to lose patience but to hold ourselves ready at all times* to profit by the favourable moment."

CHAPTER IV

THE ENGLISH BREAK INTO THE HELIGOLAND BIGHT

THE nightly cruises from the foremost patrol line by Heligoland were continued and extended. On August 12 the light cruisers *Köln* (Flagship of the First Flag Officer of the destroyer flotillas, Rear-Admiral Maass) and *Hamburg* went out with Flotilla VI; *Köln* and *Stuttgart* with Flotillas I and II on the 15th, and the light cruiser *Mainz* with the Flotilla VIII on the 16th. As no enemy was met on any of these enterprises the light cruisers *Stralsund* (Captain Harder) and *Strassburg* (Captain Retzmann) were sent out to the Hoofden against the destroyer patrol line, the existence of which had been reported by submarines.

They put to sea on the morning of August 11 with two U-boats, which stood by near Vlieland while the cruisers steamed south to about the line Lowestoft—Scheveningen. When this was reached they turned, early on the morning of the 18th. Shortly afterwards the *Strassburg* sighted three enemy submarines, distant about 100 hm. (11,000 yards). These were fired on and one of them seemed to be hit. Soon after eight destroyers were sighted in a northerly direction and a light cruiser with another eight destroyers in an easterly direction, which were in a position to cut off the retreat of our cruisers. The range, however, did not fall below 100 hm., so that no success was obtained on either side. The possibility that there might be other English forces not far off seemed to make it imperative for our ships not to lose time in manœuvring for attack, for the sixteen destroyers of the enemy had an immense preponderance of gun-power over our cruisers armed only with 10.5 cm. guns. Both cruisers returned home without trouble.

In the second half of August the number of reports of submarines sighted at the mouth of the Ems and in the Heligoland Bight increased, and very heavy demands were made on the destroyers to drive them out. On August 21 the light cruisers *Rostock* and *Strassburg* with Flotilla VI made a sweep in the direction of the Dogger Bank with a view to searching the fishing-grounds for English fishing-smacks. They also met enemy submarines, one of which fired two torpedoes at the *Rostock*, but

42

The English Break into the Heligoland Bight

both missed. On this cruise six fishing-steamers were destroyed which were found, well separated, in a circle round Heligoland, and were suspected of working with English submarines.

As all these cruises pointed to the conclusion that we could not expect to find considerable enemy forces in the southern half of the North Sea, our two mine-laying cruisers, *Albatros* (Commander West) and *Nautilus* (Commander Wilhelm Schultz) received orders to lay a minefield at the mouths of the Humber and Tyne. By day their operations were covered by a light cruiser and a half-flotilla of destroyers, as mine-layers must be kept out of action if at all possible. Both ships were able to carry out their commission undisturbed and laid their mines accurately at the places indicated. The actual work began at midnight and was favoured by thick weather. On the way back another six fishing-steamers were sunk.

The previous raids had been favoured by luck inasmuch as the forces employed, which were anything but strong, had not been located and cut off by superior forces. Their safety lay in speed alone. Before support from units lying ready in the estuaries could reach them it might easily be too late. But for that purpose it was considered inadvisable to have proper supporting forces hanging about in the Heligoland Bight on account of the submarines reported there.

August 28th brought us the first serious collision with English cruisers. The reports taken back by the English submarines as to our offensive arrangements in the Heligoland Bight must have decided the English to roll up our patrol line. As the English dispatches on the events of this day have been published, a clear idea of the course of the action can be obtained (see plan, p. 44). My own observations from Squadron II, which lay in the Elbe, are confined to the wireless messages received. About nine o'clock in the morning the first of these came in. "In squares 142 and 131 [that is 20 sea miles north-west of Heligoland] enemy cruisers and destroyers are chasing the 5th Flotilla." *

The *Stettin* and *Frauenlob* (light cruisers) were sent out to help. Two flotillas of U-boats took up station for attack. The remaining wireless messages from nine o'clock in the morning to five in the afternoon gave the following picture:

The ships which took part in the action comprised Destroyer

* Naval charts are drawn squared, to simplify the location of places according to length and breadth, in degrees and minutes. This facilitates delivery of reports or commands and the identification of places on the chart. The size of the squares, a side of which represents five or ten sea miles, is governed by the scale of the chart.

Germany's High Sea Fleet

Flotillas I and V, the light cruisers *Mainz, Strassburg, Köln, Stralsund, Ariadne, Kolberg* and *Danzig*, and two mine-sweeping divisions.

On the enemy's side were several cruisers of the "Town" class, armoured cruisers of the "Shannon" type, four battle-cruisers under the command of Admiral Beatty in *Lion*, and about thirty destroyers and eight submarines.

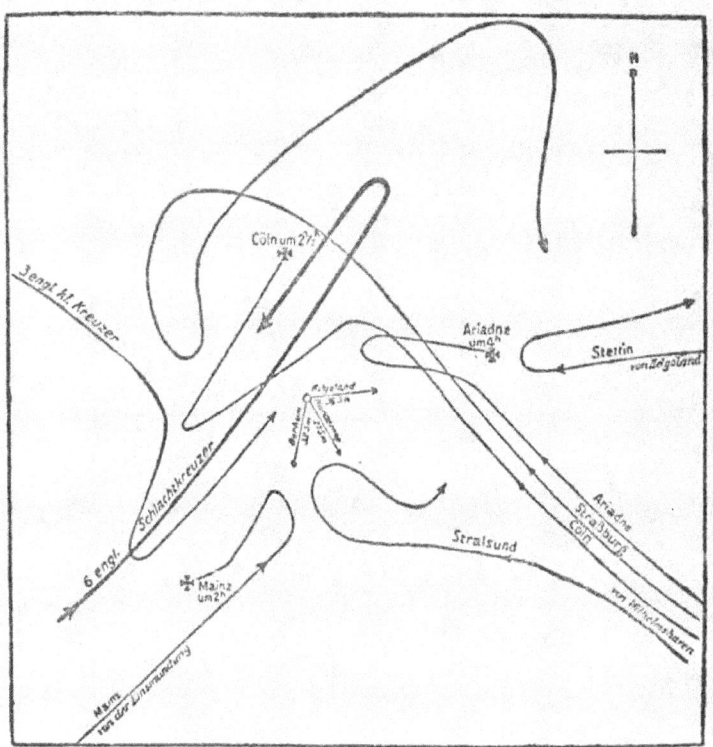

The Heligoland Bight Engagement, August 28, 1914

About six o'clock in the morning one of these submarines had fired two torpedoes, which missed, at a ship of Destroyer Flotilla I, which was retiring to the day patrol line. We had no other information on our side of the further doings of the English submarines on that day; the weather was thick, and as there was hardly any wind, visibility in the neighbourhood of Heligoland was only three to four miles. The upper part of the island was completely shrouded in mist.

44

The English Break into the Heligoland Bight

The marine artillery on the island saw nothing of the action which raged within range of the island in the morning. It was not possible for our battle-cruisers to put to sea before one o'clock owing to the state of the tide at the bar of the Outer Jade. Their intervention came too late. The orders which were issued by the Flag Officer of the German cruisers proceeded on the assumption that the same weather conditions prevailed outside as in the Jade, and the cruisers regarded the situation as such that they would be able to retire in time before a superior force. Unfortunately this was not the case. *Mainz* and *Köln*, all unsuspecting, thus came upon English battle-cruisers and fell victims to their guns. Our plan of surrounding the English forces which had penetrated by cutting off their retreat to the west with *Mainz*, which was in the Ems, while other light cruisers barred the way in the north, was actually put into execution before a general view of the whole situation had made it feasible.

Exceptionally high demands were made on the presence of mind of the Flag Officers in command when they saw themselves faced with more powerful ships than they had expected. The battle-training of our light cruisers revealed a high standard of efficiency. In spite of the serious damage to the ships and heavy losses in personnel, the gun crews served their guns and overcame the confusion of action with exemplary calm and precision. The bold intervention of the other ships and the impulse to hasten to where the thunder of the guns called and bring help, cost us, in addition to the loss of *Köln* and *Mainz*, the loss of the light cruiser *Ariadne*, which had been so damaged by fire that the men had to throw themselves overboard. The question was put whether it would have been of any avail for our big ships to come out of the estuary. They could have had no success, and this is obvious enough in view of the prevailing low visibility.

In the action between the cruisers and destroyers, the light cruiser *Ariadne* and the torpedo-boat "V 187," leader of Flotilla I, were sunk on our side. Most of the ship's company of *Ariadne* were saved by *Stralsund* and *Danzig*. Half of that of "V 187" were taken off by other ships of Flotilla I.

"Wireless communication with *Köln* and *Mainz* has stopped. They are both sunk. Two cruisers (*Strassburg* and *Stettin*) are damaged as well as the torpedo-boats D 8, V 1 and T 33. Many dead and wounded. Nothing known of English losses."

After the first news arrived Squadron II was held ready to

raise anchor in case battleships were required to go out in support. However, we received no order to intervene.

The surprise of our patrols by the two English cruisers *Arethusa* and *Fearless*, which were escorted by seventeen destroyers of the "I" class and fourteen of the "L" class (according to English reports), was a success for the enemy. The intervention of our two light cruisers *Stettin* and *Frauenlob* limited the losses on our side to one torpedo-boat, "V 187." As soon as the news of the break-through of light forces was received all other available light cruisers were sent out to meet them. In the action that now followed *Arethusa* and *Fearless* were seriously damaged and had to call in the help of the very strong English force held ready in support and not yet employed. Its intervention put our cruisers in an evil plight. Very thick weather made a survey of the whole situation difficult.

There are some who think that the way in which the light cruisers went out separately is open to criticism as a piece of temerity. With the safe withdrawal of Flotillas I and V and the driving off of the cruisers *Arethusa* and *Fearless* as the result of the prompt and resolute intervention of *Stettin* and *Frauenlob*, the English attack had lost the character of a surprise, and the plan, which involved a great show of force, had gained but a moderate success with the sinking of "V 187." On the other hand, was a baffled enemy to be allowed to withdraw from the Heligoland Bight unpursued in broad daylight?

Four weeks had passed before the first occasion had presented itself of getting to close quarters with the enemy. Were our ships to content themselves, the first time enemy light forces appeared, with hiding in the estuaries and make no attempt to deal with the enemy, who might perhaps fall into our hands if he were badly damaged? The Flag Officers and commanders would have incurred a serious reproach if they had neglected to make the attempt to get to close quarters with the enemy. If the impression of the first meeting had been a feeling of inferiority and the conviction that we could do nothing but retire before the English, it would have had an unhappy effect on the spirit of the ships' companies and the further course of the operations. The effect produced was exactly the opposite, and we were all burning to avenge the slap in the face we had received.

The disintegration of the engagement into a number of detached actions which were fought at close range, owing to the poor visi-

The English Break into the Heligoland Bight

bility, produced such remarkable examples of the presence of mind and contempt of death of our men that they deserve better than to sink into oblivion. I shall therefore give a few extracts from war diaries.

REPORT OF THE ACTION OF THE FLOTILLA LEADER OF THE DESTROYER FLOTILLA I, WALLIS—"V 187"

(Drawn up by Lieutenant Jasper)

"The Flotilla leader 'V 187' was on patrol (at 16 knots) about 24 sea miles N.W. to W. of Heligoland on a W.N.W. course. Shortly after eight o'clock the ship on our right, 'G 194' (Lieutenant-Commander Buss) reported: 'Am chased by enemy armoured cruiser.' We turned and made for 'G 194.' At 8.20 A.M. in thick weather, two destroyers came in sight in N.W. about three miles off, and were reported to S.M.S. *Köln* by wireless. The ship bore S.E. to E. and put on speed. The destroyers were kept in sight. After a short time another four destroyers or cruisers were observed. Accurate observation impossible owing to failing visibility. 'V 187' now put on full speed and altered course for Heligoland.

"Meanwhile an order from *Köln* to Flotillas I and V had been received, 'Make for shelter of Heligoland.' Simultaneously, four destroyers, which stood between us and Heligoland, emerged from the mist on our port quarter about four degrees to 50 hm. away. At about 40 hm. they opened an intermittent fire. 'V 187' turned south and replied with her after 8.8 cm. gun. The destroyers' shooting was mostly very poor. Only at regular intervals one gun fired shells which passed close over our bridge. The Commander intended to make shooting difficult by altering course and reaching the Jade or Ems at top speed. The ship ran 28 or 29 miles. The destroyers had only caught up a little and were now shooting at about 30 hm. Suddenly an enemy cruiser with four funnels appeared four points on our starboard bow. She apparently made a signal with her searchlight to 'V 187' or her own destroyers. Immediately afterwards she fired a series of salvos at 35 to 40 hm. After the third salvo the shooting was good. As escape was no longer possible the officer in command decided to close. The whole ship's company with the exception of the stokers caught hold of firearms and lifebelts. 'V 187' ported her helm and tried to cut her way through.

47

Germany's High Sea Fleet

"The running action was fought at 12 to 8 hm. The destroyers, apparently surprised, ceased fire at first, but then they subjected us to an extremely rapid fire. A shell fell close to the 8.8 cm. gun and put the crew out of action with the exception of a slightly wounded petty officer. The forward gun only fired a few rounds after that.

"Another shell fell in stokehold 4 and penetrated the bunkers. Splinters wounded the stokers, the lights went out, the steam escaped and the boiler would not fill any more.

"Simultaneously other shots and splinters fell on the bridge. I turned to starboard with a view to ramming the destroyer immediately behind us and clearing our way past it.

"Hits now followed one another with rapid succession. Shells and splinters rained down, and the ship was completely shrouded in smoke and fumes.

"The forward turbine was hit twice and stopped. Steam, mixed with black smoke, poured out of the hatches and ventilators.

"Boiler 2 was damaged and boiler 1 had also received hits.

"Some of the bridge personnel had fallen; the ship had little way on and was listing to port for no obvious reason. The officer in command, who had been seriously wounded, now gave the order to sink the ship. I took one of the four explosive charges which were on the bridge, set it and threw it in the forward turbine room. The bridge personnel put two others in the forward part of the ship.

"Meanwhile two other destroyers coming from the north had joined in the fight. After fixing up the charges I gave orders to leave the ship on the leeside of the firing.

"I jumped overboard just before (according to my calculations) the charges would take effect. The rest of the gun crew of the after gun, which had continued firing to the last (among them Lieutenant Braune), sprang simultaneously into the water. The destroyers now ceased fire and sent out boats. Several men were picked up with lines and buoys. After a few minutes' swimming about I myself was picked up by an English boat. Just as I was getting in ' V 187 ' went down by the bows. No one could be seen on deck. The boat had three other men of the ship's company of ' V 187 ' on board.

"At that moment a German light cruiser (*Stettin*) opened fire on the destroyers. The English boat's crew went on board their destroyer. I refused to go on board with my three men as I did not want to be made prisoner. The English destroyer then started off at high speed. An English sailor had let go the hawser apparently in error.

"I then hauled another sixteen survivors into my English boat.

"Another English boat, under the command of an English officer, was left behind by the destroyers in the evening. It had on board Lieutenant Braune and several survivors.

"After a considerable time a partially submerged English submarine came from the east towards us.

"It came right up and took on board the English crew of one boat and Lieutenant Braune. At first I kept away from the submarine and took off my monkey jacket lest I should be recognised as an officer and taken prisoner. The submarine, which had the mark ' E 4 ' on the bows and the number ' 84 ' (as well as ' E 4 ' again) on the conning tower, dived and disappeared, half submerged, in the west.

"Another smaller English boat, which had on board five more survivors of ' V 187, ' now came up to me. The three boats then rowed for some considerable time in an E.S.E. direction towards the German patrol line. They were subsequently picked up by ' G 4 ' and ' G 11.' The more severely wounded of our men were bandaged on board the destroyers while the boats were sunk. After the destroyers had picked up six dead and had tried to identify the spot at which ' V 187 ' went down from the remains of charts and books, they proceeded to Heligoland. From there the six dead and forty-four survivors, the latter including seven severely and about twenty slightly wounded men, were brought to Wilhelmshaven in the steamer *Arngast*."

The light cruiser *Mainz* (Captain Wilhelm Pasche) was sunk on this day. According to the record made by the First Officer, Lieutenant Tholens, who was taken as a prisoner to England, the action developed as follows:

"The order, ' *Mainz* immediately put to sea and take the reported English forces in the rear,' reached the ship at 10 A.M. in the Ems. Thanks to the previous wireless messages from the Wallis Flotilla, she had steam up in all her boilers and was ready for sea. *Mainz* could therefore put out immediately and develop full speed very quickly. A northerly course was taken at first to cut off the retreat of the enemy ships. The aeroplane at Borkum, which was placed at the ship's disposal, was sent on in the same direction. When the ship started from the Ems the weather was calm, the air clear and visibility good. The conditions for reconnaissance by the aeroplane appeared to be the best imaginable, but after a short flight it returned without any results to show. Meanwhile the *Mainz* had run into haze. This made a surprise by enemy forces possible. About half-past twelve the *Arethusa*, with eight destroyers, appeared in N.E., moving on a westerly course and distant about 70 hm. To such a degree had visibility already decreased!

"To bring the enemy under fire with the starboard guns we turned to port somewhat on a line of bearing N.N.W. Shortly after the first salvos, to which the enemy ships replied with some of his guns, the enemy turned off on a northerly course. The conditions for shooting were extremely unfavourable, as the enemy ships were very difficult to make out in the haze. All the same, several salvos were very well placed, and hits were certainly observed on two destroyers, one of which wrecked a bridge and put out of action everyone on it, including the commander. With a view to keeping the enemy in sight, *Mainz* herself gradually turned on a northerly course. At 12.45 masses of smoke were suddenly reported in N.W., and a few minutes later revealed three cruisers of the 'Birmingham' class. *Mainz* immediately turned hard to starboard, and even as she turned the salvos of the new enemy fell around her, and a few minutes later she received the first hits. The fire of *Arethusa* and the destroyers, which had now apparently passed out of sight, had been without result.

"Our own fire was now directed exclusively at the new enemy, and simultaneously the latter was reported by wireless. By 12.55 P.M. the enemy cruisers were only distinguishable by the flashes of their guns. Shortly afterwards even this had ceased, and with it the hail of enemy shells. *Mainz* ran 25 sea miles, approximately S.S.W. in the direction of the eastern Ems, and emitted large quantities of smoke. Meanwhile almost abreast on our port beam another cruiser of the 'Birmingham' class (*Fearless*) had come into sight, as well as six destroyers close together and several others by themselves. In the course of the action which now developed with these ships and in which several torpedoes were fired at the *Mainz*, the helm suddenly jammed at 10° to starboard.

"The order, 'Steer from the wheelhouse,' came through at the very same moment as the signal from the quartermaster, 'Port your helm.' The helm remained jammed, however, as the result of an explosion under the wheelhouse. The result was that although the steering gear throughout the ship was in working order, all our efforts to steer the ship were without success. We could only conclude that a hit under water had given the whole rudder a bend to starboard. The port engine was stopped.

" *Mainz* slowly turned more and more to starboard, and thus came again within range of the first three cruisers of the ' Birmingham ' class and the *Arethusa*, with her eight destroyers. At the same moment the report reached the bridge that three guns, with their crews, had been completely put out of action. In the stage of the action that followed, in which *Mainz*, with her helm jammed and going round in a circle to starboard, faced four cruisers of the ' Birmingham ' class and about twenty destroyers, our own fire was directed exclusively at the enemy destroyers. Against these only was a success worth mentioning possible. As several of the destroyers came quite close, it was possible to observe several hits upon them.

"Meanwhile casualty had followed upon casualty on the *Mainz*. About 1.20 P.M. most of the guns and gun crews were already out of action. The decks were shot to pieces. The sending up of ammunition had come to a standstill, and more than once compartments under the armoured deck had to be cleared on account of the danger from smoke and gas. The starboard engine could only go half speed.

" It was in this condition that about 1.20 P.M. the ship was struck by a torpedo amidships on the port beam. The effect of this on the conning-tower was that the whole apparatus for transmitting orders, with the exception of the speaking-tube and telephones to the central and torpedo rooms, were put out of action. The commander thereupon gave the order, ' Abandon ship, ship's company get clear with life-belts,' and left the conning-tower. This order, however, only reached the nearest action-stations, and accordingly was only carried out in part. As the result of the torpedo we had stopped firing everywhere. At this moment the First Gunnery Officer and the Torpedo Officer were in the conning-tower. The First Officer, who thought that the Commander must have fallen and knew nothing of his last order, gave orders to resume firing, and tried to launch some torpedoes. The torpedoes he fired, one from port at a light cruiser and two from starboard at destroyers, had no luck, as the enemy ships kept out of torpedo range. On the enemy's side two battle-cruisers had now intervened in the action. Whether they also tried to get in a few hits has never been definitely ascertained. In the *Mainz* only the first and fifth starboard guns were now in action."

The picture of the scene below decks after the explosion of the torpedo is amplified by the following observations of the senior surviving engineer, whose action-station was by the pumps.

"1.15 P.M.—Hit by a torpedo. The ship staggered, heeled over quite sensibly and remained thus for a considerable time. Took even longer to right herself. The emergency lights went out. All the glass which was not already broken by concussion of the bursting shells was now broken. The electric light became dim and gradually went out. In the end our electric torches were the only light we had. The engines ceased to revolve. The gauge already showed that the ship was slowly settling by the head. The efforts to ascertain where the hole was were without result, as we could no longer get a reply from any of the compartments. After a short pause we could hear that firing had been resumed, but when the firing, and shortly afterwards the hail of enemy shells ceased, we could not get into touch with any other part of the ship. The conning-tower, too, did not reply. The water that poured out of the speaking-tube showed that the water had reached the armoured deck, and therefore that the flooded compartments must be submerged.

" As the ship was bound to sink very soon, amidships was now cleared. Between-decks over the armoured deck was so full of smoke that you could not see a yard ahead. Both the companions leading up from there were shot to pieces. It was only by scrambling through the holes made by shells, over the relics of hatches and lockers, that we managed to get out. The space under the forecastle was also filled with smoke from right forward as far as over the second gun.

"As soon as the firing had ceased on all sides the English ships made the greatest efforts to pick up the survivors. At a summons from the *Mainz*, which had not listed at all until about 2 o'clock, a destroyer came alongside the stern to take the wounded on board. All the wounded whose cases did not seem perfectly hopeless were thus removed to the destroyer, assisted by everybody who had not yet left the ship. About 2.10 P.M. *Mainz* heeled over to port and sank."

For a last example I will give the report of the action prepared by Captain Seebohm, commanding the light cruiser *Ariadne*:

"On the 28th August S.M.S. *Ariadne*, flagship of the Harbour Flotilla of the Jade and Weser, was lying in the Outer Jade. On hearing the sound of guns about 9 o'clock, and more particularly on receiving a wireless from *Stettin* that cruiser support was requested, *Ariadne* set a course for Heligoland. Near the Outer Jade Lightship she met the cruiser *Köln*, flagship of Rear-Admiral Maass, which was making west at high speed. *Ariadne* then took much the same westerly course as *Köln*, which had soon disappeared in the haze. We received further wireless messages from *Mainz* and *Strassburg* that they were in action with enemy destroyers.

" Avoiding a certain area where a minefield was suspected, we steered towards the position of the ships named. Judging by her wireless reports, *Köln* appeared to be taking the same course. About 10 o'clock an enemy submarine was sighted square on our port beam. It immediately dived, and seemed at first to be manœuvring for position, but then suddenly disappeared, so that we had no chance to fire.

" Shortly afterwards gunfire was heard on our port bow, and we made straight in that direction. Shortly before 2 P.M. there emerged from the mist two ships, one of which, on our starboard bow, did not reply to our signal. It was recognised as an armoured cruiser so we immediately turned about. The second ship was *Köln*, which was being chased and would doubtless have got away if *Ariadne* had not appeared. The enemy immediately shifted his fire from *Köln* to *Ariadne*. *Ariadne* soon received a hit forward which started a fire in the coal, so that the stokehold had to be abandoned on account of the danger from smoke. Five boilers were thus put out of action and *Ariadne's* speed was reduced to fifteen knots. Behind the enemy, which, judging by its silhouette, was the English Flagship *Lion*, a second English armoured cruiser soon appeared and joined in the action, firing at *Ariadne* for about half an hour at a range of from 45 to 60 hm., at times even from 33 hm. This last distance is only an estimate, as by now all the recording instruments were out of action. *Ariadne* received many hits from heavy guns, among them a whole series aft, which was soon enveloped in flames. Such of the personnel there as made good their escape owed it entirely to luck. The fore part of the ship also received a number of serious hits, one of which penetrated the armoured deck and put the torpedo chamber out of action, while another destroyed the sick-bay and killed its personnel. Amidships and the bridge, strange to say, were almost entirely spared. It is perfectly impossible to say how many hits in all the ship received. Apparently many shells passed through the rigging and were thereby detonated. Others were observed to fall in the water without detonating. Many

53

others passed to right and left as *Ariadne* was running away from the enemy and offered but a small target.

"The English salvos followed in succession with somewhat long pauses. The shells produced their effect mainly by starting fires. All the living quarters fore and aft were immediately in flames. The tremendous flames made it impossible to extinguish a fire which had once started. Further, the fire-extinguishers on the armoured deck had been utterly destroyed.

"About 2.30 the enemy suddenly turned west. I assume that he could no longer distinguish the *Ariadne,* which was enveloped in smoke from the fires. On the *Ariadne* the undamaged guns were still being worked and independently of the fire-control, as there were no means of transmitting orders. Further, the fumes from the ship made it impossible to see anything from the bridge.

"In spite of the enemy's annihilating fire the ship's company worked with the greatest calm, as if on manœuvres. The wounded were carried down by the stretcher-bearers. All ratings tried to carry out such repairs as were possible by themselves. The First Officer was carried away by a shell while between decks with the repairing section.

"After the enemy turned away I first ordered 'all hands' to extinguish the fire. This turned out to be impossible as we could no longer get aft and the ship had to be cleared forward as well almost at once. On the order 'Flood the magazine,' the men ran to the forward magazine. It was ascertained that this was already under water. It was impossible to get to the magazine aft. A previous attempt to open the compartments 1 and 2 where some of the men were still imprisoned proved fruitless, as the deckplates had been bent by shells. The engine-room and the after boiler-room had remained uninjured throughout, and the same was true of the rudder. The telegraph apparatus failed. The cable was apparently cut by an explosion under the conning-tower.

"The heat and smoke made it more and more unpleasant to remain on the ship, and it was even worse when the ammunition piled round the guns began to go off. These explosions, however, did not do much damage. A large number of small splinters were scattered which, for example, penetrated the bridge from below.

"The ship's company assembled in perfect order on the fo'c'sle, whither the wounded also had been brought. I asked for three cheers for His Majesty and then the flag hymn and 'Deutschland,

The English Break into the Heligoland Bight

Deutschland über alles' was sung. Even the wounded joined in. One man asked for three cheers for the officers.

"Just before 3 o'clock S.M.S. *Danzig* (Captain Reiss) came up and sent boats to us. As has already been mentioned, we had not suffered so severely amidships and it was therefore possible to lower the *Ariadne's* cutters also. The first to be put in the boats were the wounded, who were lowered from the fo'c'sle with ropes. As it gradually became impossible to remain on the fo'c'sle the rest of the ship's company jumped into the sea at the word of command. Some of the stronger swimmers swam all the way to the *Danzig* and *Stralsund* (Captain Harder), which had also approached. The non-swimmers, who had lifebelts and rafts, were picked up by the boats. Meanwhile the fire on the ship—which was gutted—had died down somewhat, and the explosions were less frequent. I therefore betook myself to *Stralsund* with a few men who had returned in *Ariadne's* boat, in order to request her captain to take *Ariadne* in tow. However, just about this time *Ariadne* suddenly heeled over to port and then capsized to starboard. The keel was visible for some time above the water."

If it was already known that the Heligoland Bight was insufficiently protected, because our scouting did not extend far enough, this day brought us the knowledge that a determined raid of the enemy against our weak forward patrol must inflict loss upon us every time. By the repetition of such surprises it might gradually be worn away altogether, while the Fleet got very little value out of its patrolling operations. The continuous employment of personnel and material on patrol work in the lengthening nights weakened both and thereby prejudiced the efficiency for their main task—to fight the enemy fleet. The unmolested irruption of the enemy cruisers and destroyers and the complete freedom of movement they had enjoyed in the Heligoland Bight must be made much more difficult, as also must the perpetual harassing operations of English submarines, although the latter had not hitherto displayed any great skill in torpedo work.

Far-reaching changes were made in both directions. As regards the patrol service a large number of armed fishing steamers were secured and prepared with the utmost despatch. They had previously been employed only in the harbour flotillas, which looked after the security of the estuaries. Moreover, in the middle of September two large minefields were laid west of Heligoland, which

55

increased the danger for the enemy and offered a safe retreat for our patrols when they were hard pressed.

On September 13 an English submarine, "E 9," succeeded in torpedoing the cruiser *Hela* south of Heligoland. The ship took twenty minutes to sink, so that there was time to save the whole ship's company, and our losses were limited to three men killed where the torpedo exploded.

The minefields before Heligoland proved effective, and in conjunction with progressive defensive measures such as aeroplanes and the equipment of our patrols with weapons which could be employed offensively against submerged submarines (such weapons were wholly lacking at the beginning of the war), kept the inner area so clear that the danger from submarines came at last to be quite a rare and exceptional possibility.

CHAPTER V

THE AUTUMN AND WINTER MONTHS OF 1914

THE affair of August 28, 1915, could be regarded as the preliminary of some enterprise on a larger scale, an enterprise
in which our Fleet would start at a disadvantage if the enemy held
the initiative. He would thus be able to make full use of his
superiority while we had to undertake the difficult deployment from
the estuaries of our rivers. By choosing his own moment the
attacker had the advantage of previously sending out his submarines
in large numbers to suitable stations. As the result of their frequent visits to the Heligoland Bight, as well as their experiences
in the August action, they must have acquired sufficient data to
be employed effectively.

The defensive attitude imposed on our Fleet was a direct help to
such a plan. To anticipate it it was therefore obvious that our
High Command would desire greater freedom of movement in order
to have a chance of locating parts of the enemy's forces. This could
only be done if the light forces sent out ahead could count on timely
intervention by the whole High Sea Fleet. On the other hand, it
was not the Fleet's intention to seek battle with the English Fleet
off the enemy's coasts. The relative strength (as appeared from
a comparison of the two battle lines) made chances of success much
too improbable. Taking battleships only, the superiority on the
English side was seven compared with our total number of battleships, thirteen, and therefore more than fifty per cent. Our older
ships of Squadron II, which dated from the pre-Dreadnought period,
would be opposed to an English squadron composed of ships of
the "King Edward VII" class of equal fighting value.

The Supreme Command attached more importance to the security
of the sea front, which was entrusted to the Fleet, in this early
period of the war than to the damage which it might possibly be
able to inflict on the enemy's fleet. The restrictions imposed on
the Battle Fleet were therefore adhered to.

Germany's High Sea Fleet

The attempts to damage the enemy by guerilla operations were continued, and in addition cruiser raids against the English coast and the Skagerrak were planned. The U-boats carried their operations ever farther afield, and at last they had their first success on September 8, when "U 21" (Hersing) sank the light cruiser *Pathfinder* at the entrance to the Firth of Forth. This was followed by the great feat of Weddigen when, with "U 9," on September 22, he made a bag of the three armoured cruisers *Cressy*, *Aboukir* and *Hogue*, twenty nautical miles N.W. of the Hook of Holland.

Weddigen's name was in everyone's mouth, and for the Navy in particular his achievement meant a release from the oppressive feeling of having done so little in this war in comparison with the heroic deeds of the army. But no such victory had been required to reveal completely the value of the submarine for our war-like operations, especially after it had given such unexpectedly convincing proof of its ability to remain at sea.

Favourable news came from abroad also. The *Emden* had begun her successful operations against English merchant ships in the Gulf of Bengal, and in East Africa the light cruiser *Königsberg* had sunk the *Pegasus* and so avenged the bombardment of Dar-es-Salaam.

About the middle of September the squadron of older ships which had been newly-formed at the beginning of the war had so far progressed in its training that it could be commissioned for service in the North Sea. The ships were not themselves fit to take part in a Fleet action, but they could take over part of the duties of patrolling the estuaries and keeping these open against attempts at interruption when the Fleet was at sea. However, they were never employed on this service, for they were not kept long in commission, as their ships' companies were needed urgently elsewhere later on. However, the work spent on them had not been wasted, for they gave the Fleet well-trained men for its new ships, and their presence in the Baltic in the first weeks of the war had the effect of giving our Baltic forces much greater importance in the eyes of the Russians than was justified by the facts. This, and possibly, too, their lack of confidence in their own efficiency, may be responsible for the fact that the Russians refrained from taking the offensive.

On the other hand, the Commander-in-Chief had immediately taken the offensive himself, although all he could promise himself for a result was the intimidation of the Russian naval forces in

Autumn and Winter Months of 1914

the Baltic. In spite of the fact that at the outset he had only two light cruisers, *Augsburg* and *Magdeburg*, a few torpedo-boats and some steamers, converted into mine-layers, at his disposal, he did not wait for the Russians to attack, but, immediately after the declaration of war, put to sea and bombarded Libau. The bombardment did not do much damage, it is true, but it compelled the Russians to take a hand in the work of demolition. Moreover, mines were laid at the entrance of the Gulf of Finland.

Our purpose was completely attained and compensated for the loss of the light cruiser *Magdeburg*, which ran ashore in a haze on August 27 and had to be abandoned. On October 11 the armoured cruiser *Pallada*, which had distinguished itself by shooting at the *Magdeburg* when she was stuck fast, fell a victim to our "U 26" (Freherr von Berckheim). This success did not fail to have a paralysing effect on Russian enterprise.

Without going further into the details of the operations in the Baltic their effect on the general situation at home can be described as extremely important. Without depriving the Fleet of important forces and thereby weakening or quite paralysing it, the modest forces employed kept the Russians in check, so that there was no bombardment of the German coast from the sea, and traffic in the Baltic, which was absolutely vital for war purposes, was not interfered with. The observation and security of the southern exit of the Belt and Sound made it possible for us to use the western basin of the Baltic for the Fleet's battle practices. Without such a training area the exercising of the new units which had been formed at the beginning of the war would have been very difficult. In the same way it would have been very doubtful whether we could have carried out trial trips and the first gunnery tests of newly commissioned ships.

As the war proceeded the importance of the western Baltic as an aid to keeping the Fleet ready to strike became a matter of life and death. Without constant training of an appropriate kind the standard of gunnery and navigation would have sunk to a precarious level. When navigating on a raid in the North Sea the attention of the Flag Officers was fully taken up with the possibility of enemy counter-measures and more especially with defence against underwater attack. Half the ship's company were on watch at action stations and the engine-room complement were on watch down below, and as their duties required their whole attention it was no good thinking of carrying out useful exercises of the whole ship's company

59

under the direction of the commander. We could only expect victory in battle if we succeeded in maintaining that standard of training in which we saw our sole and overwhelming chance of beating the enemy. A suitable practice area for this purpose was the Baltic, with Kiel Haven as base. Without this area at our disposal the development which our submarine weapon subsequently underwent would have been quite unthinkable.

In view of the importance of this practice area for our operations and the valuable establishments at Kiel dockyards, especially the torpedo-establishments at Friedrichsort, on the efficiency of which the whole submarine war was later to depend, it appears incredible that the enemy made no efforts to open this vital vein. At the beginning of the war the mining by the Danes of the northern and central portions of the Great Belt was in accordance with the wishes of our Naval Staff that the safety of the Baltic should be secured. There may be some question as to whether the Danes had the right to mine these waters, for they were an international strait, but the mining was approved by the English also, apparently because it fitted in with their plan of not penetrating into the Baltic. Our Fleet regarded these mines as a great obstacle to their freedom of movement, for they deprived it of the possibility, when large ships were sent out on a distant raid in the North Sea, of bringing them back round the Skagen into the Baltic instead of keeping them on the single line of retirement to Heligoland. For political reasons the Naval Staff regarded it as unwise to demand the opening of the Great Belt by Denmark.

Of the different mine-laying enterprises of the High Sea Fleet in the autumn months of 1914 a special mention is due to a cruise which on October 17 began at the mouth of the Ems and had the south coast of England for its goal. Four ships of the 7th Half-Flotilla (Commander Thiele) "S" 115, 116, 117, 119 were employed. These older boats had been chosen with an eye to the possibilities of casualties, because they were no longer fit for other duties. The ships' companies had all volunteered for this dangerous raid. Their task consisted of laying mines at the entrance to the Downs, the Channel leading round the S.E. corner of England from Dover to the mouth of the Thames. The English Admiralty had announced that navigation of the area between Lat. 51° 15′ N. and 51° 41′ and Long. 1° 35′ E. and 3° 0′ E. (that means a strip 35 nautical miles broad from the English to the Dutch coast) was dangerous on account of mines. For this reason traffic was compelled to use the open channel close to the land. It was thus under English control, and the English found their inspection service easier. By mining the channel leading into the Thames we might expect practically a stoppage of London's supplies.

England's behaviour in laying mines in the open sea, a policy made public in this announcement, released us from the necessity of observing the limits we had hitherto imposed on ourselves of restricting mine-laying solely to the enemy's coasts, an operation which was naturally attended with greater danger to the mine-layer the nearer she approached within reach of the coastal patrol forces.

The half-flotilla had left the Ems in the early hours of the morning when it was still dark. Near Haaks Lightship, 15 miles W. of the southern point of the Island of Texel, it met the English cruiser *Undaunted* and four destroyers of the latest type, escape from which was impossible. As this was realised our ships attacked and, after a brave defence in an action which was carried on at a range of a few hundred yards, were sunk. The English saved as many of the survivors as was possible. After we received the first wireless message that action had begun, no further news of the torpedo-boats was forthcoming, and as we had therefore to assume that they had been lost, we sent out the hospital ship *Ophelia* to pick up any survivors. However, the English captured her and made her prize, charging us with having sent her out for scouting purposes, although she was obviously fitted up as a hospital ship and bore all the requisite markings.

The auxiliary cruiser *Berlin* was sent out into the North Sea the same night. Her commission was to lay mines off the most northerly point of Scotland, as we had reason to suspect a lively movement of warships there. The cruise of the *Berlin* was favoured by better luck, for it was one of her mines to which the battleship *Audacious* fell a victim about a week later. She was so damaged that she had to be abandoned in a sinking condition. The English succeeded in keeping secret for a considerable time the loss of this great battleship, a loss which was a substantial success for our efforts at equalisation. When the news leaked out at last its truth was definitely and decisively denied.

The following points deserve to be remembered in considering these two enterprises: (1) Mine-laying in the open seas. (2) The capture of a hospital ship which was engaged in the work of saving life from the best of motives and observing all the regulations. (3)

Germany's High Sea Fleet

The suppression of the news that an important ship had been lost in the case of the *Audacious*.

The behaviour of the English was inspired at all points by consideration for what would serve their military purposes, and was not troubled by respect for international agreements. But this did not prevent England from raising loud cries later on when we also considered ourselves released from our obligation and with far more justification took action against hospital ships which, under cover of the Red Cross Flag, were patently used for the transport of troops. In the case of the *Audacious* we can but approve the English attitude of not revealing a weakness to the enemy, because accurate information about the other side's strength has a decisive effect on the decisions taken.

The complete loss of the 7th Half-Flotilla was very painful, and the Commander-in-Chief has been freely criticised for having sent it out insufficiently supported. The reply to that is that it is extremely difficult to decide what "sufficient support" is. Suppose, in relation to the case under consideration, we say in the light of after events that if we had had two more cruisers we should have had a superiority, such a method of reasoning involves a knowledge beforehand of the strength of the enemy; otherwise you might have to bring up your whole fleet at every alarm if you wished to feel perfectly safe. Besides, risk is of the very essence of war. The idea is implied even in Moltke's phrase, "Think first." On the other hand, our failure revealed the importance to our operations of the base on the Flemish coast, from which enterprises of this kind were much more feasible and indeed led to a permanent threat to the English trade route in the Channel.

In October the enemy submarines outside the Ems and in the Heligoland Bight were very active. There was hardly a day on which reports were not received that enemy submarines had been sighted. Although a good many of these turned out to be false alarms, their presence was frequently confirmed by the fact that torpedoes were fired. Apart from the loss of the *Hela* on September 13, which has already been mentioned, the torpedo-boat " G 116 " was sunk by a torpedo north of Schiermonnikoog on October 6. It was possible to save most of the men. On the other hand, the torpedo-boat "G 7" and an incoming auxiliary cruiser which were attacked in the neighbourhood of Amrum had better luck, as all the torpedoes fired at them missed.

The annoyance from submarines increased our determination

Autumn and Winter Months of 1914

to master them. In October, after the English "E 3" had fallen a victim to one of our U-boats, which had been lying in wait all day for this exceptionally well-handled ship, and several other English submarines had had unpleasant experiences with our mines in the neighbourhood of Heligoland, the area of the Bight inside Heligoland was given a wider berth. Beyond the island, however, we had perpetually to deal with the watchful activities of English submarines. Moreover, during the autumn storms the neighbourhood of the coasts was particularly unfavourable for navigation. Our own submarine cruises extended farther and farther afield as the commanders continued to gain experience, and by exchanging notes these operations became increasingly effective.

On October 15 "U 16" passed Heligoland after a cruise of fifteen days, and on her return reported that she was still perfectly effective. This month also witnessed the first cruise round the British Islands. "U 20" (Lieutenant-Commander Droescher), which had been sent out against transports in the English Channel, found itself compelled, by damage to the diving apparatus, to avoid the Channel, which was closely patrolled, and therefore returned round Ireland and Scotland. The cruise took eighteen days in all.

On November 1 the English cruiser *Hermes* was sunk off Dunkirk by the U-boats which were commissioned to hinder the transport of English troops to the French ports. Unfortunately no success in this particular direction was achieved.

To assign this task of interrupting the English troopship service to the Fleet was to make a totally impossible demand, as the losses it would inevitably involve would be out of all proportion to the advantage the army would derive from the disturbance to the transport of English troops such a Fleet action might cause. Even if the presence of our Fleet in these waters held up one or more ships, the way would be open the minute our Fleet left, and nothing could be easier than to arrange for ships to put out as soon as news was received that the enemy had gone. However important a factor in the war on land England's effort might be, the best way of neutralising it would have been the occupation of the French Channel coast.

If our Fleet went into the English Channel by the Dover-Calais Straits its tactical situation would be simply hopeless. It would have no room to manoeuvre against torpedo and mine attack. Our own destroyers would not have enough fuel, as their radius of action only just reached as far, and they would then find themselves compelled to return. The Fleet would then have had to do without them or return with them. There could be no question of the former alternative on account of the danger from submarines, defence against which was the work of the destroyers, and also because the destroyers were indispensable for battle. The Fleet was therefore dependent upon the radius of action of the destroyers. The appearance of the submarine as a defensive weapon has made it a necessity in modern times to screen the approach of a fleet with destroyers. Moreover, it is so important to increase the offensive powers of a fleet which is inferior in numbers by the employment of destroyers that these cannot possibly be dispensed with. If one compares, simply on a map, the position of a fleet which ventures into the Channel from the Heligoland Bight with that of a fleet making for the Heligoland Bight from the English coast—from the Firth of Forth, for example—the advantages and disadvantages of the prospects on either side are at once apparent. One fleet is placed as if it were corked in a bottle, while the other has freedom of movement over the whole area in its rear.

At the end of October Squadron II had visited Kiel dockyard to effect certain important improvements in armament and the comfort of the ships, which had suffered very much from the removal of everything which was likely to catch fire. This was in the interests of the health of the ships' companies during the winter. The compartments throughout the ship were insulated in the same way as those in the newer ships by the use of fireproof material. Living in ships in which every noise came as a shock from one end to the other became a severe trial to the nerves as time went by, and in view of the strenuous hours on watch, was prejudicial to the short period allowed for rest. The victims will never forget those weeks of the war in which the tapping of hammers and the scraping of chisels never ceased from first thing in the morning to last thing at night, and mountains of wood and superfluous paint vanished from the ship.

This first visit of a squadron to the Baltic was also to be employed in various exercises in which cruisers and destroyers were to participate. It appeared advisable, in view of this, to take advantage of the presence of the ships for a great enterprise against Libau which might be very unpleasant as a winter base for enemy submarines, as it was the only Russian ice-free harbour. While the orders for this enterprise were being settled with the Commander-in-Chief in the Baltic and everyone was burning for the chance of at last firing his first shot, the news reached us from the North Sea that the bombardment of

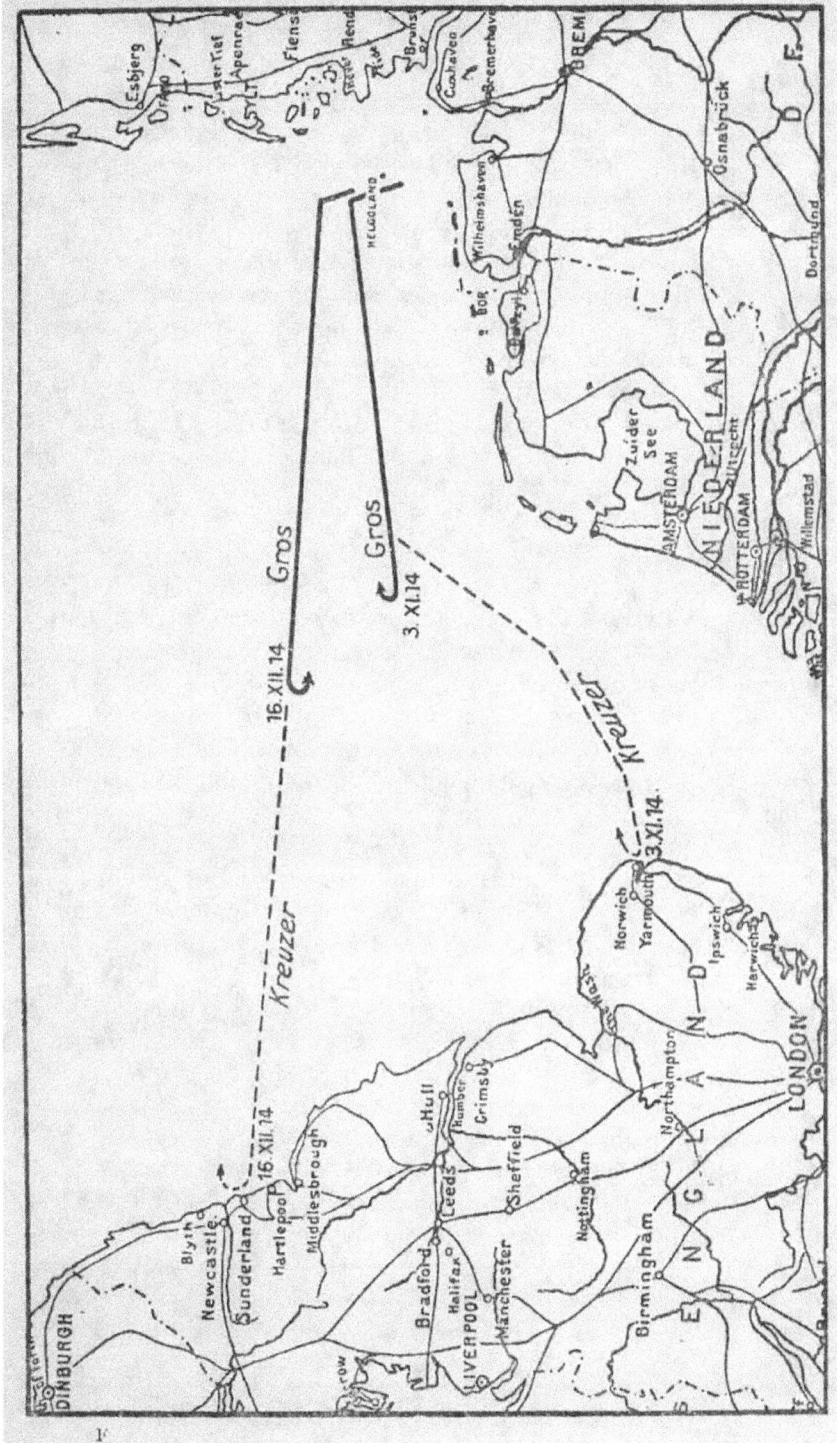

THE BOMBARDMENTS OF THE EAST COAST ON NOVEMBER 3 AND DECEMBER 16, 1914

Germany's High Sea Fleet

English coast towns had successfully been carried out on November 3. Early that morning our battle-cruisers had appeared off Yarmouth to bombard the harbour and its fortifications while mines were being laid under their protection. The absence of Squadron II had not restrained the Commander-in-Chief from taking advantage of the favourable weather and long nights for this raid, from which we could anticipate an effect on the defensive attitude of the enemy as well as the direct influence which the damage to a hostile base would have on the enemy's operations. It was not found necessary to send the Fleet out to take up an advanced station at sea in the case of the short raid to Yarmouth, because the plan was to be based entirely on surprise under cover of darkness. After returning from this raid the old armoured cruiser *Yorck* ran on a mine in a mist in the Jade and was capsized by the explosion. It was found possible to save the larger part of the crew.

The raid against Libau was cancelled at the last moment as the result of an order from the Naval Staff to Squadron II, which was already on its way. The frequent reports of the activity of English submarines in the Baltic, which had come in of late, seemed to point to the wisdom of abandoning the enterprise, as the bombardment by ships of land targets would certainly offer submarines their very best chances of attack. The submarine danger was taken very seriously because we had not yet had sufficient experience and training in the defence.

On November 6 we received the news of the victory of our cruiser squadron on November 1 off Coronel on the coast of Chile. Vice-Admiral Count von Spee had defeated in fair and open fight the English cruisers *Good Hope, Monmouth, Glasgow* and the auxiliary cruiser *Otranto* with his ships *Scharnhorst* and *Gneisenau* and the light cruisers *Leipzig* and *Dresden.* The two hostile armoured cruisers were destroyed by a superior fire, while *Glasgow* and *Otranto* escaped under cover of falling night. Great was the enthusiasm over the fact that the brave admiral had succeeded, in spite of all obstacles, in leading his ships to a victory which dealt a severe blow to the tradition of English superiority at sea. This news filled us in the Fleet with pride and confidence, and we thought in gratitude of those who, left to their resources in distant oceans, had gained immortal laurels for the German flag. Unfortunately fate was not to permit them to see their homeland again. Those who, with their leaders, rest in the ocean depths by the Falkland Islands, gave us a shining example of heroism, of devotion to duty.

CHAPTER VI

BOMBARDMENT OF SCARBOROUGH AND HARTLEPOOL, AND THE BATTLE OF THE DOGGER BANK

IN the first months of the war many efforts had been made to conduct our operations in a way that would cause the enemy such losses as would enable us to speak of a real equalisation of forces. But in vain. The results of our mine-laying were unknown, while the successes of our submarines did not weigh much in the scale, as the ships they torpedoed had no fighting value. On the other hand, raids by our cruisers were much more likely to bring considerable portions of the English Fleet out of their harbours and thus give our Fleet a favourable chance of intervening if it kept in close touch with its cruisers. For this purpose our cruisers would in any case have to go far beyond the limits of distance they had hitherto observed—not more than 100 nautical miles from Heligoland. Then only would our cruisers begin to have some real effect. Within the limits imposed upon him the Commander-in-Chief of the Fleet had described the efforts we had made—cruisers had put to sea, mine-laying was carried out continuously in spite of the losses we had suffered, submarines had done far more than was expected of them, were untiring in their efforts and had penetrated as far as the English coasts, yet for the Fleet itself these operations had proved a disappointment. Strategical reasons had made it necessary to keep our Fleet back, and this looked like a want of confidence and affected the *moral* of the men, and gradually lowered their belief in their own efficiency to a regrettable degree. An impressive recital of these facts with the request that the Commander-in-Chief of the Fleet should be allowed greater latitude was met with a decided rebuff. The grounds of this refusal, as communicated by the Naval Staff, ran somewhat as follows :

"The existence of our Fleet, ready to strike at any moment, has hitherto kept the enemy away from the North Sea and Baltic coasts and made it possible to resume trade with neutral countries in the Baltic. The Fleet has thus taken over the protection of the coast and troops required for that purpose are now available for use in the field. After even a successful battle, the ascendancy of the Fleet under the numerical superiority of the enemy would give way, and under the pressure of the enemy Fleet the attitude of the neutrals would be prejudiciously influenced. The Fleet must therefore be held back and avoid actions which might lead to heavy losses. This does not, however, prevent favourable opportunities being made use of to damage the enemy. An employment of the Fleet outside the German Bight, which the enemy tries to bring about through his movements in the Skagerrak, is not mentioned in the orders for operations as being one of the favourable opportunities. There is nothing to be said against an attempt of the big cruisers in the North Sea to damage the enemy."

These instructions served the purpose of the further enterprise against the English coast. On December 15 the big cruisers under the command of Vice-Admiral Hipper sailed under orders to bombard the fortified coast towns of Scarborough and Hartlepool and to lay mines along the coast, for there was constant traffic between the East Coast ports. Both these places, however, are 150 nautical miles nearer to the chief bases of the English Fleet in the North of the British Isles than is Yarmouth. It would, therefore, be much easier for vessels lying there or cruising at sea in the vicinity to beat off an attack, and the expedition would probably present a much greater risk, and a more urgent call for support from the Fleet.

The 2nd Scouting Division, composed of light cruisers and two torpedo-boat flotillas, was attached to the 1st Scouting Division of battle-cruisers. They left the Jade on the 15th at 3.20 A.M., followed late in the afternoon of the same day by squadrons of battleships. The hour of departure for both divisions was chosen in order to profit by the darkness and if possible put to sea unobserved. Judging from what ensued, this appears to have succeeded. A rendezvous at sea at 54° 30′ N. Lat. and 7° 42′ E. Long. was appointed for the squadrons coming from the Jade and the Elbe. In order to get there I left the anchorage at Cuxhaven with Squadron II at 4 P.M. From the meeting-place Squadron II took the course ordered by the Commander-in-Chief—W.N.W. by ½W. at a speed of 15 knots. As all the ships were most carefully darkened, nothing could be seen of the other squadrons. The navigation had therefore to be most accurate in order that the squadrons might be in their proper places the next morning. Seven to five nautical miles had been determined on as the distance between the squadrons from flagship to flagship. The sailing order of the units was: Squadrons I, III and II. To ensure the safety of the Main Fleet when under way, the two older armoured cruisers, *Prinz Heinrich* and *Roon*, were placed ahead, together with a torpedo-boat flotilla. To cover the flanks two light cruisers were utilised, each with a flotilla. The light cruiser *Stettin*, with two flotillas, covered the rear. During the night several fishing steamers were stopped by the escorting torpedo-boats but released as non-suspect.

At 5.20 A.M. a torpedo-boat in the vanguard reported four enemy destroyers in Square 105. This was at 54° 55′ N. Lat. and 2° 10′ E. Long. This spot was about 20 nautical miles north-west of the appointed meeting-place for the cruisers, to which destination the Commander-in-Chief of the Fleet was also steaming. As several hours must elapse before we could reach our destination, and no further message followed the first one, we continued on our way. An hour later there was another message from a torpedo-boat in the vanguard to the effect that ten enemy destroyers had been sighted and that flashes from guns were visible. A quarter of an hour later the same boat reported that a chase had started. Thereupon at 6.45 the Commander-in-Chief gave the signal for all the squadrons to turn into a S.E. course as it still wanted an hour and a half to daylight. By issuing that order he carried out his purpose of avoiding an encounter with the enemy torpedo-boats and denying them the opportunity to attack in the dark.

Meanwhile our vanguard had begun to fight with the enemy destroyers. At 6.58 the light cruiser *Hamburg* (Captain von Gaudecker) reported that he had sunk an enemy destroyer. At 7.10 the Fleet turned again to the E.S.E.-½E. and started on the return journey.

It had passed considerably beyond the arc from Terschelling to Horns Reef that shuts off the Bight. Having set out with the object of supporting our cruisers, there was now no possibility of carrying out that plan, seeing the great distance that lay between the two divisions. In this case, therefore, the success of the cruisers' enterprise was entirely dependent on their taking the enemy by surprise and avoiding the enemy's superior forces.

Towards daybreak, when our cruisers were approaching the English coast, the wind rose to such a pitch and the sea ran so high that the light cruiser *Strassburg* reported at 7 A.M. that, owing to heavy seas off the land, firing was no longer possible and the ship had been obliged to turn on an easterly course. As, under these conditions, the light cruisers and torpedo-boats could only be a hindrance to the big cruisers, the Commander-in-Chief decided to dispatch those vessels in the direction of the Main Fleet, with the exception of the light cruiser *Kolberg*, which was to continue laying mines at the places determined on.

The big cruisers then divided into two groups for the bombardment of the coastal towns, the northern section, the *Seydlitz*, *Moltke* and *Blücher*, making for Hartlepool. An officer of one of the U-boats who had reconnoitred the area beforehand rendered good service in locating the place. Shortly before they were off Hartlepool the cruisers were attacked by four torpedo-boat destroyers of the "River" class that ran out to sea and were brought under fire at a distance of about 50 hm. The sinking of one destroyer and heavy damage to another were observed. After firing some torpedoes without any result, they turned away. We gave up pursuing them so as not to lose time for the bombardment. The *Seydlitz* opened fire on the Cemetery Battery and scored several hits, so that at last the fire was only returned by one 15 cm. gun and one light gun from the battery. The *Moltke* was hit above the water-line, causing much damage between decks but no loss of life. From the first, the *Blücher* came under a lively fire from the land batteries; she had nine killed and three wounded by one hit alone. 15 cm. howitzers and light artillery were used on land; the *Blücher* was hit six times altogether.

The southern group, *Von der Tann* and *Derfflinger*, made for Scarborough which was easily distinguishable. The coastguard station at Scarborough and the signalling and coastguard stations at Whitby were destroyed. At the latter place the second round brought down the signalling flagstaff with the English ensign and the entire station building as well. The *Derfflinger* also bombarded trenches and barracks at Scarborough. As there was no counter-action it must be assumed that the battery at Scarborough was either not manned in proper time, or had been evacuated by the garrison.

The light cruiser *Kolberg* laid her mines at the appointed place without much difficulty, although the ship heeled over to 12 degrees and the tip apparatus (for dropping the mines overboard) drew water. At 9.45 the cruisers assembled round the *Seydlitz* and started to retire in the direction of the meeting-place agreed on with the Main Fleet. An hour later, at 10.45, a wireless message was received from the Chief of Reconnaissance with the Fleet that the task

was accomplished and that he was stationed at 54° 45′ N., 0° 30′ W.
At 12.30 noon the *Stralsund*, of the Second Scouting Division,
with Torpedo-Boat Flotilla II attached, sighted a number of enemy
cruisers and, turning in a south-westerly direction, evaded them to
try and join the large cruisers. The English cruisers were again
lost to sight, as the weather was very misty. Soon afterwards
the *Stralsund* sighted six large enemy ships which were made out to
be battleships of the "Orion" class, and therefore the Second
English Battle Squadron. The *Stralsund* kept in touch with them
and continued to report on the course and the speed of the enemy.
At 1 P.M. these groups were at 54° 20′ N. lat., 2° 0′ E. long. This
report caused our big cruisers to turn off in a north-easterly direction,
as owing to the bad visibility they were compelled to avoid an unexpected encounter with battleships of superior fighting strength
than that of our own. At that time the position of the two forces
facing each other was approximately as follows:

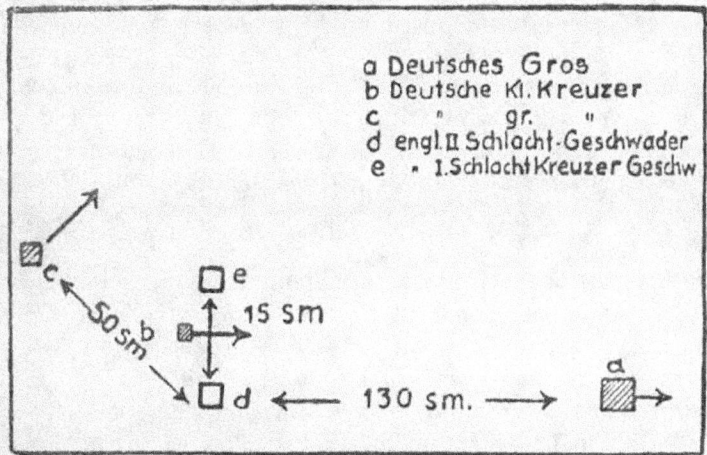

Great disappointment was caused on board my flagship by this
report. If our big cruisers had got into difficulties between the
enemy battle-squadron and other cruisers already reported and still
in the vicinity, our help would be too late. There was no longer
any possibility while it was still day of coming up with the enemy
battle-squadron, which at one o'clock was 130 nautical miles distant
from us. Our premature turning on to a E.S.E. course had robbed
us of the opportunity of meeting certain divisions of the enemy
according to the prearranged plan, which was now seen to have been

correct. At all events the restrictions imposed on the Commander-in-Chief of the Fleet brought about the failure of the bold and promising plan, owing to its not having been carried out in the proper manner. As we now know from an English source, the destroyers fired at by the *Hamburg* were about 10 nautical miles in front of the Second Battle Squadron which had come down on a southerly course—the vanguard of which had got into touch with ours between 6 and 7 A.M.; and since the position at 1 P.M., reported by the *Stralsund*, coincides exactly with the English statement, it proves that at 7 A.M. both the main fleets were only about 50 nautical miles apart. It is extremely probable that by continuing in our original direction the two courses would have crossed within sight of each other during the morning.

The advantage in a battle ensuing therefrom was distinctly on our side. The English had at their disposal on the spot the Second Battle Squadron with six ships, the First Battle-Cruiser Squadron with four ships was within attacking distance, and added to these were a few light cruisers and the Third Cruiser Squadron attached to the Second Battle Squadron.

According to his own statement, the English admiral in command did not leave Scapa Flow with the other ships till 12 noon, after receiving news of the bombardment at 9 A.M. He could not possibly have been in time; while the Third English Squadron, which had been sighted at 10 o'clock, would not have had the advantage over our Fleet.

On the part of the English, disappointment was felt that coastal towns had again been bombarded by our cruisers and that they could not succeed in stopping it, although the necessary forces chanced to be at sea and had even got into touch with our light cruisers. This, according to Admiral Jellicoe's account, may have been due to the fact that the squadrons at sea had received instructions from him how to act so as to cut off the enemy, but had also had direct orders from the English Admiralty which were totally different and which were acted upon by Sir George Warrender, in command of the Second Battle Squadron.

The weather conditions were remarkable on that day. In the east section of the North Sea—the area through which our Fleet had passed—there was a slight easterly wind, no sea running, and perfect visibility. At the 3rd deg. E. Long. there was a sharply defined spot where the weather changed. A north-westerly storm raged off the English coast and the sea was correspondingly rough, making it extremely difficult to serve the guns even on board the big cruisers. Between 9 A.M. and 2 P.M., as our Fleet withdrew, an extraordinary number of drifting mines were observed, more than 70, some of them already exploded. They must have broken loose from the big minefield at the entrance to the Canal. It was a lucky chance that we escaped damage when, on the preceding night, the ships passed through that area without being able to observe them. At 8 P.M. on December 16, Squadron II ran into the Elbe again, and the others returned to the Jade.

The impression that a specially favourable opportunity had been missed still prevailed, and the chance of another such arising could hardly be expected.

The behaviour of the English Fleet makes it obvious that our advance was a complete surprise to them, nor had they counted on our Main Fleet pushing forward to the Dogger Bank. Otherwise the English expedition would surely have comprised stronger forces than merely one battle squadron, a battle-cruiser squadron, and lighter forces. This combination certainly made them superior to our cruiser attack but not to an attack by our Fleet. The information that besides the German ships in action off the English coast a still greater number were out at sea was communicated to the English Commander-in-Chief of the Fleet at 2 P.M. by the English Admiralty.

The English received the news through their "directional stations" which they already had in use, but which were only introduced by us at a much later period. They are wireless stations for taking the directional bearings of wireless messages, and in combination are capable of indicating the direction from which intercepted wireless messages come and thus locating the signalling ship's station. The stretch of the English east coast is very favourable for the erection of these "directional stations." In possessing them the English had a very great advantage in the conduct of the war, as they were able thus to obtain quite accurate information as to the locality of the enemy as soon as any wireless signals were sent by him. In the case of a large fleet, where separate units are stationed far apart and communication between them is essential, an absolute cessation of all wireless intercourse would be fatal to any enterprise.

Towards the end of December a change was made in the squadron command. Other ships had been added to Squadron III since the declaration of war. The *König, Grosser Kurfürst* and *Markgraf* had all made their trial trips. The *Kronprinz* was very near completion and on January 2 was enrolled as the eighth ship in the squadron. I was entrusted with the command of this squadron. It was no easy matter for me to separate from Squadron II, which had been under my command for nearly two whole years, as I had learnt to value the splendid spirit of the crews, who, in spite of the inferior fighting powers of the ships, made it a point of honour never to be behindhand in anything. But personal feelings were not to be considered, and I had to look upon it as a great distinction that the command of our most powerful fighting squadron was given to me. The command of Squadron II was taken over on December 26 by Rear-Admiral Funke, whereupon I left for Wilhelmshaven to take up my position on the *Prinz Regent Luitpold*.

The ensuing time was fully occupied in learning to know the peculiarities of the new class of ship and the standard of fighting power of each individual vessel, and in judging the personality of the commanders and the corps of officers. The prevailing conditions of war made it more difficult to cultivate close relations with them than would have been the case in peace time. My chief object was so to train the unit as to make it absolutely reliable for implicit obedience to commands. I applied, therefore, to the Commander-in-Chief for an opportunity for a period of training in the Baltic towards the end of January. This was all the more necessary in view of the fact that since they were commissioned the four ships of the "König" class had had no practice in torpedo firing.

From a military point of view torpedo firing practice is an urgent necessity in the training and further development of all torpedo officers, those who are in charge of the torpedo tubes, and of those in reserve, in order to prove that the results from the use of the weapon are equal to expectations. Particular attention must be given to range practice and angle-discharging, which make a great demand on the ability of the torpedo men. During the war many ships were provided with torpedoes with all the latest improvements, without the crew having had an opportunity to fire them or become familiar with the handling of them. Experience showed that it was necessary to test every torpedo that had lain unused for more than five months to make sure that it would act when needed.

So long as enemy submarines remained in those waters the inner Bight of the North Sea was not a suitable place for gun-practice; these craft could not have had a better opportunity for firing their torpedoes. The mouths of the rivers certainly offered chances to our gunners of practice on objects passing by, but there was very little scope for gun practice at long range under fighting conditions. The necessity of combining the training period with the time required for unavoidable repairs, as also with the war activities of the Fleet which called for the participation of the highest possible number of ships, was a matter of extreme difficulty from the point of view of organisation.

Before Squadron III could sail for the Baltic there was to be another enterprise by the Fleet in the North Sea, which, owing to bad weather, was postponed from day to day. January, 1915, opened with most unfavourable weather, and one violent storm followed rapidly on another. But when, in searching for a passage for the Fleet through the minefields, it was discovered that many new ones had been laid down, both north of Amrum and west of Borkum, and also in the gap between Norderney and the safety barrier we had put down, the plan for an advance by the Fleet was abandoned. These mines would first have had to be removed, which would have been slow work owing to the bad weather. Instead of a big action by the Fleet, two light cruisers went out to lay mines and succeeded in placing a barrier 50 nautical miles from the English coast, close to the mouth of the Humber, presumably just in the enemy's outgoing course.

Towards the middle of the month the Fleet was kept at a high pitch of readiness as there was reason to believe the English were planning a blockade of our estuaries. The idea was extremely probable, as the poor visibility in winter weather offered the most favourable conditions for carrying it out. In the Jade particularly the channel for large vessels was so narrow and so shallow that the traffic was greatly hindered, especially in the case of certain vessels. There could be no warding off such an attack by a coast battery, as Wangeroog was not yet fortified. In any case, we could not afford to over-estimate the difficulty of carrying out such an undertaking; in view of the vast amount of material possessed by England for such a purpose, success in it was by no means out of the question. The fact that the Fleet would be obliged to push the undertaking to our very river mouths doubtless formed their chief reason for not making such an attempt, the success of which would have been very detrimental to the carrying out of our U-boat and mining warfare.

On the morning of January 19, an aeroplane having sighted 60 miles north-west of Heligoland numerous English ships bound

75

on an easterly course, among them several battle-cruisers and close upon 100 small craft, we made sure that their plan was to be put into execution. It is quite possible that the aeroplane was mistaken as to the number and type of the ships, although the report was confirmed from another source—two U-boats that returned from sea. However, the torpedo-boats which were sent out to reconnoitre and to attack at night if necessary saw nothing of the enemy forces, so they probably had withdrawn early. At any rate we considered the danger of a blockade to be at an end.

On January 21 Squadron III sailed for the Elbe. During the passage there was a violent snowstorm which made it very difficult to locate the mouth of the river. Owing to the rapidly falling depth of water as shown by the soundings taken, we were forced to anchor, a manœuvre carried out in exemplary fashion by the big ships, in spite of the current and the mist. It showed very clearly the difference between the navigation of a squadron of such large vessels and that of Squadron II where the ships had not half the displacement. The next morning the weather was calm and clear, and the passage through the Kaiser Wilhelm Canal was accomplished without accident. It took us only 10 hours to cover the long stretch of 100 kilometres from the lock gates at Brunnsbüttel to the entrance into the lock at Holtenau in Kiel Harbour.

While the training of Squadron III was proceeding in the Baltic, the first regular cruiser action took place in the North Sea.

BATTLE OF CRUISERS OFF THE DOGGER BANK

After a long period of inaction, the weather being apparently favourable, the commander of the scouting ships was ordered on January 23 to reconnoitre off the Dogger Bank with the cruisers of the 1st and 2nd Scouting Divisions, the First Leader of the Torpedo-boat Forces, and the Second Flotilla, and there to destroy any of the enemy's light forces to be met with. They were to set out in the evening, when darkness fell, and were expected back the following evening when it was again dark.

The speed of the advance was to be such that the cruisers, at daybreak on the 24th, would have reached the south-east edge of the Dogger Bank. It was not intended to push farther on towards the Bank while it was still dark, otherwise enemy forces might make their way unobserved in between Heligoland and the cruisers. On the way there no trade or fishing steamers were to be examined, if

Battle of Cruisers off the Dogger Bank

it could be avoided, so as not to be forced to leave any of our torpedo-boats behind; but the plan of action for the homeward run included the examination and, where necessary, seizure of all the fishing steamers encountered.

The big cruiser *Von der Tann* was missing from our cruiser squadron, being in dock for urgently needed repairs, as was also the light cruiser *Strassburg*. The fighting force, therefore, com-

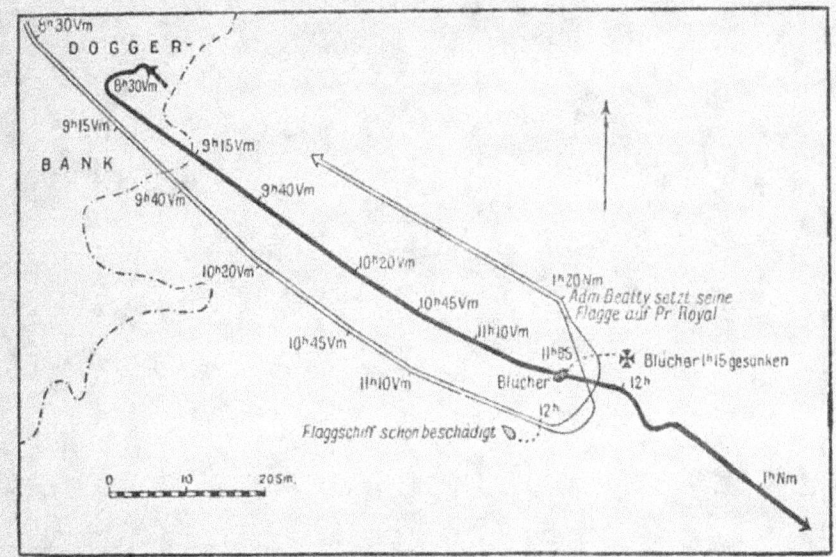

The Fight off the Dogger Bank

prised the cruisers *Seydlitz*—the flagship of Rear-Admiral Hipper—*Derfflinger*, *Moltke* and *Blücher*, the light cruisers *Graudenz*, *Stralsund*, *Kolberg* and *Rostock*, Torpedo-Boat Flotilla V, and the 2nd and 18th Half-Flotillas. The *Graudenz* and *Stralsund* formed the vanguard, the flanks were supported by the *Rostock* on the starboard and by the *Kolberg* on the larboard side. A half-flotilla was attached to each light cruiser.

At 8.15 A.M. on the 24th the *Kolberg* encountered an enemy light cruiser and destroyers. The enemy's signal of recognition was answered by the *Kolberg* turning on the searchlight and shortly afterwards opening fire, which was returned a few minutes later. The *Kolberg* was hit twice and had two men killed. At the same time she sighted thick clouds of smoke in a west-south-westerly

77

direction, and the *Stralsund* also reported to the same effect to the north-north-west.

The conclusion thus to be drawn was that other and more numerous forces were lying off the Dogger Bank.

After Admiral Hipper had picked up the *Kolberg* he assembled his group on a south-easterly course, as it was still not sufficiently

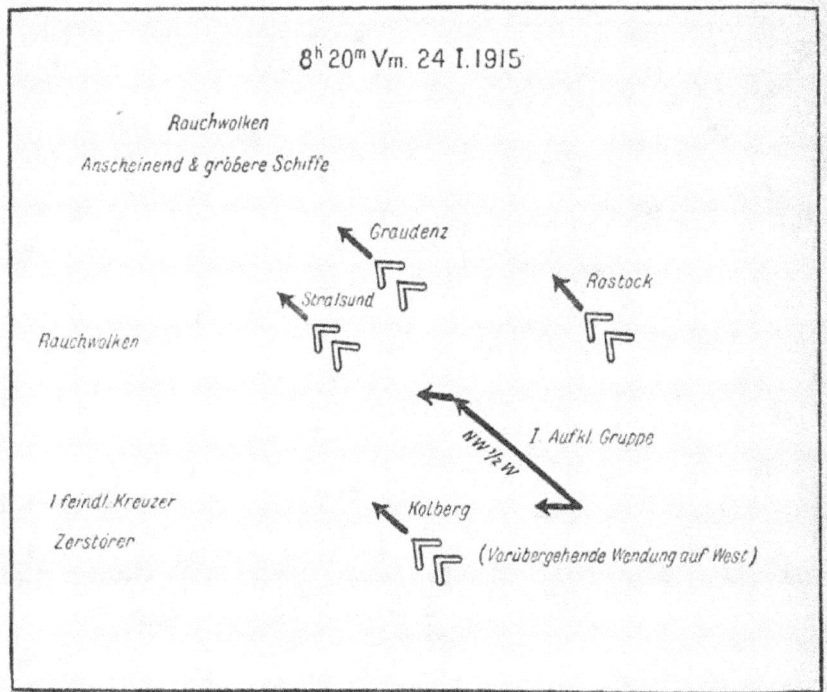

Situation at 8.20 A.M., January 24, 1915

light to make out the number and type of the enemy forces. While the ships were assembling, four cruisers of the " Town " class, three cruisers of the " Arethusa " class, and a large number of destroyers were sighted on a parallel course north of our cruisers, but out of gun range. The *Blücher* was able to count more than twenty destroyers. Further clouds of smoke could be seen in their rear, and the *Stralsund* reported that at least eight large ships were observable in a north-north-westerly direction.

Admiral Hipper was bound therefore to assume that at the rear of these numerous light forces there must be other and

Battle of Cruisers off the Dogger Bank

stronger groups of ships, and, as he could not count on any support from our own Main Fleet, he decided to push on full speed ahead in a south-easterly direction. The torpedo-boats were sent on ahead. The *Blücher*, being the rear ship, was permitted at discretion to open fire, as some of the destroyers to the north approached very near, while the light cruisers with them stood off farther to the north.

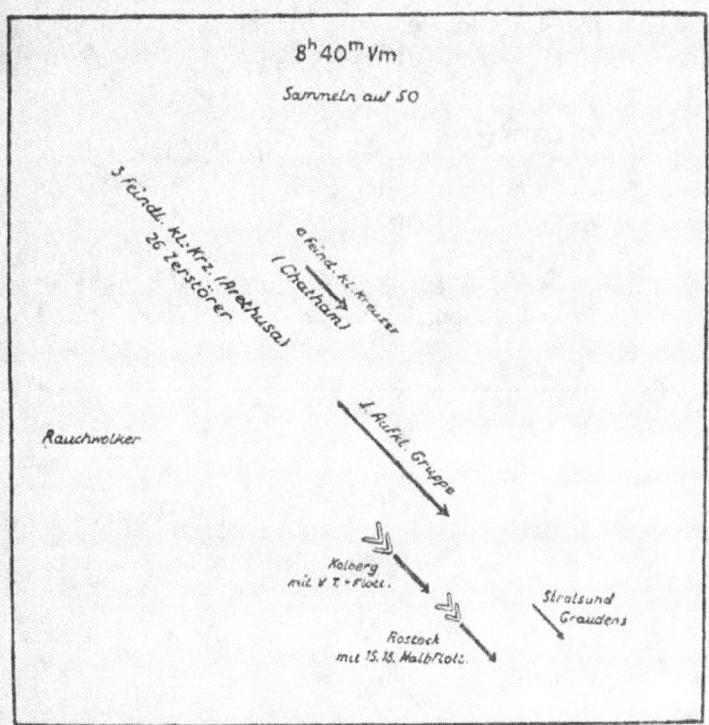

Situation at 8.40 A.M.

At 9.35 A.M., however, five thick clouds of smoke were observed from starboard in a west to west-north-west direction, which were soon made out to be from the 1st English Battle-Cruiser Squadron. They came up at full speed and opened fire at a great distance, about 200 hm., and, at first at any rate, without reaching our cruisers.

The naval command at Wilhelmshaven received the first news of the encounter of our cruisers with the enemy at 8.50 A.M., when the *Seydlitz* reported herself as being at 54° 53′ N. Lat. and 3° 30′

Germany's High Sea Fleet

E. Long., course S.E., speed 20 knots, and had sighted eight large ships, one light cruiser and twelve destroyers. The command at once issued orders for special preparation on board all ships and torpedo-boat flotillas and assembled them in the Schillig Roads. As the way to the Bight was open to our cruisers, and they were in touch with the enemy forces in the rear, it was assumed that so far our ships were not in any danger. Towards 10.30 A.M. the squadrons were all assembled in the Schillig Roads, and ran out to sea at 11.10, as a wireless message had come from the Admiral at 11 o'clock, saying he was in urgent need of support. He was then at 54° 30′ N. Lat. and 4° 35′ E. Long.

These forces were, however, not called upon to take any active part in the battle, as the further development of the fighting at that time showed it to be unnecessary.

Meanwhile the situation of the cruisers had developed as follows: At 10 A.M. our large cruisers were lying on a south-easterly course, so that all the ships could open fire from the starboard on the English

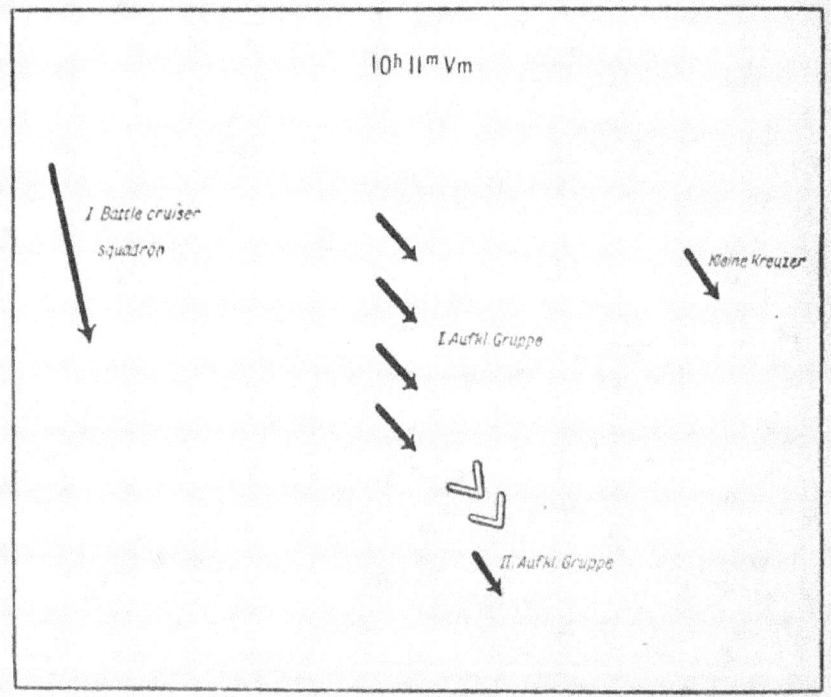

Position at 10.11 A.M

Battle of Cruisers off the Dogger Bank

large cruisers. Our light cruisers and both the flotillas were ahead of our large cruisers, slightly on the starboard side.

The enemy battle-cruisers came up very rapidly, and must have made a speed of at least 26 knots.*

Our 1st Scouting Division was not favourably situated, owing to the prevailing east-north-east wind. There was nothing for it, however, but to keep to the south-east course, leading to the Bight, as the main direction for the fighting. The chances of support from our own forces were greater there, and the farther we could succeed in drawing the enemy into the Bight the greater prospect there would be of setting torpedo-boats on him during the ensuing night. Any other course leading farther south or still farther west would not greatly have improved the smoke conditions, but would from the first have placed the enemy battle-cruisers in a frontal position. On the other hand, a north-easterly course, so as to have the wind ahead, would have carried our forces straight up against the enemy destroyers, and thus offered them a good opportunity for attack. Soon after 10 o'clock our large cruisers opened fire at 180 hm.; the enemy manœuvred so as to avoid our fire. At the same time our cruisers also turned about between E.S.E. and S.E. to a S. course. The range for the leading ship, the *Seydlitz*, varied between 180 to 145 hm. The enemy had separated and formed two groups, the leading one having three, and the other two ships.†

They were trying to keep at the farthest firing distance. Soon after the fighting began the *Seydlitz* was badly hit and both her after turrets, with their two 28-cm. guns, were put out of action, while fires were caused in them by the exploding ammunition. The gunners in both turrets were killed, and the turrets themselves jammed and put out of action. Owing to the fire, which took a long time to extinguish, the munition chamber had to be flooded.

Meanwhile some of the light cruisers and destroyers were steaming up on the larboard [port] side, so that the near ships could fire on them occasionally. In doing so *Blücher*, the last ship, hit and heavily damaged a destroyer. At 11.30 the enemy appeared to be drawing nearer; at the same time the *Blücher* reported engine trouble and dropped slowly to the rear.

---

* The English commander, Admiral Beatty, boasted in his report that his ships had achieved a speed of 28.5 knots.

† According to an English account, the *Lion, Tiger, Princess Royal, New Zealand* and *Indomitable*.

Germany's High Sea Fleet

The order "Flotilla clear for attack" was then sent to the torpedo-boat leader. At 11.45 the leading enemy ship, with a heavy list on, turned off and drew out of the line. The ship following after her passed the leader, so as to keep up the running fight. The other enemy battle-cruisers followed at irregular distances. At 12 o'clock our cruisers turned towards the enemy, and the torpedo-boats were ordered to attack. The enemy battle-cruisers then turned at once to a northerly course to evade the

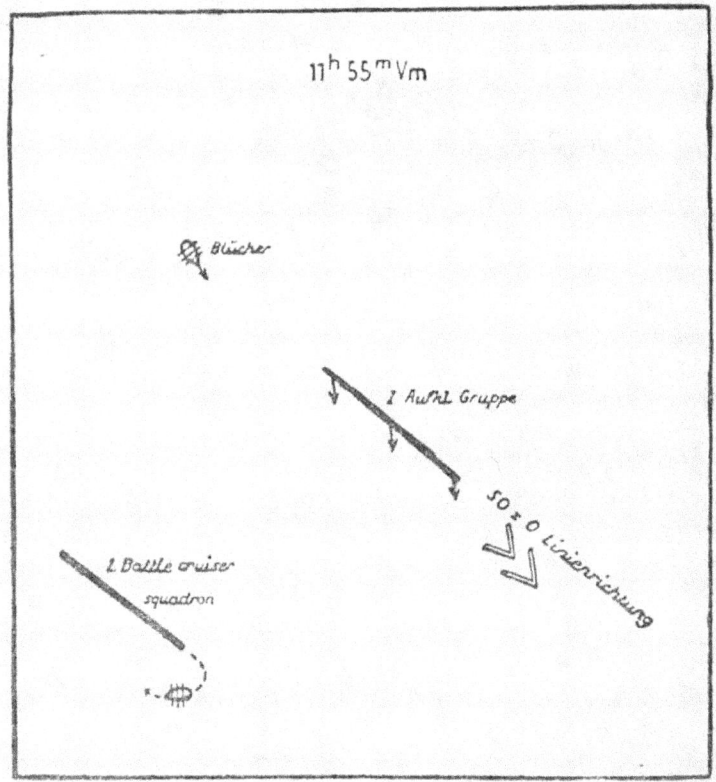

Position at 11.55 A.M.

torpedo-boats and to turn on the *Blücher*, which had been left behind. In view of this manœuvre the torpedo-boats were recalled from the attack.

Our cruisers now took up a southerly course, intending to open an encircling fight with the enemy, and if possible render help to the *Blücher*. But both turrets on board the *Seydlitz*, with two-fifths of the heavy guns, were definitely out of action, and the ship's stern was full of water which had spread to the other parts from the flooding of the munition chamber, so the Admiral of the cruisers therefore decided to profit by the increased distance caused by the enemy's manœuvre to turn again to S.E. and break off the fight. At 1.45 the enemy was lost to view, the *Seydlitz* being then 25 nautical miles north of the mouth of the Elbe.

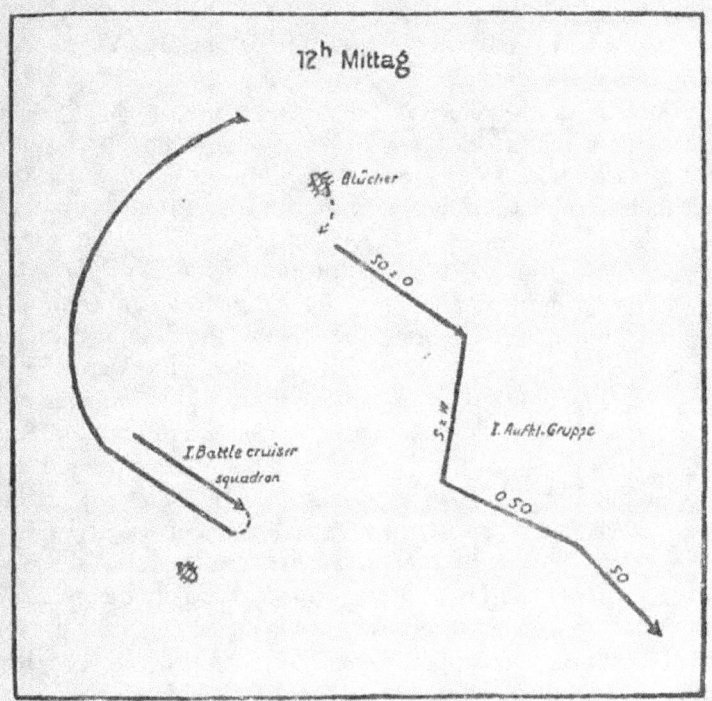

Position at 12 noon

At 3.30 P.M. the forces that had run out from the Jade joined the returning cruisers and together entered the rivers.

Besides the explosion and the list on the leading enemy ship, many other hits and a big fire on the second ship were observed. Several officers asserted positively that they had seen one of the large cruisers sink, which gave rise to the report that it was the battle-cruiser *Tiger*. Contradictory reports from an English source appeared later in the Press and confirmed the opinion that the English wished to conceal the fact. The airship "L 5," which was hovering over the spot, reported that only four large ships were seen to withdraw. The torpedo-boat "V 5," Lieut.-Commander von Eichhorn, which, after being recalled from the attack, had dropped out from between the two fighting lines, fired two torpedoes at 70 hm., and thereupon observed the withdrawal of a battle-cruiser. There seems no obvious reason why the English cruisers should so soon have stopped fighting after their leader fell out and when the number of our cruisers had already dwindled to three, unless it was because our guns had severely handled them.

On our side we deplored the loss of the *Blücher*. Very soon after her engines were damaged another shot caused an explosion and a fire amidships, apparently in the big ammunition chamber, situate in that part of the vessel. It was observed how to the very last the ship's guns on both sides fired on the battle-cruisers which concentrated their fire on that one ship, as did also the numerous enemy light cruisers and destroyers, for whom the wrecked ship was a welcome target, until at 1.7 P.M. she turned over and sank. The survivors of the crew were picked up by English destroyers and other ships that were at hand, among them being the gallant commander, Captain Erdmann, who unfortunately died afterwards while a prisoner in England of pneumonia, the result of the immersion in the cold sea after his ship had gone down. The *Derfflinger* and *Kolberg* were slightly damaged; the *Seydlitz* was badly hit a second time on her armoured belt, the plate being pressed into the ship's side and causing a leakage. The first shell that hit her had a terrible effect. It pierced right through the upper deck in the ship's stern and through the barbette-armour of the near turret, where it exploded. All parts of the stern, the officers' quarters, mess, etc., that were near where the explosion took place were totally wrecked. In the reloading chamber, where the shell penetrated, part of the charge in readiness for loading was set on fire. The flames rose high up into the turret and down into the munition chamber, and thence through a connecting door usually kept shut, by which the men from the munition chamber tried to escape into the fore turret. The flames thus made their way through to the other munition chamber, and thence again up to the second turret, and from this cause the entire gun crews of both turrets perished almost instantly. The flames rose as high as a house above the turrets.

Up to 12 noon there had been no prospect of the torpedo-boat flotillas making a successful attack; the distances were too great. The torpedo-boats would have been obliged to get within 100 hm.

of the enemy to secure an opportunity of firing. When the distances were reduced and there was an opening for attack the enemy turned away and gave up the fight. At that time Admiral Beatty, leader of the English battle-cruisers, was not in command. From information received later, it appeared he had stayed behind on the *Lion*, and had then boarded a torpedo-boat to hurry after his ships, but did not reach them till they were returning.*

The spot where the *Blücher* was sunk is at 54° 25′ N. Lat., 5° 25′ E. Long. When Admiral Hipper decided to break off the fight he, according to his report, was guided by the conviction that it would be of no avail to send help to the already sinking *Blücher*, and in view of the enemy's superior strength would only involve us in further losses. The fighting had lasted more than three hours, and the *Seydlitz* had only 200 rounds of ammunition for the guns. The Naval Command fully recognised that no objection could be raised to the conduct of the forces in the battle, or to the tactical measures adopted, and also approved of the decision, hard though it was, to abandon the *Blücher* to her fate.

If our battle-cruisers, by turning round and risking the three remaining cruisers, had approached the *Blücher*, then unnavigable, they would have entangled themselves in the most unfavourable tactical position imaginable, as their own torpedo-boats would have been astern of them, while the enemy would have had his light cruisers and destroyers directly ahead, and could have used them for a torpedo attack. The result was, therefore, more than doubtful; there would probably have been heavy casualties without corresponding loss on the other side, and the *Blücher* could not possibly have been saved.

The enemy's behaviour obviously shows that it was his intention, relying on the heavier calibre of his guns, to carry on the fighting at the greatest distance, to knock out the central guns (15-cm.) of our ships, and above all to keep themselves beyond the range of our torpedoes. It would have been easy for him to draw nearer, as was proved when he steamed up so quickly. His superior speed enabled him to select the range at his own pleasure. In spite of superior guns and the more favourable position of the English line, their

* Admiral Beatty says in his report: "I followed the squadron with the utmost speed on the destroyer *Attack*, and met them at noon as they withdrew to the north-north-west. I went on board the *Princess Royal* and hoisted my flag at 12.20 P.M., when Captain Brock informed me of what had happened after the *Lion* fell out, how the *Blücher* was sunk, and the enemy battle-cruisers very much damaged had continued their eastward course." His report does not mention any reason for their not having pursued the damaged German cruisers.

firing in the protracted running fight was not very successful when we take into consideration that three of their ships each had eight 34-cm. guns and the two others each eight 30.5-cm. guns. Opposed to them on our side were two ships each with ten 28-cm. guns, the *Blücher* with twelve 21-cm., and the *Derfflinger* with eight 30.5-cm. It is not surprising that the *Blücher* was destroyed by gun-fire; her armour plating was not very thick, and, being the last ship of our line, most of the enemy's fire was concentrated on her.

However regrettable was the great loss of life on board the *Seydlitz* through the fire spreading to the munition chamber of each turret, a valuable lesson had been learned for the future in dealing with reserve ammunition, and it was applied in subsequent actions.

The unexpected presence of the English ships on the morning of the 24th leads to the conclusion that the encounter was not a matter of chance, but that our plan in some way or other had got to the knowledge of the English. The leader of our cruisers, seeing so many ships assembled, must have considered it extremely probable that still more forces were behind. Whether there was any other reason for such a concentration cannot be maintained with certainty. It may possibly be that it was connected with the conduct of the English on the 19th, or with preparations for a new action.

As we know from the English accounts, the *Lion* was not able to reach harbour under her own steam but was taken in tow by the *Indomitable* during the afternoon, and towed to the Firth of Forth. The question as to whether our flotillas that stood by the cruisers could have kept in touch with the enemy so as to attack at night must be negatived, as they would not have had sufficient fuel. As regards the flotillas assembled in the Jade, when the news of the encounter reached them the enemy was already so far ahead as to exclude the prospect of a successful night attack.

This first serious fight with large ships which the Fleet had had the opportunity of participating in proved that the fighting preparedness of the ships as regards the training of all on board was on a very high level, that the ships were handled in a correct and reliable manner, and that the serving of the guns, the signalling, and the transmission of orders from ship to ship during the fight, as well as the measures necessitated by leakages, had all worked admirably. Everywhere the behaviour of the crews was exemplary. The case of the *Seydlitz* (Captain von Egidy), from which ship, in spite of the fierce fire raging on board, the command of the whole unit was calmly maintained, deserves special emphasis.

CHAPTER VII

THE YEAR OF THE WAR 1915

ENTERPRISES at sea are doubtless in a greater measure dependent on chance than those on land, owing largely to the lack of reliable information of the enemy's movements and the rapidity with which a situation changes. Therefore absolute liberty of action is essential to the officer in command of the operation within the prescribed limit of the objective. The general aim of our Fleet may be summarised as follows: not to seek decisive battle with the entire English Fleet, but to test its strength against separate divisions.

If, however, the burden of responsibility of the officer in command is to be complicated by suggestions and instructions restricting the operation of his plans, the possibility of a successful result will be greatly lessened. The lack of tangible results from the various enterprises by the Fleet may be attributed to that cause, and no blame can be attached to the leader, whose whole character was a guarantee that, trusting implicitly to the powers of those under him, he would make a determined use of them.

When a change was made in the chief command early in February, 1915, the entire Fleet was unanimous in its regrets that the departing chief, Admiral von Ingenohl, who was highly esteemed and respected by all his officers, had not been able to obtain any great results.

The command was given to Admiral von Pohl, the former Chief of the Naval Staff. Whilst acting in the latter capacity, Admiral von Pohl had brought about the U-boat trade-war on England, which on February 4th was notified under the form of a declaration that the waters round England were to be included in the war zone. The use of the U-boat in this connection opened up a new field for the conduct of naval warfare and might prove of the greatest importance on the issue of the war. The necessity of resorting to it arose from the nature of the English method of conducting naval warfare, and will be discussed in detail in a later section.

The action of the Fleet under Pohl's leadership coincided with

87

the views held by him when Chief of the Naval Staff—that the maintenance of the Fleet intact at that stage of the war was a necessity. His plan was by frequent and constant advances of the entire High Sea Fleet to induce the enemy to operate in the North Sea, thus either assuring incidental results or leading to a decisive battle under favourable conditions to ourselves, that is to say, so close to our own waters that, even if the actual battle were undecided, the enemy's total losses, owing to the longer route home with his damaged ships, would be much greater than ours.

He therefore determined to make such advances with the strongest possible forces at every possible opportunity. There was to be no lack of important units, such as torpedo-boats or indispensable fighting vessels, whether battleships or battle-cruisers. The advances were not to be pushed farther forward than was compatible with the plan of fighting closer to our own than to the enemy waters; therefore, they did not extend a greater distance than could be covered in a night or a day. Owing to our shortage of cruisers, our scouting was inadequate and had to be supplemented by aerial means. Before any encounter with the enemy, and during an advance, every precaution had to be taken to prevent our being exposed to any damage from submarines; moreover, a careful search for mines, the driving away of enemy submarines from our coastal waters, precautions to be observed by torpedo-boats against submarine attacks, and the highest possible speed while under way, were all matters of the greatest importance. Besides the preparedness of the Fleet, fine weather was a primary necessity to the fulfilment of all plans of this kind, and thus it was not always possible to make a forward movement.

During the months of February and March, therefore, only two advances were made, while in the more favourable period of April and May there were four. But in none of these enterprises was there any encounter with the enemy. They were carried out in a westerly to north-westerly direction from Heligoland at a distance of about 100 to 120 nautical miles, thus presenting a considerably wider area for our airships, but they failed to locate the enemy. On May 18, during one of these advances, the light cruiser *Danzig*, when forty-five nautical miles from Heligoland, ran into a minefield, but was able to reach dock under her own steam.

The Year of the War 1915

Whenever the news of our putting to sea reached the enemy, as we gathered from his wireless messages and certain other means, he began to make a move, but he never left the northern part of the North Sea. The enemy thus left to us that area of the sea in which our movements took place, and we observed a similar method of procedure with regard to him, so that a meeting between the two Fleets seemed very improbable. If it was the enemy's object to entice us nearer to his coasts, he failed to achieve it; we did not favour him by adapting our course of action to suit his pleasure. Admiral von Pohl considered that a big surplus of forces was necessary for an offensive of that kind, and if it was available for the enemy it certainly was not for us.

Although there seemed little prospect of an advance for our Fleet, the Commander-in-Chief, in spite of the danger from submarines that it involved, never ceased in his reconnoitring efforts, for only by such means could efficiency in the navigation of the ships be secured and familiarity gained with the dangers of the submarines and mines.

The Commander-in-Chief was of opinion that the enemy would suffer most from the U-boat and mine-warfare; but after the U-boat trade-war was started, very few of those boats could be told off to seek out the English Grand Fleet. An advance of mine-laying steamers to the English bases in the north could only lead to needless sacrifice of the boats.

The auxiliary cruiser *Meteor*, under the command of Captain von Knorr, certainly made two successful trips, but the ship was lost on the second. At her christening she had received the name of the gunboat commanded by that officer's father, Admiral von Knorr, of many years' service, which in 1870 had distinguished herself in a fight with the French cruiser *Bouvet*, off Havana. On May 30 this new *Meteor* had gone to the White Sea, returning thence to Kiel on June 20, bringing in several prizes. In August a fresh expedition was made to the Moray Firth with the object of laying mines off this English naval station. Just as she had completed the greater part of this task, the *Meteor* was seen by an English guardship, the *Ramsay*, which was at once torpedoed and sunk. Captain von Knorr rescued four officers and thirty-nine of the crew, and then started to return. The sinking ship had managed to call up help, so that in the course of the next day the *Meteor* found herself encircled by English cruisers. The captain, with his crew and his prisoners, transferred just in time to a Swedish sailing-vessel and sank his ship. The enemy, on arriving, took no notice of the Swedish ship, and when a Norwegian vessel came along Knorr handed over the prisoners to her, as it would have been impossible for so many to remain on board the Swede for any length of time. The crew from the *Meteor* were taken on board a ship sent to meet them off Sylt.

The different advances made during the summer months did not impress the Fleet with the idea that any serious effort was contemplated of getting at closer quarters with the enemy and challenging him to action, although the addition made to the forces of the now complete Squadron III rendered us more than ever capable of chancing it. Even among the Naval Staff under Admiral Bachmann the opinion prevailed that the policy of holding back the Fleet was being carried too far. But no one there would issue an order that would involve greater risk than the Commander-in-Chief of the Fleet, guided by his acquired convictions, was himself inclined to run.

The restrictions enforced by the previous Command, not to expose the Fleet to serious losses in the gaining of a prescribed objective, had meanwhile been swept away. The Fleet Command had merely been notified that the necessary caution, by means of reconnaissance, must be observed in all enterprises, and that action should be broken off if unfavourable conditions arose.

So far as the Fleet was concerned, the general situation of the war had altered very much to our advantage through the successes achieved by the Army on the Eastern Front. For the Fleet the only object in the war lay now in fighting English power at sea, for there was no longer any question of a Russian landing on our Baltic coast.

The situation, indeed, had veered round directly opposite, and the question was whether we should threaten the Russians with a landing. Our squadron IV was therefore detached and sent to the Baltic at the beginning of June. It was composed of the ships of the "Wittelsbach" class, under the command of Vice-Admiral Schmidt. Scouting Division IV and Torpedo-Boat Flotilla VIII were also detached from the Fleet for the Baltic, and placed at the disposal of the Commander-in-Chief there. Any important naval action with a view to defeating Russia was, as already pointed out, quite purposeless. On account of the enormous area of that Empire, the cutting off of imports by sea could not inflict any mortal injury. Any maritime enterprises would be in the

The Year of the War 1915

nature of support to the operations of the army, by ensuring safety in the use of the more suitable sea route for the transport of troops and war material to the Gulf of Riga, or when the town itself was taken by the army, then to protect it against attacks from the sea.

The Russians, who had always shown great skill in their use of mines, had laid down masses of them in the Gulf of Riga. The removal of such a minefield to enable a division of our ships to enter the Gulf was a difficult undertaking. Meanwhile the occupation of Libau afforded a very desirable point of support. The actual forcing of the Gulf of Riga began early in August.

An opportunity was thus provided of proving whether England was willing to attempt an entry into the Baltic in order to assist her Allies. In that case we would be compelled to move our forces stationed in the east to the west portion of the Baltic. In anticipation of the necessity of quickly transferring large divisions of the Fleet to the Baltic, Squadron III was moved to the Elbe, whither the Commander-in-Chief of the Fleet, on board the *Friedrich der Grosse*, had also betaken himself so as to be ready if required to assume command in the Baltic.

But the English had no intention of altering their line of action; they continued to rely on the effectiveness of their barriers. They had withdrawn the line of guardships at the north entrance to the North Sea in the direction of the Faröe Islands, as a permanent patrol of the line from the Shetland Isles to the Norwegian coast was considered too dangerous owing to our U-boats. The loss of the cruiser *Hawke* and the attack on the *Theseus*, carried out by "U 29," under command of Lieut.-Commander Weddigen—afterwards killed—induced them to change their guard system and to depend chiefly on auxiliary cruisers. They had also succeeded in forcing neutral shipping to submit to examination at their naval base in the Orkney Islands.

The U-boat trade-war, which was thought to be our most effective counter-measure against the blockade, had started bravely, but owing to America's protests soon took on a very modest form. The obligation imposed on the U-boats, first to make sure whether they were dealing with neutral steamers or not, was bound inevitably to lead to many casualties on account of the misuse of flags by the English.

In the middle of July two more valuable boats, "U 23" and "U 36," were lost. The only survivor from the latter was a petty officer of the name of Lamm; he had been entrusted with the task of bringing in as prize to the Elbe the American ship *Pass of Balmaha,* bound for Archangel with cotton, which had been captured when going round Scotland. He succeeded in achieving this purpose, although on arriving at Cuxhaven we discovered, to our great surprise, that an English officer and four men were on board. They had been kept secure during the voyage by this one German petty officer, and were then handed over as prisoners.

Although in this instance the prize was brought in successfully, there was no general possibility of the U-boats being able to spare any of their crew to bring larger ships into a home port. On the west coast of the British Isles the U-boat trade-war entirely ceased from the middle of September.

In August our torpedo-boats notified a success during a night attack. On August 18, Flotilla II, Captain Schuur, returning from a reconnaissance trip, encountered north of Horns Reef an English flotilla consisting of one light cruiser and eight destroyers. The visibility conditions on our side were so excellent that, apparently unobserved, our craft approached the enemy to within 3,000 metres. Three torpedoes fired from the leading boat hit and sank the English cruiser and the destroyer next to her. The other destroyers made off, probably thinking they had got into a minefield.

A raid on London carried out the same night by the new airships "L" 10, 11 and 14, and the favourable news received from the Baltic theatre of war of the successful bombardment of the *Slava* in the Gulf of Riga and the destruction of several gunboats and torpedo-boats, added greatly to the day's success.

The Fleet's Baltic enterprise was broken off at the end of August, as at that time the Army had no troops available to support the entrance of the Fleet into the Gulf of Riga, and no importance was attached then to the possession of the town. Besides the desired opportunity of confronting the enemy with the Fleet, the investigation of conditions to be reckoned with in the conduct of the war, in relation to the conquest two years later of the Baltic Islands, was of importance. Apart from the loss of some of our lighter craft while engaged in searching for mines, the battle-cruiser *Moltke* was the only vessel to be damaged. She was hit in her bow by a torpedo. About 450 tons of water poured into the vessel, which, however, was able to pass through the Kaiser-Wilhelm Canal and make for a repairing dock at Hamburg, where the damage was made good in a few weeks' time.

The Year of the War 1915

Owing to a difference of opinion between the heads of the Navy and the Government concerning the conduct of the U-boat warfare, a change in the post of Chief of the Naval Staff was effected early in September. Admiral Bachmann returned to his former post as Chief of the station at Kiel, and was replaced by Admiral von Holtzendorff.

The latter, in January, 1913, resigned the command of the Fleet, which he had held during the previous three years. A brilliant officer, of a very active mind, and possessed of great eloquence and personal charm, he was a splendid seaman, who, by virtue of the varied positions he had held, could look back upon an unusually long spell of foreign service. He was a chivalrous and amiable personality in whose character courtesy was a prevailing quality. It was a known fact, however, that he was not on friendly terms with the State Secretary, Grand-Admiral von Tirpitz. Strained relations between the two persons who were called upon to control the Navy in time of war could not serve to further the cause, in spite of the best intentions of both parties, which one was bound to assume. The association, however, involved no change in the fundamental views respecting the Fleet's duties in the war.

In September the Fleet again advanced in the direction of the Hoofden, and at the same time mines were laid by the light cruisers. Fresh minefields were discovered which in part were noticeable from having mines attached near the surface.* These new mines lay in the centre of the arc from Horns Reef to Borkum. Taken in connexion with previously discovered minefields closer inland off the North and East Frisian Islands, the conclusion was arrived at that the English purpose was to encircle that part of the Bight with mines. According to reports from steamers, a number of English ships, five large ones among them, had been met with the day before in that district.

The opening of the great Anglo-French autumn offensive on the Western Front, together with the news that the English Fleet was also taking part, kept our Fleet in a perpetual state of tension, although no opportunity was offered for any action, as the report proved to be untrue.

In September our airships "L" 1, 11, 13, 14, 15 and 16 were able

---

* These are mines which through some error in reckoning the depth of the water, instead of reaching the desired depth below the surface, have floated up again sufficiently to be seen, and are therefore the more easily avoided.

to carry out a very effective raid, when all the airships reached London and returned safely in spite of very strong counter-action.

Greater activity on the part of the Fleet during the autumn months was prevented owing to the ships of Squadron III being forced to remain for a long time in dock in order to be fitted with range-finders and because, on most of the ships, the bearings of the screw shafts had to be renewed. Owing to the ships lying so constantly in the sandy waters of the Jade basin they had suffered far more than was the case in peace time, when they were either out in the open sea or were lying in the clear and calm waters of the harbour.

In October the Fleet attempted an advance in the usual way and in a northerly direction, but did not get beyond the latitude of Horns Reef, where the wind rose so high as to make aerial reconnaissance and the use of torpedo-boats doubtful, and the enterprise was broken off.

During the months of November and December the separate units in turn were given opportunity for gun practice in the Baltic. This break in the monotonous outpost duty on the Jade was a very welcome one for the crews, although it by no means signified a lessening of daily duty, as the time at their disposal had to be used to the utmost advantage so as not to prolong unnecessarily their absence from the North Sea. Whilst at Christmas a short frost set in, opening a prospect of carrying out the Commander-in-Chief's plan of an enterprise into the North Sea, the weather soon changed again, and lasting well into January there ensued a period of bad weather that prevented expeditions of any kind, even searches for mines.

On January 8, 1916, Admiral von Pohl was taken seriously ill and transferred to a hospital ship, whence he was later conveyed to Berlin for an operation. He never recovered and died on February 23. In him the Navy lost an officer of quite exceptional steadfastness and devotion to duty; one who exacted much from himself and who was entirely wrapped up in his calling. As commander and squadron chief he distinguished himself by sea-manlike assurance in manœuvres and a correct grasp of the tactical situation, so that under his leadership in battle the best results might have been expected. His highest ambition was to live to see it, but it was not to be granted to him.

Thanks to the confidence in me of the All-Highest War-Lord, I was appointed to deputise for Admiral von Pohl, and represented

The Year of the War 1915

him until my formal appointment as Commander-in-Chief of the High Sea forces was issued on January 18.

In the first place I procured information respecting enterprises that were proceeding and those in prospect. The bad weather prevented their being carried out, with the exception that one steamer bound for East Africa undertook, without any escort, to set out on the voyage, hoping to get through best under those conditions.

The auxiliary cruiser *Moewe*, commanded by Count zu Dohna, had shortly before, with U-boat escort, safely reached the North Atlantic, but we had had no further news of her.

I begged to have Captain von Trotha, commander of the battleship *Kaiser*, as my Chief of Staff, and Captain von Levetzow, commander of the large cruiser *Moltke*, as Chief of the Operating Division. I had previously served for a long time with both those officers, and they both declared themselves ready to take over their new duties. It appeared to me an important matter to make a change in these posts and thus show the Fleet my complete independence in that and all matters. All other members of the Fleet Staff retained their positions. They comprised: Captain Hans Quaet-Faslem, Captain Dietrich Meyer, Lieut.-Commander Heuinger von Waldegg as Admiral's Staff Officer of the Operating Division; Captain Paul Reymann, Admiral's Staff Officer for torpedo- and U-boats; Captain Walther Franz, Admiral's Staff Officer for Artillery; Captain Wilke, Fleet Navigation Officer; Captain Bindseil, Flag-Lieutenant for Wireless Telegraphy; Lieut. Commander Weizsacker, Flag-Lieutenant; Chief Naval Engineer Schutzler, Fleet Engineer; Chief Naval Surgeon Dr. Cudden, Fleet Surgeon; Chief Naval Chaplin Klein, Catholic Priest to the Navy; Stollhof, Chief Councillor to the Navy; Cöster, Staff Paymaster to the Navy; Paul Wulff, Secretary to the Fleet. I feel deeply indebted to all these gentlemen for the devoted and untiring assistance they rendered to the Fleet and to myself in their respective posts.

My very special gratitude is due to Rear-Admiral von Trotha, my Chief of Staff, on whose prudent and circumspect judgment I invariably relied. He supplemented in the happiest manner the keen and eager leader of the Operating Division, Captain von Levetzow. They were both upright men with independent views based on much learning, who stood by their opinions, were closely linked in faithful comradeship, and formed a circle to which I look back with pride and gratitude.

CHAPTER VIII

PREPARATIONS FOR INCREASED FLEET ACTIVITY

AFTER taking over command of the Fleet my first and most important task was to draw up a plan for the future tactics of the High Sea Fleet and to work out a programme of operations. The success hitherto of the conduct of naval warfare lay in the effect produced by the existence of the Fleet, the coastal defence, the influence exercised on neutrals, and the support given to the Army. The conviction that English maritime power was a serious menace to our capability of resistance seemed to make it imperative that, if a successful issue of the war were to be expected, it must be waged far more energetically against that adversary. There was no question of England giving in unless she was made to feel the pressure of war at home much more forcibly than hitherto had been the case. After having carried out her transport of troops from overseas on a much larger scale than was anticipated, thereby imposing on the country great sacrifices both of money and men, her determination for war was bound to increase, in order to reap the benefits of those efforts and to compensate for the blunders made, such as the surrender of Antwerp and the abandonment of the Dardanelles enterprise. So far, the war, for England, was merely a question of money and men. There was no lack of either, thanks to the support from the Colonies, the systematic manner in which volunteers were pressed into Kitchener's Army, and the ruthless employment of coloured auxiliaries.

Thus England was better able to stand the war for a lengthened period than we were, if the hunger blockade were to continue to oppress the country. The English public never thought of urging the Fleet to more active warfare; its object was achieved without its being weakened or being forced to make undue sacrifices. The nation readily understood this, especially when it was made clear to them that the Fleet had succeeded in keeping open those overseas communications on which the country was so dependent. This fact was specially brought into prominence by the destruction of our cruiser squadron off the Falkland Islands.

Preparations for Increased Fleet Activity

The danger from the U-boat warfare, which at first appeared serious, was reduced to slight importance owing to the mutual interests with America. But when the danger really was recognised, England prepared to ward it off, and did splendid work in this connection. On our part, the conduct of naval warfare in 1915 was less satisfactory. Even though we succeeded in preventing the neutrals from joining our opponents, it always remained an open question to what cause it could be attributed. If the utility of our High Sea Fleet were not made more distinctly manifest, then its deeds were not sufficient to justify its existence and the vast sums exacted from the resources of our people for its maintenance. The principal task stood out clearly defined—to punish England in such a way as to deprive her speedily and thoroughly of the inclination to continue the war. That might be expected if success could be achieved either by a blow at her sea power centred in her Navy, or at her financial life—preferably both.

The continued numerical superiority of the English Fleet from the beginning of the war kept us at a disadvantage; but, from a purely tactical point of view, our Battle Fleet, by the addition of four ships of the "König" class, was very differently organised from formerly, when Squadron II had to form part of the fighting line and was confronted in battle with "Dreadnoughts" with which it could not possibly cope. From the beginning of 1915 we also had had a double squadron of "Dreadnought" ships at our disposal (Squadrons I and III) and were therefore better able to avoid bringing the ships of Squadron II into a situation in battle where they must inevitably have suffered losses. Certainly the English had added greatly to their fighting powers by ships of the "Queen Elizabeth" class which must have been ready early in 1915. They carried guns of 38-c.m. calibre, were strongly armour-plated, and had a speed of 25 knots, in all of which they were nominally the same as our battle-cruisers, whereas in the strength of their attack they appeared to be vastly superior to all our vessels.

The then prevailing conditions of strength kept us from seeking a decisive battle with the enemy. Our conduct of the naval war was rather aimed at preventing a decisive battle being forced on us by the enemy. This might perhaps occur if our tactics began to be so troublesome to him that he would try at all costs to get rid of the German Fleet. It might, for instance, become necessary, if the U-boat war succeeded again in seriously threatening English economic life. Should the English thus manœuvre for a decisive battle, they could fix the time so as to allow the full use of their vast superiority, whereas some of our ships would be either under repair or otherwise unfit for service, or absent in the Baltic for exercises, of which the enemy would be well informed.

But for us to get into touch with the English Fleet, a definite and systematic operation would have had to be carried out with the view of compelling the enemy to give up his waiting tactics and send out forces which would provide us with favourable conditions of attack. The methods previously employed had failed. Either they were undertaken with inferior forces—in the case of an advance by cruisers the Main Fleet could not intervene in time to be of use—or else, as in most of the 1915 enterprises, they had not pushed on far enough for an encounter with important units of the enemy Fleet.

If we wished to attempt an effective and far-reaching offensive, it was necessary that we should be masters in our own house. The waters of our coast must be so controlled that we could be left free to develop and have no fear of being surprised and called out against our will. With the exception of the Fleet, we had nothing that the enemy could attack, as unfortunately our maritime trade had been put down from the beginning. The enemy, however, was still vulnerable in so many places that it was surely possible to find ample opportunity to make him feel the gravity of the war!

The ways and means of effecting this were the U-boat trade-war, mines, trade-war in the North and on the open seas, aerial warfare and aggressive action of the High Sea forces in the North Sea. The U-boat and aerial warfare had already started; the three other factors were to be operated in combination. The activities in the near future were laid down in a programme of operations submitted to the Naval Staff and their general sanction obtained. Above all things, every leader, as well as each commander, was to be told his part, so as to facilitate and encourage independent action in accordance with the combined plan.

The first and most important task was the safety of the German Bight. Fresh rules were laid down dealing with the action of the Fleet when in the Bight, and instructions issued concerning protection and outpost duty. Arrangements were also made as to action under an enemy attack which would save waiting for lengthy orders in an urgent emergency, and would render it possible for all subordinate officers to play the part expected of

Preparations for Increased Fleet Activity

them in such an event. The aim of the organisation was to keep the Bight clear by means of aeroplanes, outpost flotillas, mine-sweeping formations, and barrier-breakers, and regular reconnaissance, guard, and mine-searching service was established. The outpost-boats were to form a support for the active protective craft in the North Sea, be sufficiently strong to meet a surprise hostile attack and always ready to pick up at sea any forces returning to harbour. The command of the protective services was, as hitherto, retained by the Chief of the Reconnaissance. The actual aerial reconnaissance in the vicinity was undertaken by aeroplanes and airships from the stations at List—on the island of Sylt—Heligoland, and Borkum. The North Sea Outpost Flotilla, the Coastal Defence Flotilla from the Ems, and boats of the Harbour Flotilla, were ready for guard service; their duties consisted chiefly in driving away enemy submarines. As a rule, the following positions were occupied:—

The List Group: The waters off List (to keep neutral fishing boats out of German coastal waters).

The North Group: The Amrum Bank passage.

Line 1: The Heligoland—Hever Line.

Line 2: The Heligoland—Outer Jade Line.

The Outer Group: The Jade—Norderney line and the barrier opening at Norderney.

Heligoland boats: North and West of the island.

Jade boats: Off the Jade.

S. Group: Three boats (chiefly intended to fight enemy submarines, or for other special duties, as, for instance, the cutting of cables).

The chief object of these outpost boats was to search the Inner German Bight for enemy submarines, for which purpose they set out every day in groups from the lines where they were stationed. The service of the outpost boats, some eighty fishing steamers, was so arranged that half were on duty for three days and then had three days off. The Ems Coastal Defence Flotilla had the guarding of the waters off East and West Ems, and westward to about 6 degrees E. Longitude. The Harbour Flotilla boats joined in when there was a chase after submarines which might have shown themselves at the mouth of the river. A torpedo-boat flotilla stationed on the Ems also did duty when required, and helped further to ensure the safety of the sea area off the Ems.

Germany's High Sea Fleet

The regular mine-sweeping service consisted of two Mine-sweeping Divisions and one Auxiliary Flotilla. The latter was composed of vessels that had only been requisitioned for the purpose since the war, and were mostly trawlers. They were specially suitable for the North Sea, owing to their sea-going qualities, but we lost several of them because of their deep draught. When a mine-sweeping division was off duty the crews were billeted at Cuxhaven, while the men on leave from the half-flotilla of the Auxiliary Flotilla were quartered at Wilhelmshaven. For their mine-sweeping duties all those boats were armed in order that they might be prepared to fight submarines. As soon as hostile submarines were sighted within their area they had to stop their mine-sweeping and take part in the chase.

The barrier-breaker service consisted of three groups of barrier-breakers to every four steamers, internally constructed so as to enable them to keep afloat should they strike a mine. At the outbreak of war we had no apparatus whatever for sweeping up mines and protecting the steamers. Every effort was made, however, to invent such an apparatus, which, as soon as it had been tested, was at once supplied to the barrier-breakers and the mine-searching division.

Special credit is due to Captain Walter Krah, Chief of the Auxiliary Mine-Flotilla, who profited by much practical experience and was successful in his efforts to avert unnecessary losses in his flotilla. It was the duty of the barrier-breaker group to protect the navigation of certain channels, chiefly those where our mine-laying divisions had been at work, and to make sure that no mines had been laid by the enemy in the interval. The activity of our mine-layers could not be entirely concealed from the enemy. When they were working in the inner section of the Bight, the enemy submarines had every opportunity for their observations, and the farther the mine belt was pushed out in the North Sea the nearer it drew to the area of English observation planes. The English were in advance of us with the Curtis plane, a hydroplane which, even with a considerable sea running, was able to keep on the water and so husband its strength.

The search for, and the chasing of, hostile submarines was principally the business of the torpedo-boats. The outpost boats had, of course, to keep a look-out for submarines, and had to follow up on any occasion when there was a chance of fighting them; but they had not the speed, nor were their numbers sufficient to carry out a systematic search and pursuit. The torpedo-boats of the outpost service were told off for that purpose. The same flotillas were employed when there was a question of warding off the submarines from any ships or units engaged in special enterprises. The safety of the German Bight at night—to ensure which the guardships were too far apart besides being inadequate in armament—was further ensured by torpedo-boat patrols taken from the outpost service which cruised along the line of guardships and the shores of the German Bight.

In order not to keep the entire Fleet constantly under steam and thus overtire both men and machinery and use up material to no purpose, and yet provide that they should be ready with considerable forces for any enemy enterprises, an outpost service was organised. In the Jade there lay always in readiness a squadron of battleships, two battle-cruisers, a cruiser-leader of torpedo-boats, and a torpedo-boat flotilla; a scouting division of light cruisers were in the Jade and the Weser, a torpedo-boat flotilla in Heligoland harbour, half the ships of Squadron II in the Cuxhaven Roads, at Altenbruch; and, if sufficient torpedo-boats were available, another torpedo-boat half-flotilla was stationed on the Ems or in List Deep (at Sylt)—constituting approximately half the total forces of the High Sea Fleet. The ships were kept clear to put to sea from their station three-quarters of an hour after the order reached them. The torpedo-boat flotilla at Heligoland was held ready to run out immediately, and the flotilla on the Jade three-quarters of an hour after receipt of an order.

The outpost forces were under the command of the Senior Naval Commander stationed on the Jade. In the event of a sudden enemy attack, his duty was to arrange independently for the necessary measures of defence, and to exercise the command. If the High Sea Command had any duty to assign to him, it was restricted to a suggestion or general directions, as, for instance, to station torpedo-boats at such and such a place at daybreak, leaving the rest to be carried out by the Chief of the U-boat forces.

The other ships, not belonging to the outpost service, lay, half of them in harbour (about a fourth part of the fleet), the other half remaining on the inner roads at Wilhelmshaven or Brunsbuttel. The torpedo-boats off outpost duty were always allowed to enter harbours. The ships off duty had to seize that opportunity to carry out such necessary repairs as could be done by their own men; ships that had been long in dock for refitting and repairs were regarded as being in the same category. The usual preparedness of ships lying in the inner roads and in harbour was fixed at three hours. But whenever news came which seemed to necessitate the calling out of the ships, orders were issued to hurry the preparations, the entire crew remained on board, and the ships kept ready, on receipt of further orders, to weigh anchor at once.

These far-reaching measures for the protection of the German Bight were, above all, intended to ensure that the Fleet should be able to take up a position in line if it was deemed advisable to pass out in expectation of an enemy attack. General regulations were issued for two eventualities; the one, in case information and messages were received announcing an impending hostile attack, and the other, in case the enemy came entirely unexpectedly. It was not long before there was an opportunity to test them in practice.

Finally the defence resources of the German Bight were improved by adding to the already existing minefields, partly by laying them adjacent to those laid by the enemy, which he was forced to avoid. The intention of establishing a safe area for assembling within the line Horns Reef—Terschelling, was soon carried out, as the enemy laid his minefields in still further concentric rings outside that line.

Constant navigation and firing exercises by separate units as well as by the assembled Fleet, were carried on within this zone, and they were very rarely interrupted by an alarm of submarines. The dispatch of units for practice in the Baltic was no longer so necessary, and the readiness of the Fleet for action was perceptibly improved.

Heligoland, which at the beginning of the war was our advanced outpost, had thus assumed the character of a point of support in the rear, from which radiated a free zone extending over a radius of 120 nautical miles. Unfortunately the island never had occasion to use her excellent armament on the enemy. But the newly constructed harbour was of great service to the light forces of the Fleet, besides which the possession of the island was indispensable in order that a fleet might be able to leave our estuaries.

Even though security from enemy attacks was necessary and called for immediate action, nevertheless a still more important duty was that of attacking and injuring the enemy. To this end various enterprises were started. Foremost among these were nocturnal advances by light forces in the boundary area of the German Bight in order to destroy enemy forces stationed there, the

Preparations for Increased Fleet Activity

holding up of suspicious craft, and readiness to afford help to airships raiding England, which always took place at night. These advances were carried out by several flotillas led by an escorting cruiser. They were supported by a scouting division of light cruisers sent either to the Ems or to a certain quadrant in the North Sea. The battle-cruisers were told off to the Schillig Roads, or deployed in line at sea; all other outpost ships were held in strictest readiness, and all measures were taken to ensure the speedy intervention of vessels lying in the Roads. In this way, the entire Fleet was kept in a certain state of tension, and unvarying alertness in view of eventualities at sea was maintained in order to be prepared at once to take part in the proceedings.

A further system of enterprise was to prolong these nocturnal sallies till daybreak in order to patrol a more extended area, in which case the entire Fleet had to be at sea as a support. The furthest advanced flotillas received support from Scouting Divisions I and II which, reinforced by one or two flotillas, followed them at a suitable distance. The extension of such enterprises was designed to reach to the Skagerrak and the Hoofden.

Finally, other important enterprises were planned, such as the bombardment of coastal towns to exercise a still greater pressure on the enemy and induce him to take counter-measures which would afford us an opportunity to engage part or the whole of his Fleet in battle under conditions favourable to ourselves.

In all these enterprises the co-operation of the Naval Corps in Flanders was desirable by stationing their U-boats along the nearest stretch of coast and thereby supporting the Fleet. This was carried out regularly, and with the greatest readiness.

The employment of our U-boats was of fundamental importance for our warfare against England. They could be used directly against English trade or against the English naval forces. The decision in the matter influenced the operations very considerably. It was not advisable to embark on both methods simultaneously, as most probably neither would then achieve success. Also the poor success resulting from our U-boat action on English warships in the North Sea seemed to point to a decided preference for trade-war. In military circles, there was no doubt that success in trade-war could only be looked for if the U-boat were empowered to act according to its own special methods; any restrictions in that respect would greatly reduce the chances of success. The decision in the matter lay in the political zone. It was therefore necessary that the political leaders should recognise what we were compelled to do to achieve our war aim. Hitherto our politicians, out of anxiety with regard to America or in order not to exasperate England to the utmost, had not been able to decide on energetic action against England in the naval war. The naval authorities, however, should have known what they had to reckon with in order to be able to beat down England's resistance. It was also their duty to protest against enterprises specially unsuited to the U-boat, which inevitably led to useless sacrifices.

The restricted form of U-boat warfare against English mercantile ships, adopted in the course of 1915, was extremely unsatisfactory. The damage caused thereby to her trade could be borne by England, and, on the other hand, the only result to us was vexation and disappointment, for our Fleet could obtain no support for its own enterprises from the U-boats. Co-operation with separate units or with the entire Fleet could not be sufficiently well organised to prove dependable for certain operations. First of all, only temporary co-operation was possible in the case of enterprises by the Fleet and attacks by the U-boats when each unit had a special duty, to be mutually supplemented but without exacting any tactical union. If, for instance, there was the intention to bombard a certain coastal town, it might be assumed that English fighting forces would at once rush out from different harbours where they were lying to drive off or capture the disturbers of their peace. If U-boats had been stationed off such towns, where it was presumed there were enemy ships, they would probably have had a chance of attacking.

Tactical co-operation would have been understood to mean that on the Fleet putting out to sea with the possibility of encountering the enemy, having the fixed intention of leading up to such an encounter, numbers of U-boats would be present from the beginning in order to be able to join in the battle. Even as certain rules have been evolved for the employment of cruisers and torpedo-boats in a daylight battle to support the activity of the battleship fleet, so might an opportunity have been found for the tactical employment of the U-boats. But no preliminary work had been done in that respect, and it would have been a very risky experiment to take U-boats into a battle without a thorough trial. The two principal drawbacks are their inadequate speed and the possibility of their not distinguishing between friend and foe.

The first-mentioned method, however, offered the most varied possibilities, and consideration was given as to what would be the most desirable way to station U-boats off enemy harbours; how they could be used in the form of movable mine-barriers, as flank protection, or otherwise render assistance.

In order to gain assurance in the use of U-boats and secure a basis for the activity of the Fleet, I went, in February, to Berlin to a conference with the Chief of the Naval Staff, in which Prince Henry of Prussia, Commander-in-Chief of the Baltic forces, also took part.

The result of this conference was the decision to come to close grips with England. Our chief maritime elements were to be centred absolutely in the North Sea, and the greatest restriction put on all active measures in the Baltic. Shortly afterwards an unrestricted U-boat warfare was to be instituted and the Naval Command was to make the necessary preparations. March 1 was the date on which it was intended to begin, as General von Falkenhayn, Chief of the General Staff, recognising the importance of England's contribution to the hostile resisting forces, had given up his previous scruples concerning the unrestricted U-boat warfare.

On January 31, nine airships set out for an attack on England: "L" 11, 13, 14, 15, 16, 17, 19, 20, and 21. On this occasion Liverpool was reached for the first time, where doubtless large quantities of war material from America were stored. Several other large factory towns in central England were also bombed, which they hardly expected although they were of great importance on account of their output of munitions and war material. A cruiser on the Humber was hit and badly damaged, which, according to subsequent information, was the new light cruiser *Caroline* (3,810 tons). The farther such air-raids spread over the country the greater would be the efforts made to defend the most important places, involving the withdrawal from the principal scene of war of guns, airmen, gunners and munitions to protect England from danger from the air.

Although the chief objective of every air-raid was London, where the Admiralty controlled the whole Naval war, and where the docks and the mouth of the Thames represented many other important objectives, to destroy which was highly necessary for the continuation of the war, still wind and weather did not always allow of its being attained. Sometimes during the flight of the airships they would be obliged to deviate from their plan of attack for other reasons than wind and weather. Therefore all airships that went up were given a general order to attack England in the south, centre and north. "South" signified the Thames, "centre," the Humber, and "north," the Firth of Forth. These three estuaries were the main points of support for the English Fleet, and were amply provided with all kinds of naval and mercantile shipbuilding works. The direction of the attack, whether south, centre or north, was determined by the wind, as the airships usually had the wind against them in going, in order, on the return journey, to have it behind in case they had to cope with damage or engine trouble.

Commander Odo Lowe, "L 19," never returned from an attack made during the night of January 31—February 1, 1916. On the return journey the airship, owing to fog, found itself over Dutch territory, and was fired at, not being at a very great altitude. Owing to the damage done, when it again came over the water it was unable to rise on account of a strong northerly wind, and so was forced to come down at about 100 nautical miles from the English coast, in a line with Grimsby. It was seen there in a sinking condition by a steam trawler (*King Stephen*) which, although within hailing distance, allowed the helpless crew to perish in the waves. This shameful deed was publicly acclaimed by an English bishop— a strange manifestation of his Christian principles! The behaviour of that bishop is so typical of English mentality that it is worth while adding a short comment on it. Two points are invariably and entirely lacking in English views on the war: they never admit the "necessity of war" for their opponent and never recognise the difference between unavoidable severity and deliberate brutality. The Englishman thinks it quite justifiable to establish a blockade in the North Sea which exposes his naval forces to a mere minimum of danger, and pays no heed to the rules of International Law. That the consequence of the blockade was to bring starvation on the entire German nation—the step indeed was taken with that avowed purpose—does not in the least affect his feelings for humanity. He employs the means that serve his war aims, and no objection could be raised did he allow the same to hold good for the enemy. But instead—whether in conscious or unconscious hypocrisy is an open question—he raises indignant opposition to all counter-measures. Our air-raids caused injury to civilians. It was inevitable, when institutions serving war purposes were so close to populous districts—perhaps with a view to secure protection for them. To the Englishman, it was of no moment that the airship crews exposed themselves to the greatest personal danger in thus fighting for their suffering Fatherland. Accustomed as he was to carrying on a war with hirelings, and mostly abroad, he considered any personal encroachment on his comfort as a crime against humanity and made a terrible ado to increase favour for his cause. English behaviour in the mine-war is an example of this.

At the second Peace Conference at the Hague, Satow,* an English delegate raised a violent protest against a decision authorising the laying of mines in the open sea, in view of the danger to neutral shipping. In spite of this, an extensive area at the eastern egress from the Channel was mined by the English. Success in war, in their view, stood higher than their former principles and professed consideration for the neutrals. Our mine-warfare along the English coast was cried down as a terrible crime, although it is distinctly allowed by the Hague regulations. The same hypocrisy concerning what the rights of war conceded to the one or the other belligerents is also prevalent in English professional literature. According to English ideas it is quite right and correct that the English Fleet, in spite of its double numerical superiority, does not consider it necessary to advance to the German coast. But when the weaker German Fleet refrained from committing what obviously would have been military errors, it was ascribed to lack of courage. When the English Fleet, as we see later, had in battle, in spite of twofold superiority, twice as many losses as the weaker adversary, it was still termed an English victory! What has become of all common sense? After this digression we must resume.

One of the first enterprises on the newly-drawn-up programme of operations was an encounter during the night of February 10—11 with English guardships off the Dogger Bank; they were in all probability stationed there in connection with our airship raids, either to give warning of their approach or to give chase on their way back. Torpedo-Boat Flotillas II, VI and IX, led by Captain Hartog, the First Leader of torpedo-boats, while patrolling at night came across a new type of English vessel which they at first took to be a cruiser, but finally decided it was a new vessel of the "Arabis" class. After a brief exchange of shots, the vessel was sunk by a torpedo; the commander, some officers and 28 of the crew were saved and taken prisoners. A second ship was also hit by a torpedo and observed to sink. The ships had only recently been built, were of 1,600 tons, had a crew of 78, slight draught, and a speed of 16 knots.

On February 11 an order from the Chief of the Naval Staff

* Right Hon. Sir E. Satow.

was sent to the Fleet regarding the action to be observed towards armed enemy merchantmen. At the same time a "Note from the Imperial German Government on the Treatment of Armed Merchantmen" was published in the Press. This Note contained uncontestable proof, gathered from instructions issued by the English Government, and various other sources, that the armed English merchantmen had official orders whenever they saw and were close to German U-boats, maliciously to attack and wage ruthless war on them. The Note concluded with the following notice:

Berlin, February 8, 1916.

"1. Under the circumstances now prevailing, enemy merchantmen carrying guns have no longer the right to be considered as such. German naval forces will therefore, after a short respite in the interests of the neutrals, treat such ships as belligerents.

"2. The German Government notifies the neutral Powers of the conditions in order that they may warn their subjects to desist from entrusting their persons and property to armed merchantmen belonging to the Powers at war with the German Empire."

The order to the naval forces, which out of consideration for the neutrals was not to come into force until February 29, was as follows: "Enemy merchantmen carrying guns are to be considered as warships and destroyed by all possible means. The officers must bear in mind that mistaken identity will lead to a breach with the neutrals and that the destruction of a merchantman because she is armed, must only be effected when the guns are clearly distinguished."

This new announcement from the Government, in which the Chief of the Naval Staff evidently had a share judging from his order to the Fleet, came as a surprise to me and appeared as though it were a reversal of the policy of unrestricted U-boat warfare which on February 1 had been promised for certain within a month, and which it now seemed doubtful would be carried out. The order imposed upon the officers that they should "distinguish" the guns made action very difficult for the U-boat officers, and it was they who were chiefly concerned. For at the distance necessary to secure this evidence the enemy, if he vindictively opened fire, could hardly miss. But if a U-boat when submerged were in a position to attack a steamer, and could only fire a torpedo when there

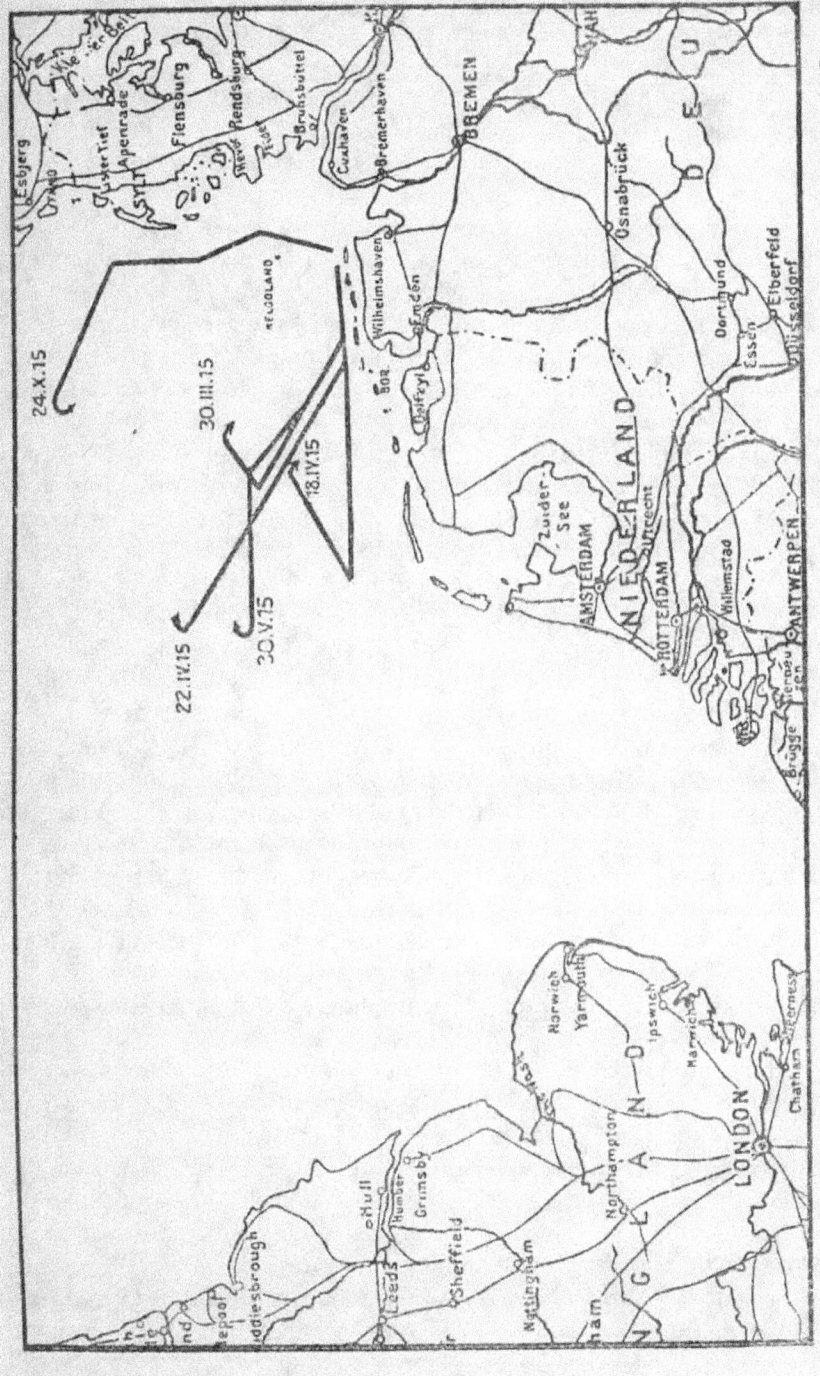

OPERATIONS OF THE HIGH SEA FLEET IN THE NORTH SEA DURING MARCH, APRIL, MAY AND OCTOBER, 1915

was no doubt that she carried guns, the opportunity would almost always be lost.

I made known my objections to the order, both verbally and in writing, when I had occasion in the course of the month to go to Berlin; a violent north-westerly storm, which set in on February 16—17 had stopped all operations in the Fleet. I was informed that the intention shortly to open the unrestricted U-boat warfare still held good. An order to that effect sanctioned by the Emperor was already drawn up; it merely remained to fill in the date of starting. This appeared to me of the greatest importance, and, as meanwhile the Emperor had announced his intention of visiting the Fleet on February 23, I took that to mean that I need no longer entertain any doubts.

On the appointed day, at 10 A.M. His Majesty went on board the flagship *Friedrich der Grosse* lying in dock at Wilhelmshaven ready to put to sea. Besides his own personal suite, he was accompanied by Grand Admiral Prince Henry of Prussia, the Secretary of State of the Imperial Naval Department, Admiral von Tirpitz, and the Chief of the Naval Staff, Admiral von Holtzendorff. It was the second time during the war that the Emperor had visited the Fleet. A little more than a year before the Emperor had introduced to the Fleet my predecessor in command. I had occasion to give a long report on the situation in the North Sea and also to express my opinion on the conduct of the war, for which I took as basis the matter of the unrestricted U-boat warfare. The Emperor agreed with my statements and interpreted them to a meeting of admirals and officers, when he spoke in laudatory terms of the activities and deeds of the Navy during the previous year and gave an explanation of the orders that had caused the Fleet to be held back. His Majesty then took the opportunity to remark that he fully approved of the order of procedure submitted to him by the Commander-in-Chief of the Fleet.

This announcement was of great value to me, as thereby, in the presence of all the officers, I was invested with authority which gave me liberty of action to an extent I myself had defined. The intentions of the Commander-in-Chief of the Fleet were thoroughly understood in this circle, as I had discussed in detail the programme of operations and had handed it in writing to those whom it concerned. The date for beginning the unrestricted U-boat warfare was, however, still uncertain. When I put the question to him, the Emperor remarked that he could not be influenced by the military suggestions only, though he recognised that they were justified, as besides his position as Chief War Lord, he was also responsible as Head of the State. Were he now to order the unrestricted U-boat war, it would probably meet with approval in the widest circles, but he must be careful that the entry into war of America on the side of the enemy did not give rise to consequences that might outweigh the advantages of unrestricted U-boat warfare.

When convinced that this decision could not then be altered, and not knowing what the political counter reasons were, and since it was the business of the Naval Staff to come to an understanding with the Imperial Administration, I selected two U-boats to test the effect of the war under the new conditions in the war zone off the west coast of England, in order that a judgment might be formed for further plans. The commanders of the boats, "U 32" (Baron von Spiegal) and "U 22" (Hoppe), gave me a verbal report on their return on March 18. "U 22" had sunk four steamers, with about 10,000 tons of cargo, three times as many neutral ships, but had been forced to let two passenger steamers get through. Owing to bad weather and damages "U 22" had no success. Meanwhile other U-boats were on the way to operate with the same intent. The success of their activities had not then been reported.

A wireless message on March 3 from the auxiliary cruiser *Moewe* was a surprising and joyful piece of news. She reported being stationed south-west of the Norwegian coast, and asked to be enrolled in the High Sea forces. This opportunity of practically testing the newly established outpost service was most opportune. It was a point of honour for the Fleet to preserve the intrepid and successful raider from a disastrous end off a home harbour. But great was the anxiety, as the *Moewe* had reported clouds of smoke sighted in clear weather, and evidently, from the appearance, belonging to a group of warships, but the distance was then too great to proceed to her assistance. The enemy, however, did not turn his attention to our cruiser, which endeavoured to give off as little smoke as possible, and when night fell had not been molested. Acting on the warnings given her, she happily escaped the further danger of striking any of the numerous English mines, that were unknown to her, between Horns Reef and Amrum Bank. Such ample protection was afforded her by a dense fog that she passed our first outposts unnoticed. But the fog lifted at the right moment to allow the ships sent out to meet her to escort her in triumph into the Jade, where she received a splendid welcome. The prudent and resolute behaviour of the commander, Count zu Dohna, his firm belief in the success of his undertaking—lightly called "luck" by some, though really based on the intrepid courage of the man, which spread to the entire crew—did not fail to make a deep impression on all of us who, on the first evening after his return, listened to the vivid description of his adventures. From January 1 to February 25, 1916, the *Moewe* captured 15 steamers, of a total tonnage of 57,835 tons. The first news received of her activities was the arrival of the prize *Appam*, under Lieut. Berg, of the Naval Defence, at Norfolk (Virginia), carrying the crews of the seven steamers sunk up to then. Further news of the *Moewe's* activities was the announcement that the steamer *Westburn*, under command of Badswick, had been taken into Teneriffe. After taking his 200 prisoners into port, he sank his ship the next day before the eyes of the English armoured cruiser *Sutlej* that was lying in wait, so that the prize might not fall into her hands. And now the *Moewe* had got back to us in safety! We considered her most important success to be the sinking of the *King Edward*, the flagship of the 3rd English Battle Squadron, which, on January 3, struck a mine laid by the *Moewe*, and owing to the damage caused, sank between Cape Wrath and the west ingress to the Pentland Firth.

This encouraged our hopes that the *Greif*, an auxiliary cruiser sent out a few days previously under Captain Tieze, would have an equally successful trip. Unfortunately, news came to hand a few weeks later that she had been held up on the English guard-ship line between the Shetlands and Norway, and after a fierce fight had succumbed, but not until she had torpedoed and sunk the auxiliary cruiser *Alcantara*, a vessel three times her size. This first encounter, in which the *Greif* had already suffered severely, attracted a second auxiliary cruiser, the *Andes*, and the light cruiser *Comus*, which came up with two destroyers and joined in the fight. Faced by such superior forces, Tieze, after a fierce fight lasting two hours, left the ship, with the surviving members of the crew, and sank her. While the English at first took part in the rescue of the crew, the *Comus*, according to the statements of prisoners since returned home, again opened fire on the life-boats and rafts, asserting that a U-boat had been sighted. The result was that several others were killed, the commander among the number. Commander Nevetzky, First Lieut. Weddigen, and Lieut. Tiemann, had already been killed in the battle. About two-thirds of the crew were taken prisoners by the English.

CHAPTER IX

ENTERPRISES IN THE HOOFDEN, AND BOMBARDMENT
OF YARMOUTH AND LOWESTOFT

ON March 5, the day after the return of the *Moewe*, the High Sea Fleet, under my command, carried out the first of its greater enterprises, partaking of the nature of a more extended advance. The idea prompting this move was to attack the enemy light forces that were constantly reported in the Hoofden, and thus attract support from the English harbours to the south, and if possible force them between the pincers formed by our advanced cruisers and the Main Fleet following in the rear. At daybreak the distance between the battleships and the cruisers was approximately 30 nautical miles. The cruisers were then to advance from a position Terschelling Bank Lightship S.S.E. 15 nautical miles to the Hoofden, and push on to the northern boundary of the English minefield. The battleships were to follow the course of the cruisers up to 10 A.M., when they would have reached latitude 53" 30', provided that in the meantime our action had not been checked by intervening circumstances. Squadron II (the older battleships) did not form part of this expedition, but was held ready, with a mine-sweeping division, to secure the safety of the Bight, in order to keep the return route open for the Fleet. Two flotillas accompanied the cruisers; the others were with the Main Fleet. To ensure the safety of the proceeding an airship was allotted to the Chief of Reconnaissance, while other airships were to reconnoitre early the following day in the sector north-west of Heligoland as far distant as 200 nautical miles, to protect the flank and rear of the Fleet. Should the weather the preceding night be favourable for airships, advantage was to be taken of it. This was carried out, and led to a very effective bombing of the important naval yards at Hull, on the Humber.

A graphic picture of the attack is given in the description by one of the airship commanders who took part, Captain Victor Schulze, on board the "L 11," but who has since died the death of a hero. He writes:

Germany's High Sea Fleet

"Our orders were: 'March 5, in morning, "L 11," with "L 13" and "L 14," to attack England in the north.' At noon (on 4th) an ascent was made with the object of attacking the naval yard at Rosyth. In consequence, however, of an ever-increasing strong north-north-west wind, bringing heavy snow and hailstorms with it, it was decided on the way to seek out the munition factories at Middlesbrough instead of proceeding to Rosyth. The only shipping traffic noticeable was limited to a few fishing-boats off the Dogger Bank. Following the throwing out of benzine casks, the airship, towards 10 P.M., was fired at through a thin bank of clouds, but without success, and the incident was not worth noticing. At 10.45 P.M. the English coast was crossed between Flamborough Head and Spurn Head at Hornsea, which led to the belief that the north wind was stronger. The ship now steered northwards over the distinctly visible snow-clad coast. Wherever the landscape was not hidden from view by heavy snow-clouds visibility was good. The upper line of cloud was at a height of 2,000-3,000 m.; above, the sky was bright and starry. Violent hailstorms again came on, the ship became coated with ice, and although all water was discharged and the temperature of the air was 16 degrees, she could not rise above 2,000 m. Not until just before the attack, and after a further discharge of benzine, did she achieve a somewhat risky 2,300 m. The antennæ and the ends of the metal props in gondola and corridor glistened through the snow and hail with balls of light—St. Elmo's fire. The gondola and platform were thickly covered with snow. When the weather cleared at 1 A.M., it appeared from the position of the ship that to steer further north would be fruitless with the velocity of the wind at 12 doms (doms—2 metres per second).

"Meanwhile the course of the Humber was now distinguishable in the snowy landscape further south, offering a very favourable chance of attack. The town of Hull was well darkened, but from where 'L 11' was stationed we could easily make out the dropping of bombs from 'L 14'. Fresh snow-clouds then interfered with the view, but I had time to spare and remained at my post until the clouds cleared away in an hour's time. At 2 A.M. 'L 11' opened the attack and first dropped some bombs on Hull to induce the defence batteries and searchlights to disclose their position, for if that failed the ship could not have attacked against so strong a wind. The town remained quiet and dark, but at that moment the clouds cleared away and disclosed the following picture. The town

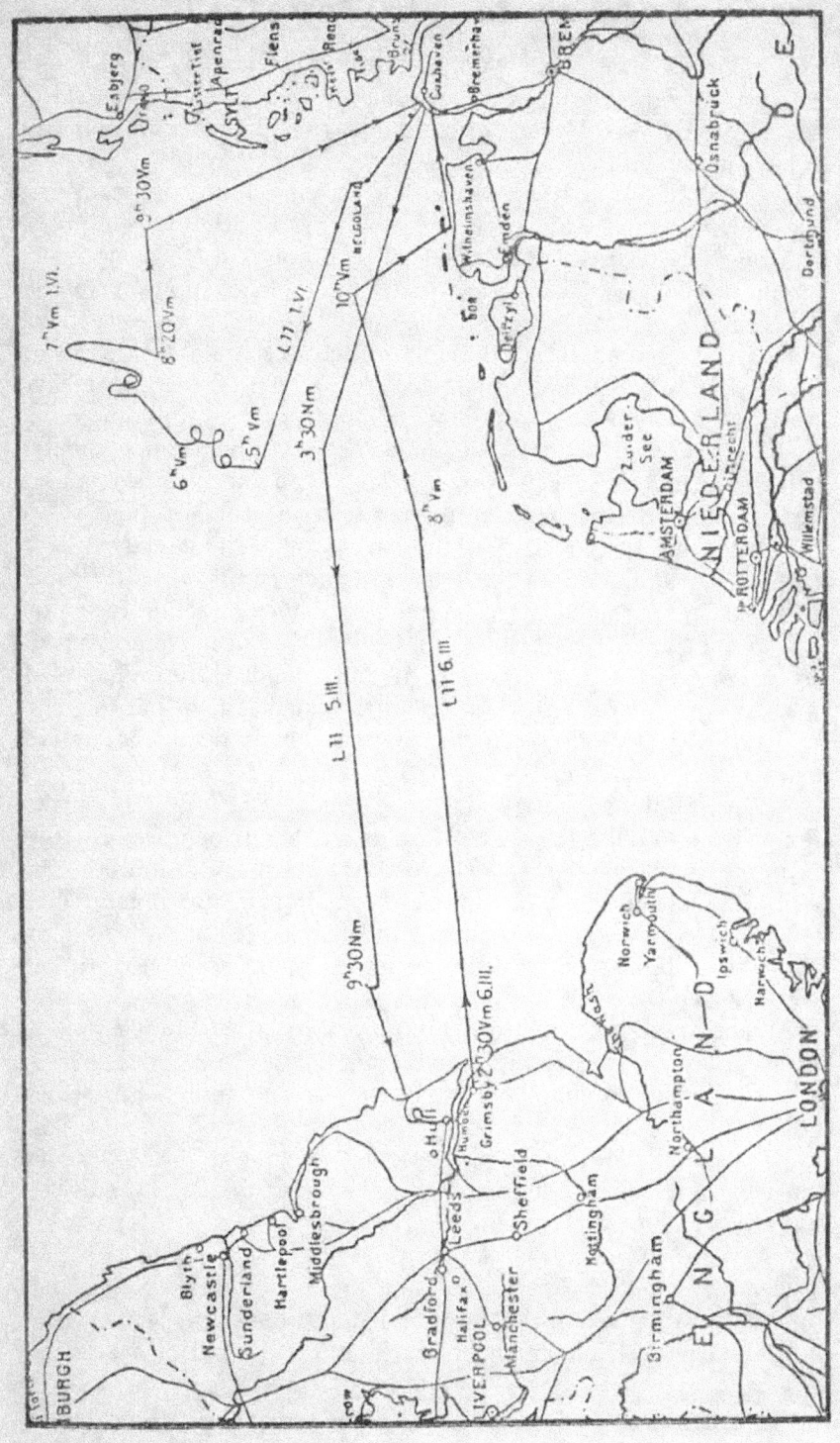

THE ZEPPELIN RAID ON MARCH 5, 1916, BY "L 11" ON THE HUMBER, AND SCOUTING CRUISE OF "L 11" IN NORTH SEA ON JUNE 1

and environs were white with the freshly-fallen snow. Although plunged in darkness the town lay sharply defined under the starlit sky with its streets, blocks of houses, quays and harbour basins just below the airship. A few lights moved in the streets. The ship, taking a northerly course and with all her engines at high pressure, was poised over her objective and stood by. For twenty minutes, following my instructions, bombs were dropped quite composedly on the harbour and docks and the effect of each bomb carefully watched. The first bomb hit the quay, knocking a great piece out of it; a second hit the middle of the dock-gate of a harbour basin. The bomb fell so accurately on the gate that it might have been taken for a shot deliberately aimed at it. Buildings fell down like so many houses built of cards. One bomb in particular had a tremendous effect. Near the spot where it exploded houses kept falling on each side until at last a huge black hole stood out on the snowy ground near the harbour. A similar large black spot in the neighbourhood was apparently caused by ' L 14 '. People were seen through the telescope running hither and thither in the light of the flames. Ships that had been hit began moving about in the harbour. All counter-action at and round Hull was limited to a few weak searchlights that failed to find the ship, and to some isolated firing. While the bombs were being dropped, the airship ventured up to 2,700 m.

"When fully convinced of the excellent effect of the bombs dropped on Hull, I decided to drop the remainder on the fortifications at Immingham which, as I had already noticed, had been heavily bombed by ' L 14 '. The airship made for Immingham with the last five explosive bombs and was received at once by four strong searchlights and very lively gun-fire. The searchlights tried in vain to find the ship through the light clouds just passing over, for although it was brightly lit up by them they always moved farther away. South of the searchlights on the bank the batteries were using much ammunition. From 40 to 50 fiery lights or fire-balls were scattered round the ship on all sides, above and below. The height these missiles reached was reckoned at 3,000 m. or more. The first explosive bomb that fell among the searchlights extinguished first one and then all the others. No other results were observed. Towards the end of the attack on Hull the fore engine was put permanently out of action by the stoppage of the water gauge and the consequent freezing of the oil and water pipes at a temperature of 19 degrees; at Immingham the aft engine was out of action for half an hour. The coast was crossed at 2.40 A.M. on the return journey. Again heavy snow and hailstorms accompanied by electric disturbances were encountered. In the space of three minutes a sudden squall carried the ship upwards from 2,400 to 3,200 m., 250 m. above our previous highest altitude. Coming down shortly afterwards the elevating gear got out of order, but the ship was worked by the crew as well as could be until the damage was repaired, though we were forced to rise again to 3,200 m. At 5 A.M. the rear engine gave way again and, owing to the freezing of oil and water, stopped altogether shortly before we landed. At 7 A.M. we met the First Scouting Division of Squadrons I and II 30 nautical miles north-north-west of the Terschelling-Bank Lightship. At 2 P.M. we landed safely at Nordholz. The ship was quite able to fly again."

The airship did the trip in 26 hours; it must be mentioned in this connection that the crew of a raider is so limited that all the men have to be on duty the whole time.

The Naval Corps in Flanders supported the Fleet's enterprise by stationing 12 U-boats off the English south-east coast. In spite of good visibility, there was no encounter with the enemy. The expedition, therefore, was only useful for the purpose of practising unity of command, and the handling of individual ships under circumstances likely to arise during an offensive engagement of any big unit. The return voyage was made an occasion for different exercises in manœuvring the Fleet in fighting formation until we were compelled to withdraw, alarmed by the sighting of enemy submarines, for which the Fleet at Terschelling would have presented a good target. After our return all opportunity for further operations was put a stop to for a time owing to the bad weather, and to high east and north-east winds which our airships only just succeeded in escaping.

The dismissal of Grand Admiral von Tirpitz as Secretary of State of the Imperial Admiralty which was announced to the Fleet Command on March 18 aroused great sympathy, not only in view of his services in connection with the many-sided development of the Fleet through long years and in all branches of maritime service, but because, in these critical times for the country, much anxiety was aroused at the thought of being deprived of the services of a man who had shown himself to be a genial personality and of unwavering energy. This change in the conduct of the Naval

Germany's High Sea Fleet

Department, in particular, gave rise to grave fears as to the prompt carrying out of resolute and adequate U-boat warfare.

At the beginning of March the decision in this connection had again been postponed for four weeks. The Fleet was therefore bound all the more to aim at active action against the enemy, and every attention was given to that purpose by the new Fleet Command.

Meanwhile the English, by an unexpected attack, provided us with the opportunity of testing our preparations. The repeated air-raids, and particularly a very big and successful one on London on February 1, roused them to make an effort to seek out and destroy these troublesome raiders in their own homes. The hangars at Tondern were the nearest. There had been no further attack on this group since the first unsuccessful one on Christmas Day, 1914. On March 25, in very unfavourable weather for flying—so much so that our own scouting machines did not go up owing to fog and snowstorms—an attack was made at 9.30 A.M. by some torpedo-boat destroyers on our outpost group at List. They sank two fishing steamers that could have reported the attack, but were themselves obliged to withdraw before our aeroplanes which had gone up from List, and which dropped bombs on the enemy, hitting the destroyer *Medusa*. She was abandoned later on in a sinking condition. The English report gave out that the loss of the *Medusa* was owing to a collision with the destroyer *Laverock*.

Various reports were made by our aeroplanes, from which we gathered that an aerial attack had started from two vessels carrying aeroplanes, which were supported by battle-cruisers, light cruisers and destroyers. We were not able at once to determine what the intention was—whether there was to be a simultaneous attack from the west on the hangars at Hage (south of Norderney), or whether we were to expect an encircling movement of the enemy against our forces sent northwards, in an effort to force them to come out. The counter-action of our aeroplanes and the bad weather compelled all the five English airmen to come down. Two of them were picked up by one of their own torpedo-boats; the other three were taken by our aeroplanes. They did not succeed in doing any damage.

The English attack caused great commotion among our outpost forces, as well as among all the other ships, which at once got ready to put to sea, until the further purpose of the enemy was revealed. Our cruisers and several flotillas went in pursuit of the retiring enemy, who evidently did not attach much importance to the rescue of the airmen; but the weather becoming still worse, we were unable to get near the ships. There was an encounter that night between our torpedo-boats and English light cruisers, when the English cruiser *Cleopatra* succeeded in ramming and sinking one of our torpedo-boats, " G 194," which had crossed her bows by mistake.

During these nocturnal proceedings another torpedo-boat, "S 22," Commander Karl Galster, struck a mine 55″ 45″ North Lat., and 5″ 10″ East Long. This boat broke in two at once; the fore-part sank quickly, the remainder floated for about five minutes and then suddenly went to the bottom. The hurrahs from the crew, led by the commander, proved that they stood firm at their posts to the very last. Torpedo-boat "S 18" immediately tried to render assistance, but the wind and the high sea running made it impossible, and in spite of every effort only ten petty officers and seven of the crew were saved. We learned from the English report that the same night the *Cleopatra* was also run into by the English cruiser *Undaunted,* the latter receiving such heavy damage that she had to be towed into harbour. An English wireless intercepted during the night stated that a warship, together with destroyers, had tried to take a damaged English destroyer in tow, and it might be presumed that the ships would proceed northwards by night and return again at daybreak when there would be the possibility of encountering and capturing certain units of the enemy. Squadron II and Scouting Divisions I and IV were ordered to proceed to 55″ 10′ N. Lat. and 6″ 0′ E. Long., whither Squadrons I and II would follow; the flagship was with Squadron III. At 6.30 A.M. the cruisers reported that the sea was so rough that an engagement was impossible; the push was therefore given up as hopeless. We heard later that the same reason had induced the English to abandon the destroyer *Medusa* and return home very much battered by the storm.

At the end of March, by way of reprisal for the attempt to injure our aerial fleet, our airships enjoyed a very successful series of expeditions which, aided by a combination of favourable weather and dark nights, resulted in five successive attacks. It is difficult for airships to bring back an exact statement of their successes owing to the great altitude at which they fly, and also to the darkness and their exposure to anti-aircraft defences. The reports issued by the English official censor were, therefore, the only means of ascertaining the extent of the damage done, which was often represented as being of little importance in order to calm the fears of the population. But it is certain that at the time the uninterruptedly recurring raids caused a great feeling of panic, as the destruction in London itself surpassed anything ever before known. On our side, for the first time, we had to deplore the loss of an airship, brought down by enemy anti-aircraft guns. "L 15," Commander Breithaupt, was forced down on to the water at the mouth of the Thames after the airship's gondola had been repeatedly hit by shells. The crew, two officers and sixteen men, were rescued by English boats and taken prisoner; they did not, however, succeed in towing in the airship, the destruction of which had been provided for. It is worthy of note that in the night of the 2nd-3rd the Firth of Forth was reached for the first time, and ships lying there and buildings along the Firth were attacked. Bad weather set in again on April 6 and put an end to this exceptionally successful period. "L 11" took part more than once in the attacks and its commander has given the following description:

"Order: 'L 11,' together with 'L 14,' to attack England south or centre on morning of April 1st.' At 12 noon an ascent was made for the purpose of attacking England in the south, but owing to the wind soon veering round to north-west, the centre of the coast was made for. There was lively traffic among steam-trawlers off the Dogger Bank, and the English wireless was distinctly heard at work. In spite of throwing out two casks of benzine, the temperature of the air did not allow of the ship rising above 2,200 m.; at 10 P.M. the English coast was reached south of the Tyne. While trying to bomb the docks on the Tyne and cross the coast, the ship was greeted with violent firing, which came from the whole coastal area north and south of the river. To draw back and seek the required weather side for the attack would have occupied several hours with the prevailing wind (W.N.W., 5-7 doms.) I decided not to cross the batteries on account of not being very high in relation to the firing, and also because of slow progress against the wind and the absolutely clear atmosphere up above. I fixed, therefore, on the town of Sunderland, with its extensive docks and the blast furnaces north-west of the town. Keeping on the weather side, the airships dropped explosive bombs on some works where one blast-furnace was blown up with a terrible detonation, sending out flames and smoke. The factories and dock buildings of Sunderland, now brightly illuminated, were then bombed with good results. The effect was grand; blocks of houses and rows of streets collapsed entirely; large fires broke out in places and a dense black cloud, from which bright sparks flew high, was caused by one bomb. A second explosive bomb was at once dropped at the same spot; judging from the situation, it may have been a railway station. While over Sunderland, the airship was caught by a powerful searchlight and was pelted with shrapnel and fire-balls, but to no purpose. The concussion from a shell bursting near the airship was felt as though she had been hit. After leaving the town, two other searchlights tried to get the ship, but only with partial success. Then followed slight firing, apparently with machine-guns. The last explosive bombs were dropped with good aim on two blast-furnace works in the neighbourhood of Middlesbrough. On returning, we again saw numbers of steam trawlers off the Dogger Bank. At 10 A.M., April 2nd, we landed at Nordholz."

The day following "L 11" again set out for a raid on England in company with "L 17," and reported as follows:

"Owing to the expected warm temperature of the air, only five mechanics and forty-five bombs were carried; the spare parts were limited; two machine-guns and a landing-rope were left behind, and the supply of benzine very sparingly measured out, as both going and coming back the wind was expected to be behind us. The ascent was made at 2.30 P.M. The flight was so rapid that the last bearings* taken showed that the English coast would be reached near Sheringham at about 10.30 P.M. As the atmosphere was becoming still thicker, it was impossible to distinguish anything beyond a few dim lights. As the coast could not be made out at the expected time, I turned by degrees out of my previous course W. ½ S. to S.W. to S., presuming that the wind would have gone further south on land. Finally, however, the bearings taken at 1.10 A.M. revealed the surprising fact that the slight W.S.W. wind blowing had risen to 8-10 doms. When, therefore, we found at 2.45 A.M. that the ship was over the land, a further advance towards London became purposeless. Moreover, on ascertaining the exact position, it was too late, and in view of the

* To ascertain its position wireless signals are sent out from the airship, picked up at two different stations, and registered on the map. The position fixed is then transmitted to the airship by wireless. The whole proceeding occupies the shortest space of time, but when several airships are on an expedition together the wireless must be worked most carefully to avoid mutual misunderstandings and mistakes.

strength and direction of the wind, to turn off towards the mouth of the Humber. So long as the darkness lasted, I determined to try for some objective in the county of Norfolk. Norwich, which was in complete darkness, could not be made out.

"Towards 3.55 A.M., after 'L 11' had crossed the coast close to the west of Yarmouth, violent gun-fire was observed through the mist in the rear. A turn was made, and altogether thirteen well-aimed bombs were dropped on the place where the firing came from. We had to give up the idea of staying longer on the coast as at the altitude of the airship day was already beginning to break. The return journey, as was anticipated, was favoured at a high altitude by a fresh W.S.W. wind. At 10 A.M. we landed at Nordholz."

On April 5, "L 11" with "L 13" and "16" again went up for a raid on the English Midlands. An account of this expedition will serve to give the reader some idea of the strain and exertions to which our airships' crews were exposed on such occasions.

"At 9.45 p.m. the airship crossed the distinctly visible coast south of Flamborough Head and took a course for Sheffield. When over Hull to the north the airship was found to be over several newly-erected batteries with four very strong searchlights, which caught up the ship easily in the very clear air; whereupon, from 10.10 to 10.30 an unusually heavy firing with shells and shrapnel was kept up. The aim was good; many shells burst quite close to the airship, causing the frame to shake violently. The next battery was at once attacked and silenced by explosive bombs. Being at the low altitude of 2,300 metres and in such clear air it was not considered advisable to continue to pass over the numerous other batteries, so we turned round intending to take a southerly course outside the coast in order to rise higher when the moon had gone down and to proceed inland. In setting off, the rear engine was put out of order through worn-out crank bearings. The commander decided, therefore, to put himself on the lee side of the north-north-east wind and look for Hartlepool. The line of the coast and the course of the rivers were just as plainly visible as on the map. North and south of Flamborough Head there was much shipping activity. Several neutral vessels were distinguishable by the bright lights above their neutrality marking.

"At 2 A.M., just off Hartlepool, the fore engine gave out. The attack on the town was abandoned, and it was decided on the way back to destroy a large iron factory at Whitby. Even from the high altitude of the airship, the factory appeared to be a very extensive establishment with many brightly illuminated blast furnaces and numerous buildings. It was situated on the shore and had steam extinguishing apparatus. The airship hovered sufficiently long over this factory to drop carefully aimed bombs. The distinctly visible result consisted not only in the utter destruction of the furnaces and buildings through fire and explosion, but there were also heavy explosions in the darker sections of the factory, which led to the conclusion that the entire establishment had been destroyed. At 10.30 A.M. the airship got into a dense fog on the return journey, and with a view to safer navigation went over land and made good her way at 50 m. altitude, landing at Nordholz at 3 P.M. in clear weather."

No ships were sent up by Captain Strasser, the Commander of Airships, on April 6. His estimate of the weather conditions proved quite correct, for in the course of the afternoon the slight northeasterly wind veered round to the east and when night came a regular storm was blowing. While the air raids of the previous night were proceeding several torpedo-boats started out from Horns Reef in a north-west and westerly direction and kept the outpost forces in constant activity. It led, however, to no engagement with the enemy.

From April 13 to April 19 the Fleet was kept in constant expectation of an English attack, news having been received that one was pending. But the enemy did not show himself.

BOMBARDMENT OF YARMOUTH AND LOWESTOFT

On April 24, Easter Monday, the Fleet put out on an important enterprise which, like that in the beginning of March, was directed towards the Hoofden, but was to be extended farther so as to force the enemy out of port. I expected to achieve this by bombarding coastal towns and carrying out air raids on England the night the Fleet went out. Both these actions would probably result in counter measures being taken by the enemy that would give our forces an opportunity to attack. On the occasion of the advance of March 5—6 the enemy preferred withdrawing all his forces into port, as we learnt afterwards from intercepted wireless messages, as soon as he had news of our advance, either through agents or from submarines in the North Sea.

Germany's High Sea Fleet

The news we obtained from the enemy had repeatedly announced strong enemy forces in the northern section of the North Sea under the Norwegian coast; forces had also been sighted in the Hoofden and harbours on the south-east coast of England so that an opportunity would probably occur for our Fleet to push in between those two divisions of the enemy Fleet and attack with equal strength that section which should first present itself. It was, therefore, obvious that the most suitable direction for attack would be towards the south-east counties of England. If the enemy then wished to cut off our return he would have to move into the neighbourhood of Terschelling Bank, where the waters were favourable for offering battle. With luck we might even succeed in attacking the enemy advancing from the Hoofden on both sides; on the south with the forces told off to bombard the coast and on the north with the Main Fleet.

Lowestoft and Yarmouth were the only coastal towns it was intended to bombard. Both were fortified and were important military points of support for the enemy—Lowestoft for mine-laying and sweeping; Yarmouth as a base for the submarines whence they started on their expeditions to the Bight. The destruction, therefore, of the harbours and other military establishments of both these coastal towns was a matter of great military importance, apart from the object of the bombardment in calling out the enemy. Simultaneous air-raids on southern England would offer the advantages of mutual support for the airships and the sea forces. The airships would reconnoitre for the forces afloat on their way to and fro, while the latter would be able to rescue the airships should they meet disaster. It was also hoped there might be an opportunity for trade-war under prize conditions.

All the available High Sea forces were assembled, including Squadron II, and the Chief Command of the Naval Corps in Flanders was enjoined to keep his available U-boats in readiness. The Naval Corps also offered to station two U-boats east of Lowestoft to facilitate the advance; they did excellent service in assisting the bombardment. The U-boats at the disposal of the High Sea Command were placed in a position to attack the Firth of Forth and the southern egress from the Firth was closed by a U-minelayer.

Eight of the newer airships were selected for the raid and three older ones were ordered to hold themselves in readiness on the second day in the rear of the fleet for reconnoitring. If at all possible, the bombardment was to take the towns by surprise at daybreak, in order to prevent counter-measures by the enemy, such as calling up sub-

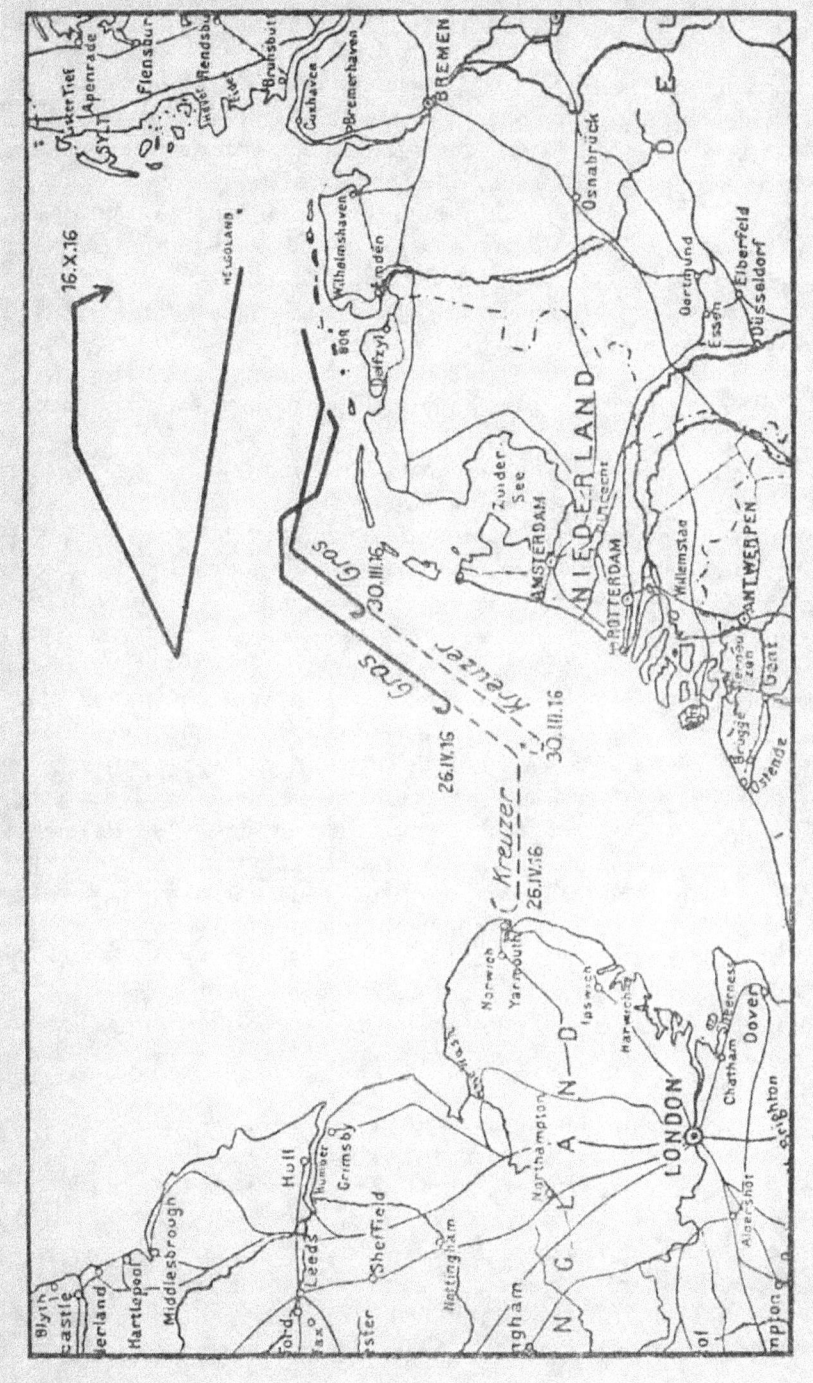

OPERATIONS OF THE HIGH SEA FLEET DURING MARCH, APRIL AND OCTOBER, 1916

marines from Yarmouth to protect the coast. The forces intended to accompany the cruisers had to endeavour to keep, not actually in the Hoofden, but in the open waters west and north of Terschelling Bank in case it should come to a fight, as that was the only position where liberty of action in all eventual developments could be ensured. The bombardment of both the coastal towns was entrusted to the battle-cruisers. They were supported by Scouting Division II and two fast torpedo-boat flotillas (VI and IX). The Main Fleet, consisting of Squadrons I, II and III, Scouting Division IV, and the remainder of the torpedo flotillas was to accompany the battle-cruisers to the Hoofden until the bombardment was over, in order, if necessary, to protect them against superior enemy forces.

At noon on the 24th all the forces, including the airships, started. The course led first through the south opening in the barrier at Norderney and then north, round a minefield laid down by the English out of sight of the Dutch coast, and into the Hoofden where the bombardment was to open at daybreak and last for about thirty minutes. At 4 P.M. the movement received an unwelcome set-back owing to a message from Rear-Admiral Bödicker, leader of the reconnaissance ships, that his flagship, the battle-cruiser *Seydlitz* had struck a mine and her forward torpedo compartment was damaged. The ship was thus debarred from taking part in the expedition; she was still able to do 15 knots and returned to harbour under her own steam. The leader was, therefore, obliged to hoist his flag in another cruiser. The route on which the ship had struck the mine had been searched and swept last on the night of the 22nd and 23rd and had been constantly used by light forces on their night patrols.

Owing to this occurrence the battle-cruisers behind the *Seydlitz* stopped and turned according to agreement, awaiting further orders in case they, too, should come across mines. As the *Seydlitz* turned to follow them in order to transfer the admiral to the *Lützow*, two of the ships simultaneously reported the track of a torpedo and submarines. With that danger so near it would not have been advisable to attempt to stop the ship and transfer the admiral to another; and as the cruiser was already badly damaged, it would have been dangerous to expose her to still further injury. The *Seydlitz* continued, therefore, on her westerly course and the Chief of Reconnaissance on board a torpedo-boat reached the *Lützow* later and there resumed his duties. The *Seydlitz* was escorted on her homeward way by two torpedo-boats and "L 7" and reached the harbour without further misadventure.

In consequence of this incident, the Fleet Command thought fit to alter the intended course, and the only alternative was to take the route along the coast of East Friesland. The weather being so very clear, it would have to be borne in mind that in following that route the ships could be observed from the islands of Rottum and Schiermonnikoog and the news probably dispatched farther. Unfortunately this lessened the chance of carrying out a surprise bombardment of Lowestoft and Yarmouth, but there was no reason on that account to give it up altogether. Relying on aerial reconnaissance, further developments might be expected and the enterprise was continued.

Towards 8 P.M. a wireless message from the Naval Staff confirmed what the Naval Corps in Flanders had already reported at noon, that since 6 A.M. numerous enemy forces had been assembled off the Belgian coast, at the mouth of the Scheldt; it was not possible to divine their intention, but very probably it was connected with a bombardment of the coast of Flanders. It was welcome news for our Fleet to hear of the assemblage of enemy forces there. Another wireless announced that on the morning of the 23rd large squadrons of English warships of all types had been sighted off Lindesnaes, the south-west point of Norway. I could count, therefore, on my presumption that the English Fleet was divided into two sections being correct.

At 9.30 P.M. a message was sent us from Bruges that according to an intercepted English wireless all patrol boats had been ordered back to port. This showed that the meeting of our battle-cruisers with English submarines during the afternoon had resulted in their sending news of our movements.

Shortly before daybreak reports were received from the airships of the results of their attack. They were obliged to fight against unfavourable wind conditions, and bad visibility over the land; they also met with strong counter-action. The six air-ships taking part had raided Norwich, Lincoln, Harwich and Ipswich and had been engaged with outpost ships. None had been damaged and they were then in the act of returning home. At 5 A.M. our large cruisers approached the coast off Lowestoft. Good support was afforded them by the U-boats placed in position by the Naval Corps. The light cruiser *Rostock*, which formed the flank cover for the battle-cruisers, reported enemy ships and destroyers in a west-south-west direction. But as the light was not good enough to open fire, Admiral Bödicker proceeded to bombard the towns. This was carried out at a distance of from 100—130 hm. Excellent results were observed in the harbour and the answering fire was weak. A north-west course was then taken to proceed with the bombardment of Great Yarmouth and to engage the ships reported by the *Rostock*.

Meanwhile the *Rostock*, supported by the light cruiser *Elbing*, had kept in touch with the enemy forces in the endeavour to bring them nearer to the battle-cruisers. The ships in question were four modern light cruisers and about twelve destroyers. As soon, however, as they caught sight of our battle-cruisers they turned at full speed southwards. We opened fire on them at a distance of 130 hm. until they were beyond our range. Many hits were observed, and on one of the cruisers a big fire was plainly visible. The high speed kept up by the enemy made pursuit useless. The cruisers then shaped their course in the direction of our Main Fleet and reported that their task was accomplished.

During the bombardment of the two coastal towns, the light cruiser *Frankfurt* sank an armed patrol steamer by gunfire. A second one was sunk by the leader of Torpedo-boat Flotilla VI, "G 41," the crew of which were rescued. From what the latter stated she was the *King Stephen*, of evil repute, which had allowed the crew of the airship "L 19" to perish. These men certainly denied most emphatically that they were on the trawler then, and laid the blame on a former crew. They contradicted themselves so constantly, however, that the captain and the engineer grew very suspicious, and as the steamer had been used for war service the crew were made prisoners.

At 5.30 A.M. "L 9" reported being chased by aeroplanes in a south-westerly direction. When the fleet was sighted the airmen departed, probably to announce the approach of our ships, which at that time were steaming on a south-westerly course to meet the cruisers. At the same moment "L 11" and "L 23" came in sight; they had not been able to discover the enemy. At 6 A.M., therefore, when the cruisers had reported the conclusion of the bombardment, Terschelling Bank was made for according to plan. Towards 7.30 A.M. the Naval Corps in Flanders reported that the English ships assembled there had been ordered by wireless, which was intercepted, to return. The English destroyers were to finish coaling and then move on to Dunkirk.

An approach, therefore, on the part of the enemy was not to be looked for from that quarter. The only hope now left was that enemy forces might be encountered off Terschelling. As we drew near to that zone, the Fleet was constantly obliged to evade submarine attacks, but no other enemy forces were met.

The return trip passed without further incident. Two neutral steamers, as well as some smaller vessels were stopped and searched for contraband goods. The enemy, hearing of the advance of our forces, withdrew all his ships from the Belgian coast and made no effort to locate us. It appears from subsequent English statements that the English Fleet had put to sea the day before for one of the usual North Sea expeditions, and it would be interesting to find out whether it could have had an opportunity of crossing our path in the Bight.

When the *Seydlitz* was docked a hole of 90 sq. m. was found in her, through which about 1,400 tons of water had poured into the ship. Eleven of the men had been killed at their post in the torpedo chamber. In spite of the considerable quantity of ammunition stored there, no further explosion occurred or the disaster would have been far greater.

Early in May the weather conditions were such as to allow of a resumption of the air raids on England. But this favourable phase in the weather was not of so long duration as in the preceding month, which was quite exceptional. Two raids were carried out in which eight airships took part. "L 20" was lost in the second raid as a strong south-westerly wind had arisen, and the airship, owing to engine trouble, was unable to reach the home coast. The captain, Commander Stabbert, made, therefore, for the Norwegian coast, where he came down with his damaged airship in the neighbourhood of Jäderen, where the crew alighted and were interned. Then ensued a period of short nights which caused a cessation in the airships' raiding activities as the hours of darkness were not enough to afford them sufficient protection, and it was also obvious that latterly the defensive measures had become much more effectual. But the Fleet made good use of the airships for all reconnoitring purposes in connection with important enterprises, which gained in value through co-operation with the U-boats and on which all the more energy had to be expended since the trade-war by the U-boats had been stopped since the end of April.

Just as we were proceeding to Lowestoft a wireless message was received from the Chief of the Naval Staff, to the effect that trade-war by U-boats was only to be carried out now in accordance with prize regulations. This was the result of the American protest in the case of the *Sussex* disaster. As I could not expect the U-boats to carry on a war of that description owing to the heavy casualties that might be expected, I had called back by wireless all the boats engaged in the trade-war, and subsequently received approval of this action in high quarters. It was left to me until further notice to employ the U-boats in purely military enterprises. This helped still further to protect the German Bight, as definite areas could now be continuously occupied and we could expect early reports of enemy movements; we also hoped to find opportunities to attack the enemy submarines employed as guard-ships. The experiences of our U-boats confirmed the danger caused by the enemy submarines, which, appearing unexpectedly, had come to be very unpleasant adversaries, and we intended, therefore, to make use of our boats for defence purposes.

PART II

From the Battle of the Skagerrak to the
Unrestricted U-boat Warfare

CHAPTER X

THE BATTLE OF THE SKAGERRAK

THE bombardment of April 25 had not failed to make an impression in England. The expectation that the fleet was bound to succeed in warding off all German attacks on British shores had repeatedly been disappointed. On each occasion the English main fleet had arrived too late—in December, 1914; in January, 1915; and now again this year—so that, to the great annoyance of the English, the German "raiders" got away each time unpunished. Wherefore Mr. Balfour, the First Lord of the Admiralty, felt called upon to announce publicly that should the German ships again venture to show themselves off the British coast, measures had been taken to ensure their being severely punished. However, we were ready to take our chance.

The question was whether it would be advisable to include Squadron II in an advance which in all probability would involve us in a serious battle. Early in May I ordered the squadron temporarily into the Jade Basin that I might have an opportunity of discussing with the Squadron Commander the action to be observed in battle under the most varied conditions. Military reasons entered into the question as to whether the squadron should be taken out or left behind, as well as consideration for the honour and feeling of the crews, who would not hear of being reduced, themselves and their ships, to the second class. For battleships to have their activity limited absolutely to guarding the German Bight without any prospect of getting into touch with the enemy— to which they had been looking forward for a year and a half—would have caused bitter disappointment; on the other hand, however, was the responsibility of sending the ships into an unequal fight where the enemy would make use of his very best material. I cannot deny that in addition to the eloquent intercession of Rear-Admiral Mauve, the Squadron Commander, my own former connection with Squadron II also induced me not to disappoint it by leaving it behind. And thus it happened that the squadron played its part on May 31, and in so helpful a manner that I never had cause to regret my decision.

133

Germany's High Sea Fleet

The repairs on the *Seydlitz*, damaged on April 24, were not completed until the end of May, as the reconstruction of the mine-shattered torpedo tubes necessitated very heavy work. I had no intention, however, of doing without that battle-cruiser, although Vice-Admiral Hipper, Chief of the Reconnaissance Forces, had meanwhile hoisted his flag in the newly repaired battle-cruiser *Lützow* (Captain Harder, formerly on the *Stralsund*). The vessels belonging to Squadron III were also having their condensers repaired, as on their last trip there had been seven cases of damaged machinery in that squadron. The advantage of having three engines, as had each of these ships, was proved by the fact that two engines alone were able to keep up steam almost at full speed; at the same time, very faulty construction in the position of the engines was apparent, which unfortunately could not be rectified owing to limited space. Thus it happened that when a condenser went wrong it was impossible to conduct the steam from the engine with which it was connected to one of the other two condensers, and thus keep the engine itself working. It was an uncomfortable feeling to know that this weakness existed in the strongest unit at the disposal of the Fleet, and how easily a bad accident might result in leakages in two different condensers and thus incapacitate one vessel in the group!

The object of the next undertaking was a bombardment of the fortifications and works of the harbour at Sunderland which, situated about the middle of the East coast of England, would be certain to call out a display of English fighting forces as promised by Mr. Balfour. The order issued on May 18 in this connection was as follows:

"The bombardment of Sunderland by our cruisers is intended to compel the enemy to send out forces against us. For the attack on the advancing enemy the High Sea Fleet forces to be south of the Dogger Bank, and the U-boats to be stationed for attack off the East coast of England. The enemy's ports of sortie will be closed by mines. The Naval Corps will support the undertaking with their U-boats. If time and circumstances permit, tradewar will be carried on during proceedings."

The squadrons of men-of-war had made over the command of prizes to the torpedo-boat flotillas, as torpedo-boats are the best adapted for the examination of vessels, but have not a crew large enough to enable them to bring the captured vessels into our ports. The First and Second Scouting Divisions were placed at the

The Battle of the Skagerrak

disposal of the Chief of Reconnaissance, and the Second Leader of the torpedo-boats with Flotillas II, VI, and IX. Scouting Division IV* and the remainder of the flotillas were with the Main Fleet. Sixteen of our U-boats were told off for the positions of attack, with six to eight of the Flanders boats. On May 15 they started to reconnoitre in the North Sea, and from May 23 to June 1 inclusive were to remain at the posts assigned to them, observe the movements of the English forces, and gain any information that might be of use to the Fleet in their advance; at the same time they were also to seize every opportunity to attack. Provision was also made for the largest possible number of our airships to assist the enterprise by reconnaissance from the air. The fact that the U-boats could only remain out for a certain period put a limit to the execution of the plan. If reconnaissance from the air proved impossible, it was arranged to make use of the U-boats, and so dispense with aerial reconnaissance.

As the weather each day continued to be unfavourable and the airship commander could only report that it was impossible to send up any airships, the plan was so far changed, though without altering other preparations, that it was decided to embark on a campaign against cruisers and merchantmen outside and in the Skagerrak, with the expectation that the news of the appearance of our cruisers in those waters would be made known to the enemy. With this object in view, they had been told to keep in sight of the coast of Norway, so that the enemy might be notified. In further describing the course of this undertaking, which led to the Battle of the Skagerrak, I shall keep strictly to the official report I sent in.

In judging the proceedings it must be borne in mind that at sea a leader adapts his action to the events taking place around him. It may possibly reveal errors which can only be accounted for later by reports from his own ships or valuable information from enemy statements. The art of leadership consists in securing an approximately correct picture from the impression of the moment, and then acting in accordance with it. The writer of history can then form a tactical inference where obvious mistakes were made, or where a better grasp of the situation would have led to a more advantageous decision. In this event a certain reticence should be

* The Third Scouting Division, which contained the oldest armoured cruisers, *Prinz Adalbert, Prinz Heinrich,* and *Roon,* had long since been handed over to the commander of the Baltic forces, as, owing to their lack of speed and inferior armour-plating, the vessels were not suitable for use in the North Sea.

observed in making definite assertions that a different movement would have been more successful, for armed efficiency plays the chief part in success and cannot be determined with mathematical precision. I have in mind one hit that did so much damage to our battle-cruiser *Seydlitz* on January 24, 1915, that one almost came to the conclusion that such ships could not stand many shots of such heavy calibre, and yet the following battle proved the contrary. At all events, a good hit can seal the fate of a ship, even one of the strongest. A naval battle may be open to criticism as to why it happened thus, but anyone who asserted that it might have happened otherwise would be in danger of losing his case.

I

THE ADVANCE

On May 30, as the possibility of a long-distance aerial reconnaissance was still considered uncertain, I decided on an advance in the direction of the Skagerrak, as the vicinity of the Jutland coast offered a certain cover against surprise. An extensive aerial reconnaissance was an imperative necessity for an advance on Sunderland in the north-west, as it would lead into waters where we could not allow ourselves to be forced into giving battle. As, however, on the course now to be adopted, the distance from the enemy points of support was considerably greater, aerial reconnaissance was desirable, though not absolutely necessary. As already stated, our U-boats were in position, some of them in fact facing Scapa Flow, one boat off Moray Firth, a large number off the Firth of Forth, several off the Humber and the remainder, north of the Terschelling Bank, in order to be able to operate against enemy forces that might chance to come from a south-westerly direction. The combination of our total forces taking part was as follows:

A list of warships which on May 30 to June 1, 1916, took part in the Battle of the Skagerrak and the operations connected therewith:

Chief of the Fleet: Vice-Admiral Scheer in *Friedrich der Grosse*.
Chief of Staff: Captain von Trotha (Adolf).
Chief of the Operating Section: Captain von Levetzow.
Admiralty Staff Officer: Captain Quaet-Faslem (Hans).
Commander of "Friedrich der Grosse": Captain Fuchs (Theodor).

The Battle of the Skagerrak

SQUADRON I

Chief of Squadron: Vice-Admiral Ehrhard Schmidt, *Ostfriesland.*
Admiralty Staff Officer: Captain Wegener (Wolfgang).
Admiral: Rear-Admiral Engelhardt, *Posen.*

Ostfriesland: Captain von Natzmer.
Thüringen: Captain Küsel (Hans).
Helgoland: Captain von Kameke.
Oldenberg: Captain Höpfner.
Posen: Captain Lange.
Rheinland: Captain Rohardt.
Nassau: Captain Klappenbach (Hans).
Westfalen: Captain Redlich.

SQUADRON II

Chief of Squadron: Rear-Admiral Mauve, *Deutschland.*
Admiralty Staff Officer: Captain Kahlert.
Admiral: Rear-Admiral Baron von Dalwigk zu Lichtenfels, *Hanover.*

Deutschland: Captain Meurer (Hugo).
Pommern: Captain Bölken.
Schlesien: Captain Behncke (Fr.).
Schleswig-Holstein: Captain Barrentrapp.
Hannover: Captain Heine (Wilh.).
Hessen: Captain Bartels (Rudolf).

SQUADRON III

Chief of Squadron: Rear-Admiral Behncke, *König.*
Admiralty Staff Officer: Captain Baron von Gagern.
Admiral: Rear-Admiral Nordmann, *Kaiser.*

König: Captain Brüninghaus.
Grosser Kurfürst: Captain Goette (Ernst).
Markgraf: Captain Seiferling.
Kronprinz: Captain Feldt (Constanz).
Kaiser: Captain Baron von Kayserling.
Prinz Regent Luitpold: Captain Heuser (Karl).
Kaiserin: Captain Sievers.

Germany's High Sea Fleet

Chief of the Reconnaissance Forces: Vice-Admiral Hipper, *Lützow.*
Admiralty Staff Officer: Captain Raeder (Erick).

Scouting Division I

Seydlitz: Captain von Egidy (Moritz).
Moltke: Captain von Karps.
Derfflinger: Captain Hartog.
Lützow: Captain Harder.
Von der Tann: Captain Zenker.

Leader of Scouting Division II: Rear-Admiral Bödicker, *Frankfurt.*
Admiralty Staff Officer: Commander Stapenhorst.

Scouting Division II

Pillau: Captain Mommsen.
Elbing: Captain Madlung.
Frankfurt: Captain von Trotha (Thilo).
Wiesbaden: Captain Reiss.
Rostock: Captain Feldmann (Otto).
Regensburg: Captain Neuberer.

Leader of Scouting Division IV: Commodore von Reuter, *Stettin.*
Admiralty Staff Officer: Captain Weber (Heinrich).

Scouting Division IV

Stettin: Captain Rebensburg (Friedrich).
München: Captain Böcker (Oskar).
Frauenlob: Captain Hoffmann (Georg).
Stuttgart: Captain Hagedorn.
Hamburg: Captain von Gaudecker.

Torpedo-Boat Flotillas

First Leader of the Torpedo-Boat Forces: Commodore Michelsen,
    Rostock.
Admiralty Staff Officer: Captain Junkermann.
Second Leader of the Torpedo-Boat Forces: Commodore Heinrich,
    Regensburg.
Chief of Flotilla I: Commander Conrad Albrecht, "G 39."
Chief of 1st Half-Flotilla: Commander Conrad Albrecht, "G 39."
Chief of Flotilla II: Captain Schuur, "B 98."

The Battle of the Skagerrak

Chief of 3rd Half-Flotilla: Captain Boest, "G 101."
Chief of 4th Half-Flotilla: Captain Dittamar (Adolf), "B 109."
Chief of Flotilla III: Captain Hollmann, "S 53."
Chief of 5th Half-Flotilla: Commander Gautier, "V 71."
Chief of 6th Half-Flotilla: Commander Karlowa, "S 54."
Chief of Flotilla V: Captain Heinecke, "G 11."
Chief of 9th Half-Flotilla: Commander Hoefer, "V 2."
Chief of 10th Half-Flotilla: Commander Klein (Friedrich), "G 8."
Chief of Flotilla VI: Captain Max Schultz, "G 41."
Chief of 11th Half-Flotilla: Commander Rümann, "V 44."
Chief of 12th Half-Flotilla: Commander Laks, "V 69."
Chief of Flotilla VII: Captain von Koch, "S 24."
Chief of 13th Half-Flotilla: Commander von Zitzewitz (Gerhard), "S 15."
Chief of 14th Half-Flotilla: Captain Cordes (Hermann), "S 19."
Chief of Flotilla IX: Captain Goehle, "V 28."
Chief of 17th Half-Flotilla: Commander Ehrhardt, "V 27."
Chief of 18th Half-Flotilla: Captain Tillessen (Werner), "V 30."

SUBMARINES

Leader of the Submarines: Captain Bauer, *Hamburg.*
Admiralty Staff Officer: Captain Lützow (Friedrich).

"U 24"—Commander: Lieut. Schneider (Rudolf).
"U 32"—Commander: Lieut. Baron Spiegel von und zu Peckelsheim.
"U 63"—Commander: Lieut. Schultze (Otto).
"U 66"—Commander: Lieut. von Bothmer.
"U 70"—Commander: Lieut. Wünsche.
"U 43"—Commander: Lieut. Jürst.
"U 44"—Commander: Lieut. Wagenführ.
"U 52"—Commander: Lieut. Walther (Hans).
"U 47"—Commander: Lieut. Metzger.
"U 45"—Commander: Lieut. Hillebrand (Leo).
"U 22"—Commander: Lieut. Hoppe.
"U 19"—Commander: Lieut. Weizbach (Raimund).
"U B 22"—Commander: Lieut. Putzier.
"U B 21"—Commander: Lieut. Hashagen.
"U 53"—Commander: Lieut. Rose.
"U 64"—Commander: Lieut. Morath (Robert).

Germany's High Sea Fleet

AIRSHIPS

"L 11"—Commander: Captain Schütze (Viktor).
"L 17"—Commander: Lieut. Ehrlich (Herbert).
"L 14"—Commander: Lieut. Böcker.
"L 21"—Commander: Lieut. Dietrich (Max).
"L 23"—Commander: Lieut. von Schubert.
"L 16"—Commander: Lieut. Sommerfeldt.
"L 13"—Commander: Lieut. Prölt.
"L 9"—Commander: Captain Stelling.
"L 22"—Commander: Lieut. Dietrich (Martin).
"L 24"—Commander: Lieut. Koch (Robert).

Vice-Admiral Hipper, Chief of the Reconnaissance Forces was ordered to leave the Jade Basin with his forces at 4 a.m., May 31, to advance towards the Skagerrak out of sight of Horns Reef, and the Danish coast, to show himself off the Norwegian coast before dark, to cruise in the Skagerrak during the night, and at noon the next day to join up with the Main Fleet. The ships under his command comprised the Scouting Division I and II. To the latter was attached the light cruiser *Regensburg*, flagship of the Second Leader of the torpedo-boats; under his command were the Flotillas II, VI, and IX. The Main Fleet, consisting of Squadrons I, II, and III, of Scouting Division IV, the First Leader of torpedo-boats, in the *Rostock*, and Torpedo-Boat Flotillas I, II, V, and VII, were to follow at 4.30 A.M. to cover the reconnaissance forces during the enterprise and take action on June 1. The sailing order of the battleships was as follows: Squadron III in van, Squadron I following, and Squadron II in the rear.

The *König Albert* was absent from Squadron III, having been incapacitated a few days previously through condenser trouble. Notwithstanding the loss of this important unit, I could not bring myself further to postpone the enterprise, and preferred to do without the ship. Squadron II was without the *Preussen*, which had been placed at the disposal of the Commander-in-Chief of the Baltic forces to act as guard-ship at the south egress from the Sound. *Lothringen* was deemed unfit for service. Scouting Division IV, and the Leader of Torpedo-Boats in the light cruiser *Rostock*, together with the Torpedo-Boat Flotillas I, II, V, and VII, were attached to the battleships.

To the west of the Amrum Bank a passage had been cleared

through the enemy minefields which led the High Sea forces safely to the open sea. Visibility was good, with a light north-westerly wind, and there was no sea on. At 7.30 A.M. "U 32" reported at about 70 miles east of the Firth of Forth, two battleships, two cruisers, and several torpedo-boats taking a south-easterly course. At 8.30 a second wireless was received stating that she had intercepted English wireless messages to the effect that two large battleships and groups of destroyers had run out from Scapa Flow. At 8.48 A.M. a third message came through from "U 66" that about 60 nautical miles east of Kinnairel [sic., ? Kinnaird Head], eight enemy battleships, light cruisers, and torpedo-boats had been sighted on a north-easterly course.

These reports gave no enlightenment as to the enemy's purpose. But the varied forces of the separate divisions of the fleet, and their diverging courses did not seem to suggest either combined action or an advance on the German Bight or any connection with our enterprise, but showed a possibility that our hope of meeting with separate enemy divisions was likely to be fulfilled. We were, therefore, all the more determined to keep to our plan. Between 2 and 3 P.M. "L" 9, 14, 16, 21 and 23 ascended for long-distance reconnaissance in the sector north to west of Heligoland. They took no part in the battle that so soon was to follow, neither did they see anything of their own Main Fleet, nor of the enemy, nor hear anything of the battle.

2

THE FIRST PHASE OF THE BATTLE: CRUISER ENGAGEMENT

At 4.28 P.M. the leading boat of the 4th Torpedo-Boat Half-Flotilla, "B 109," reported that *Elbing*, the west wing cruiser on the Chief of Reconnaissance's line, had been sent to examine a steamer about 90 nautical miles west of Bovbjerg, and had sighted some enemy forces. It was thanks to that steamer that the engagement took place; our course might have carried us past the English cruisers had the torpedo-boat not proceeded to the steamer and thus sighted the smoke from the enemy in the west.

As soon as the enemy, comprising eight light cruisers of the "Caroline" type, sighted our forces, he turned off to the north. Admiral Bödicker gave chase with his cruisers. At 5.20 P.M. the Chief of the Reconnaissance then sighted in a westerly direction two columns of large vessels taking an easterly course. These soon showed themselves to be six battle-cruisers, three of the "Lion" class, one "Tiger," and two "Indefatigables," besides numbers of lighter forces. The Chief of Reconnaissance called back Scouting Division II, which he had sent to give chase in the north, and prepared to attack. The enemy deployed to the south in fighting line. It was Vice-Admiral Beatty with the First and Second English Battle-Cruiser Squadrons, consisting of the *Lion, Princess Royal, Queen Mary, Tiger, New Zealand,* and *Indefatigable.* That the enemy deployed to the south was a very welcome fact for us, as it offered the possibility of inducing the enemy to fall back on his own main fleet. The Chief of Reconnaissance therefore followed

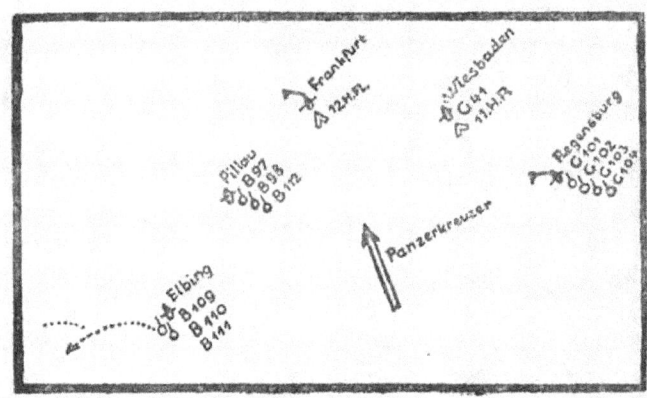

Order of Sailing

the movement, manœuvred to get within effective firing range, and opened fire at 5.49 P.M., at a range of about 130 hm.*

The fighting proceeded on a south-easterly course. The Chief of Reconnaissance kept the enemy at effective distance. The batteries fixed their aim well; hits were observed on all the enemy ships. Already at 6.13 P.M., the battle-cruiser *Indefatigable,* the last in the line of the enemy cruisers, sank with a terrible explosion caused by the guns of the *Von der Tann.* Superiority in firing and tactical advantages of position were decidedly on our side up to 6.19 P.M., when a new unit of four or five ships of the "Queen Elizabeth" type, with a considerable surplus of speed, drew up from a north-westerly direction, and beginning at a range of 200 hm., joined the fight-

* Earl Beatty gives the range at about 18,500 yards.

ing. It was the Fifth English Battle Squadron.* This made the
situation critical for our cruisers. The new enemy fired with extraordinary rapidity and accuracy, with the greater ease as regards the
latter that he met with almost no opposition, as our battle-cruisers
were fully engaged with Admiral Beatty's ships.

At 6.20 P.M. the fighting distance between the battle-cruisers
on both sides was about 120 hm., while between our battle-cruisers
and those with *Queen Elizabeth* the distance was something like

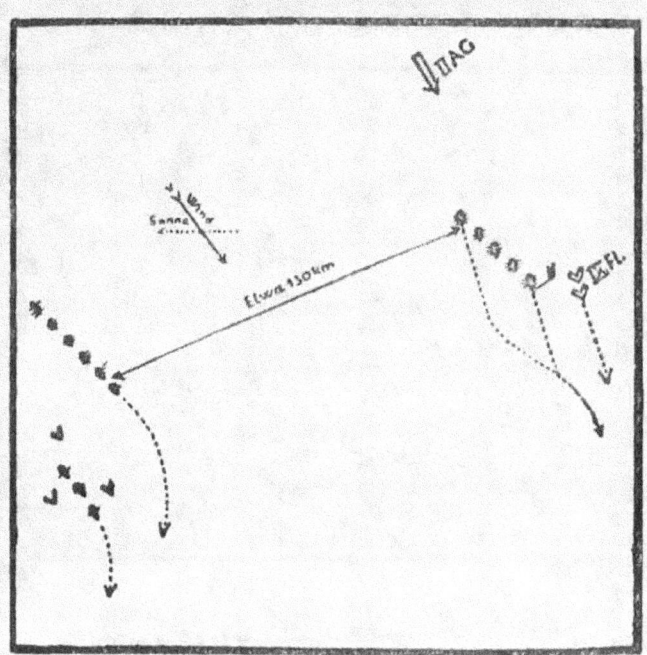

Position at 5.49 P.M.

180 hm. At this stage Torpedo-Boat Flotilla IX was the only one of
the flotillas under the Chief of Reconnaissance that was in a position to attack. The Second Leader of Torpedo-Boats, Commodore
Heinrich, on board the *Regensburg*, and some few boats belonging
to Torpedo Flotilla II, were getting up steam with all speed in a
diagonal line from the Chief of Reconnaissance's furthest point.

* According to English accounts, it comprised the *Barham, Warspite, Valiant,*
and *Malaya.* Mention is made of four ships only. According to various observations
on our side (by Squadron III and the leader of Scouting Division II), there were
five ships. If *Queen Elizabeth,* or a similar type of ship, was not in the unit,
it is possible that another recently built man-of-war replaced her.

Germany's High Sea Fleet

The cruisers of Scouting Division II, together with the remaining torpedo flotillas, were forced by the "Queen Elizabeths" to withdraw to the east to escape their fire and had, therefore, in spite of working their engines to the utmost, not been able to arrive in position at the head of the battle-cruisers.

In view of the situation, the Second Leader of the Torpedo-Boats ordered Torpedo Flotilla IX (whose chief, Captain Goehle, had already decided on his own initiative to prepare to attack) to advance to the relief of the battle-cruisers.

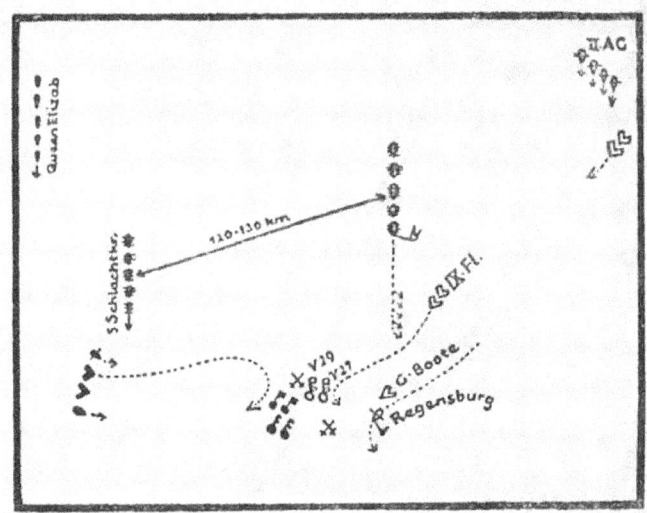

Position at 6.20. P.M.

At about 6.30 P.M. Torpedo Flotilla IX proceeded to attack, running through heavy enemy firing. Twelve torpedoes were fired on the enemy lines at distances ranging between 95—80 hm. It was impossible to push the attack closer on the enemy, as at the same time that Flotilla IX got to work, eighteen to twenty English destroyers, covered by light cruisers, appeared on the scene to counter-attack and beat off our torpedo-boats. The result was a torpedo-boat fight at close range (1,000—1,500 m.). The *Regensburg*, together with the boats of Torpedo-Boat Flotilla II that were with her, and the centrally situated guns on the battle-cruisers, then joined in the fight. After about ten minutes the enemy turned away. On our side "V 27" and "V 29" were sunk, hit by shots from heavy calibre guns. The crews of both the boats were rescued in spite of enemy fire, by "V 26" and "S 35." On the enemy side two, or perhaps three, destroyers were sunk, and two others so badly damaged that they could not get away, and fell later into the hands of our advancing Main Fleet. The enemy made no attempt to rescue the crews of these boats.

During the attack by the torpedo-boats, the English battle-cruisers were effectively held in check by the Scouting Division I with heavy artillery, which at the same time manœuvred so successfully that none of the numerous enemy torpedoes observed by Torpedo-Boat Flotilla IX hit their objectives. Towards 6.30 P.M. a powerful explosion was observed on board the third enemy cruiser— the *Queen Mary*. When the smoke from the explosion cleared away the cruiser had disappeared. Whether the destruction was the result of artillery action or was caused by a torpedo from the battle-cruisers or by a torpedo from Torpedo-Boat Flotilla IX can never be ascertained for certain, but most probably it was due to artillery action which caused an explosion of ammunition or oil on board the enemy vessel. It was not until night that I heard of the destruction of the two battle-cruisers.

The attack by Flotilla IX had at all events been successful in so far that for a time it checked the enemy's fire. Admiral Hipper took advantage of this to divert the cruisers to a north-westerly course and thus secure for himself the lead at the head of the cruisers in the new phase of the fight. Immediately following on the attack by the torpedo-boats, the German Main Fleet appeared on the scene of battle just in the nick of time to help the reconnaissance forces in their fight against considerably superior numbers.

3
THE SECOND PHASE OF THE FIGHT: THE PURSUIT

At 4.28 P.M.* about 50 nautical miles west of Lyngoig, on the Jutland coast, the first news of the sighting of enemy light forces was reported to the Main Fleet proceeding in the following order:

Squadrons III, I, II, the flagship at the head of Squadron I, on a northerly course, speed 14 knots—distance between the vessels, 7 hm., distance between the squadrons, 35 hm., the torpedo-boats

---

* In comparing the time given in the German and English accounts it must be remembered that there is a difference of two hours, for the reason that we reckon according to summer-time in Central Europe, while the difference between ordinary Central Europe and Greenwich time is one hour. Therefore 4.28 German time corresponds to 2.28 English time.

as U-boat escort for the squadrons, the light cruisers of Scouting Division IV allotted to the Main Fleet to protect their course.

At 5.35 the first report was sent that heavy forces had been sighted. The distance between the Chief of Reconnaissance and the Main Fleet was at that time about 50 nautical miles. On receipt of this message, the fighting line was opened (that is, the distance between the squadrons was reduced to 1,000 m., and between the vessels to 500 m.), and the order was given to clear the ships for action.

In the fighting line the Commander-in-Chief of the Fleet is not tied to any fixed position. When there is a question of leading several squadrons it is not advisable to take up a position at the head of the line, as it is not possible from there to watch the direction in which the fight develops, as that greatly depends on the movements of the enemy. Being bound to any such position might lead to the Commander-in-Chief finding himself at the rear instead of at the head of his assembled line. A position in the centre or at a third of the line (according to the number of units) is more advantageous. In the course of events the place of the eighth ship in the line for the flagship has been tested and approved of.

During the whole time that fighting was going on I had a clear look-out over the whole line and was able to signal with great rapidity in both directions. As the fighting line of the warships was more than 10 km. long, I should not have been able to overlook my entire line from the wing, especially under such heavy enemy firing.

The message received at 5.45 P.M. from the Chief of Reconnaissance that he was engaged with six enemy battle-cruisers on a south-easterly course showed that he had succeeded in meeting the enemy, and as he fought was drawing him closer to our Main Fleet. The duty of the Main Fleet was now to hasten as quickly as possible to support the battle-cruisers, which were inferior as to material, and to endeavour to hinder the premature retreat of the enemy. At 6.5, therefore, I took a north-westerly course at a speed of 15 knots, and a quarter of an hour later altered it to a westerly course in order to place the enemy between two fires, as he, on his southerly course, would have to push through between our line and that of the battle-cruisers. While the Main Fleet was still altering course, a message came from Scouting Division II that an English unit of warships, five ships (*not four!*) had joined in the fight.

The Battle of the Skagerrak

The situation thus was becoming critical for Scouting Division I, confronted as they were by six battle-cruisers and five battleships. Naturally, therefore, everything possible had to be done to get into touch with them, and a change was made back to a northerly course. The weather was extremely clear, the sky cloudless, a light breeze from N.W., and a calm sea. At 6.30 P.M. the fighting lines were sighted. At 6.45 P.M. Squadrons I and III opened fire, while the Chief of Reconnaissance, with the forces allotted to him, placed himself at the head of the Main Fleet.

The light enemy forces veered at once to the west, and as soon as they were out of firing range turned northwards. Whether the fire from our warships had damaged them during the short bombardment was doubtful, but their vague and purposeless hurrying to and fro led one to think that our fire had reached them and that the action of our warships had so surprised them that they did not know which way to turn next.

The English battle-cruisers turned to a north-westerly course; *Queen Elizabeth* and the ships with her followed in their wake, and thereby played the part of cover for the badly damaged cruisers. In so doing, however, they came very much nearer to our Main Fleet, and we came on at a firing distance of 17 km. or less. While both the English units passed by each other and provided mutual cover, Captain Max Schultz, Chief of Torpedo-Boat Flotilla VI, attacked at 6.49 P.M., with the Eleventh Torpedo-Boat Half-Flotilla. The result could not be seen.

The fighting which now ensued developed into a stern chase;

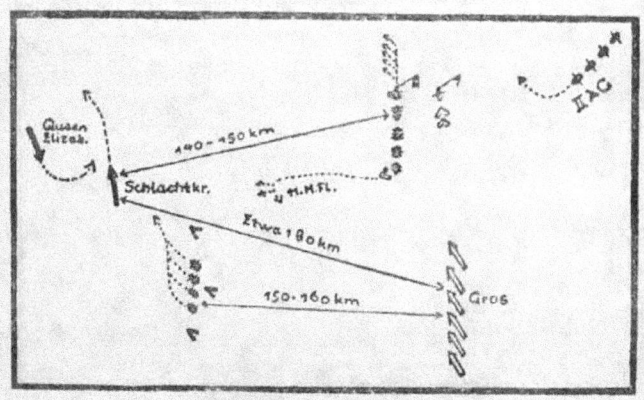

Position at 6.55 P.M.

147

our reconnaissance forces pressed on the heels of the enemy battle-cruisers, and our Main Fleet gave chase to the *Queen Elizabeth* and the ships with her. Our ships in Squadron III attained a speed of over 20 knots, which was also kept up on board the *Kaiserin*. Just before fire was opened she had succeeded in repairing damage to one of her condensers. By the *Friedrich der Grosse*, the Fleet Flagship, 20 knots was achieved and maintained. In spite of this, the enemy battle-cruisers succeeded soon after 7 o'clock in escaping from the fire of Scouting Division I. The *Queen Elizabeth* and her sister ships also made such good way that they were only under fire from the ships of Scouting Division I and of the Fifth Division (First Half of Squadron III). The hope that one of the ships pursued would be so damaged as to fall a prey to our Main Fleet was not fulfilled, although our firing was effective, and at 7.30 P.M. it was seen that a ship of the "Queen Elizabeth" type after she had been hit repeatedly, drew slowly out of the fighting line with a heavy list to leeward. Two modern destroyers, the *Nestor* and *Nomad,* were all that fell to the share of the Main Fleet;

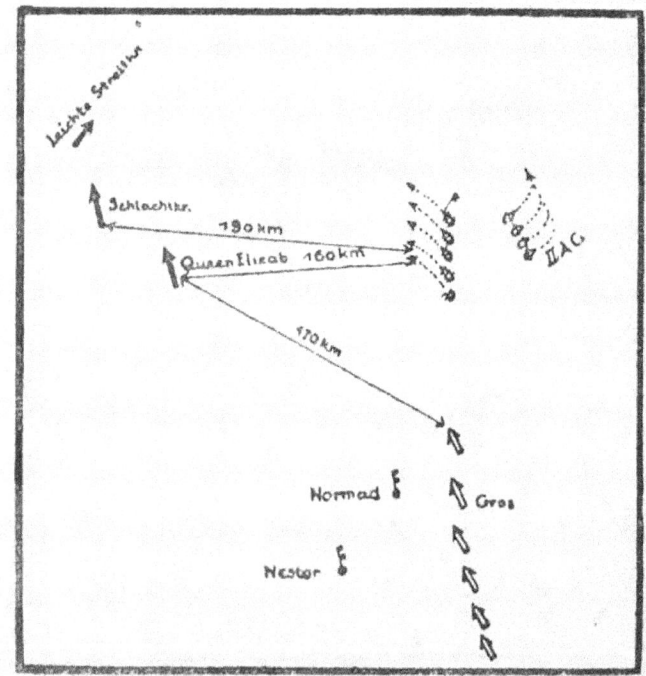

Position at 7.15 P.M.

they were hit and badly damaged in the attack by Torpedo-Boat Flotilla IX, and were overtaken and sunk by us; the crews were taken prisoner.

At 7.20 P.M., when the fire from Scouting Division I and from the ships of the Fifth Division appeared to grow weaker, the leaders of the Fleet were under the impression that the enemy was succeeding in getting away, and gave orders to the Chief of Reconnaissance and to all the fighting forces "to give chase." Meanwhile, the previously clear weather had become less clear; the wind had changed from N.W. to S.W. Powder fumes and smoke from the funnels hung over the sea and cut off all view from north and east. Only now and then could we see our own reconnaissance forces. Owing to the superior speed of Beatty's cruisers, our own, when the order came to give chase, were already out-distanced by the enemy battle-cruisers and light craft, and were thus forced, in order not to lose touch, to follow on the inner circle and adopt the enemy's course. Both lines of cruisers swung by degrees in concentric circles by the north to a north-easterly direction. A message which was to have been sent by the Chief of Reconnaissance could not be dispatched owing to damage done to the principal and reserve wireless stations on his flagship. The cessation of firing at the head of the line could only be ascribed to the increasing difficulty of observation with the sun so low on the horizon, until finally it became impossible. When, therefore, enemy light forces began a torpedo attack on our battle-cruisers at 7.40 P.M., the Chief or Reconnaissance had no alternative but to manœuvre and finally bring the unit round to S.W. in an endeavour to close up with the Main Fleet, as it was impossible to return the enemy's fire to any purpose.

4

THE THIRD PHASE OF THE FIGHTING: THE BATTLE

I observed almost simultaneously that the admiral at the head of our squadron of battleships began to veer round to starboard in an easterly direction. This was in accordance with the instructions signalled to keep up the pursuit. As the Fleet was still divided in columns, steering a north-westerly course as directed, the order "Leader in Front" was signalled along the line at 7.45 P.M., and the speed temporarily reduced to 15 knots, so as to make it possible for the divisions ahead, which had pushed on at high pressure, to get into position again.

Germany's High Sea Fleet

As long as the pursuit was kept up, the movements of the English gave us the direction, consequently our line by degrees veered round to the east. During these proceedings in the Main Fleet, Scouting Division II, under Rear-Admiral Bödicker, when engaged with a light cruiser of the "Calliope" class,* which was set on fire, sighted several light cruisers of the "Town" class, and several big ships, presumably battleships, of which the *Agincourt* was one. Owing to the mist that hung over the water,

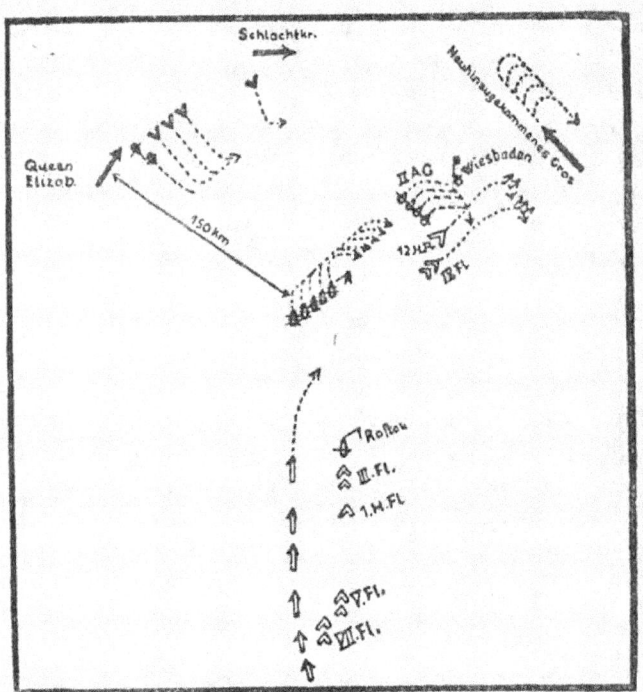

Position from 7.43 to 8 P.M.

it was impossible to ascertain the entire strength of the enemy. The group was at once heavily fired on, returned the fire, discharged torpedoes, and turned in the direction of their own Main Fleet. No result could be observed, as artificial smoke† was at once

* According to English accounts the light cruiser *Chester* was badly damaged. Her casualties were 31 killed and 50 wounded, and she had four holes just above the water-line.

† Artificial fog or smoke, prepared by a special process at the largest dye-works, and supplied to all the lighter forces to enable them to withdraw from the fire of superior forces.

The Battle of the Skagerrak

employed to protect the cruisers. In spite of the fog the *Wiesbaden* and *Pillau* were both badly hit. The *Wiesbaden* (Captain Reiss) lay in the thick of the enemy fire, incapable of action.

The Chiefs of the 12th and 9th Torpedo-Boat Half-Flotillas who were stationed behind the cruisers, recognising the gravity of the situation, came to the front. Both came under fire from a line of numbers of big ships on a N.W. course, and fired their torpedoes from within 60 hm. of the enemy. Here, too, it was impossible to observe what success was achieved, as dense clouds of smoke hid the enemy from view directly they veered round. But both the above-mentioned commanders reckon that they met with success, having attacked under favourable conditions.

While this encounter with the advance guard of the English Main Fleet was taking place, we, on our flagship were occupied debating how much longer to continue the pursuit in view of the advanced time. There was no longer any question of a cruiser campaign against merchantmen in the Skagerrak, as the meeting with the English fighting forces which was to result from such action had already taken place. But we were bound to take into consideration that the English Fleet, if at sea, which was obvious from the ships we had encountered, would offer battle the next day. Some steps would also have to be taken to shake off the English light forces before darkness fell in order to avoid any loss to our Main Fleet from nocturnal torpedo-boat attacks.

A message was then received from the leader of Scouting Division II that he had been fired on by some newly arrived large ships. At 8.2 p.m. came a wireless: "*Wiesbaden* incapable of action." On receipt of the message I turned with the Fleet two points to larboard [port] so as to draw nearer to the group and render assistance to the *Wiesbaden*. From 8.20 onwards there was heavy fighting round the damaged *Wiesbaden*, and good use was made of the ship's torpedoes. Coming from a north-north-westerly direction, the "Queen Elizabeth" ships and also probably Beatty's battle-cruisers attacked (prisoners, however, stated that after 7.0 P.M. the latter took no part in the fight).

A fresh unit of cruisers (three "Invincibles" and four "Warriors") bore down from the north, besides light cruisers and destroyers. A further message from the torpedo-boat flotillas which had gone to support Scouting Division II, stated that they had sighted more than twenty enemy battleships following a south-easterly course. It was now quite obvious that we were confronted

Germany's High Sea Fleet

by a large portion of the English Fleet and a few minutes later their presence was notified on the horizon directly ahead of us by rounds of firing from guns of heavy calibre. The entire arc stretching from north to east was a sea of fire. The flash from the muzzles of the guns was distinctly seen through the mist and smoke on the horizon, though the ships themselves were not distinguishable. This was the beginning of the main phase of the battle.

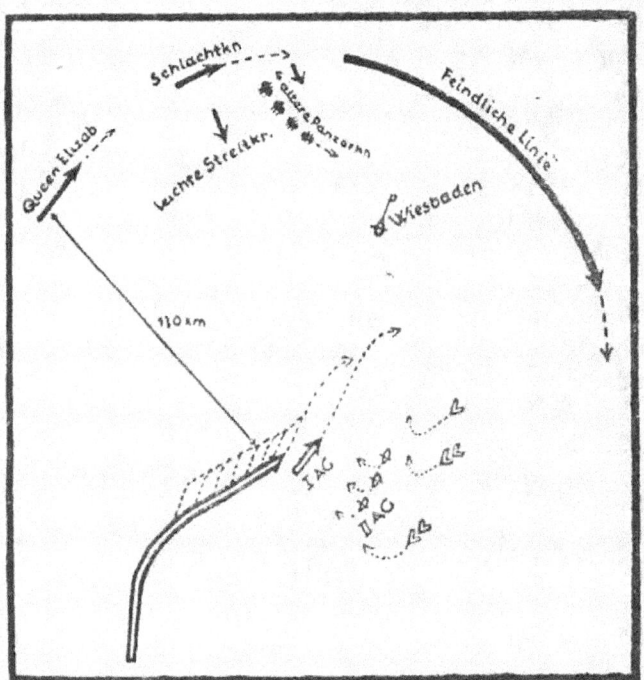

Position at 6.16 P.M.

There was never any question of our line veering round to avoid an encounter. The resolve to do battle with the enemy stood firm from the first. The leaders of our battleship squadrons, the Fifth Division turned at once for a running fight, carried on at about 13,000 m. The other divisions followed this movement on orders signalled from the flagship. By this time more than a hundred heavy guns had joined in the fight on the enemy's side, directing fire chiefly at our battle-cruisers and the ships of the Fifth Division (the "König" class). The position of the English line (whose centre we must have faced) to our leading point brought fire on us from three

152

The Battle of the Skagerrak

sides. The "Queen Elizabeths" fired diagonally from larboard [port]; the ships of the Main Fleet, which Jellicoe had brought up, from the forecastle starboard. Many shots were aimed at the *Friedrich der Grosse*, but the ship was never hit.

During this stage of the fight the cruisers *Defence*, *Black Prince*, and *Warrior* came up from the north, but were all destroyed by the fire from our battleships and our battle-cruisers. Fire from the *Friedrich der Grosse* was aimed at one of the three, which in a huge white cloud of steam was blown into the air, at 3,000 m. distance. I observed several enemy hits and consequent explosions on the ships at our leading point. Following the movement of the enemy they had made a bend which hindered free action of our Torpedo-Boat Flotilla II stationed there.

I could see nothing of our cruisers, which were still farther forward. Owing to the turning aside that was inevitable in drawing nearer, they found themselves between the fire of both lines. For this reason I decided to turn our line and bring it on to an opposite course. Otherwise an awkward situation would have arisen round the pivot which the enemy line by degrees was passing, as long-distance shots from the enemy would certainly have hit our rear ships. As regards the effectiveness of the artillery, the enemy was more favourably situated, as our ships stood out against the clear western horizon, whereas his own ships were hidden by the smoke and mist of the battle. A running artillery fight on a southerly course would therefore not have been advantageous to us. The swing round was carried out in excellent style. At our peace manœuvres great importance was always attached to their being carried out on a curved line and every means employed to ensure the working of the signals. The trouble spent was now well repaid; the cruisers were liberated from their cramped position and enabled to steam away south and appeared, as soon as the two lines were separated, in view of the flagship. The torpedo-boats, too, on the leeside of the fire had room to move to the attack and advanced.

While the veering round of the line was proceeding, two boats of Torpedo-Boat Flotilla III ("G 88" and "V 73") and the leading boat of Torpedo-Boat Flotilla I ("S 32") had attacked. The remaining boats of Torpedo-Boat Flotilla III had ceased the attack on an order to retire from the leader. The weakening of the enemy fire had induced the First Leader to give the order, being persuaded that the enemy had turned away and that the flotilla, which would be urgently

needed in the further development of the battle, would find itself
without support. Owing to the shortening of the line at the head,
the boats of the other flotillas were not able to attack. One division
(Torpedo-Boat Flotillas IX and VI) had just returned from the
8 o'clock attack. The enemy line did not follow our veer round.
In the position it was to our leading point, it should have remained
on, and could have held us still further surrounded if by a simultaneous turn to a westerly course it had kept firmly to our line.

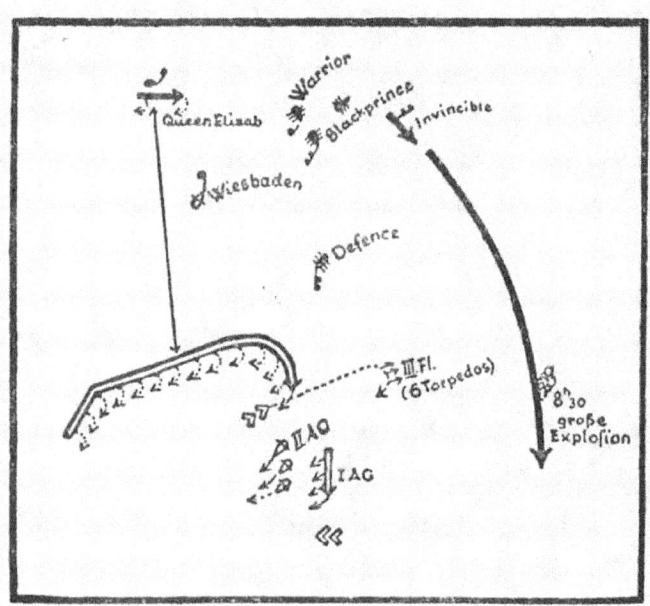

Position at 8.35 P.M.

It may be that the leader did not grasp the situation, and was afraid
to come any nearer for fear of torpedo attacks. Neither did any of
the other officers on the enemy side think of holding firmly to our
line, which would have greatly impeded our movements and
rendered a fresh attack on the enemy line extremely difficult.

Immediately after the line was turned the enemy fire ceased
temporarily, partly because the artificial smoke sent out by the
torpedo-boats to protect the line—the battle-cruisers in particular—
greatly impeded the enemy's view, but chiefly no doubt on account
of the severe losses the enemy had suffered.

Losses that were observed for certain as sunk were: a ship of
the "Queen Elizabeth" class (name unknown), a battle-cruiser

154

The Battle of the Skagerrak

(*Invincible*), two armoured cruisers (*Black Prince* and *Defence*), the light cruiser *Shark*, and one marked "O 24." Heavily damaged and partially set on fire were : One cruiser (*Warrior*, sunk later), three light cruisers, three destroyers (of which the *Acasta* was one).

On our side "V 48" was the only destroyer sunk, the *Wiesbaden* was rendered incapable, and the *Lützow* so badly damaged that the Chief of Reconnaissance was subsequently compelled at 9 P.M. to leave the ship under the enemy's fire, and transfer to the *Moltke*. The leadership of Scouting Division I was thus made over to the *Derfflinger* (Captain Hartog) until 11 P.M. The other battle-cruisers and the leading ships of Squadron III had also suffered, but kept their place in the line. No one reported inability to do so; I was, therefore, able to reckon on their being fully prepared to fight. After the enemy was forced to cease firing on our line steering S.W., he flung himself on the already heavily damaged *Wiesbaden*. The ship put up a gallant fight against the overwhelmingly superior forces, which was clearly to be seen as she had emerged from out of the clouds of smoke and was distinctly visible.

It was still too early for a nocturnal move. If the enemy followed us our action in retaining the direction taken after turning the line would partake of the nature of a retreat, and in the event of any damage to our ships in the rear the Fleet would be compelled to sacrifice them or else to decide on a line of action enforced by enemy pressure, and not adopted voluntarily, and would therefore be detrimental to us from the very outset. Still less was it feasible to strive at detaching oneself from the enemy, leaving it to him to decide when he would elect to meet us the next morning. There was but one way of averting this—to force the enemy into a second battle by another determined advance, and forcibly compel his torpedo-boats to attack. The success of the turning of the line while fighting encouraged me to make the attempt, and decided me to make still further use of the facility of movement. The manœuvre would be bound to surprise the enemy, to upset his plans for the rest of the day, and if the blow fell heavily it would facilitate the breaking loose at night. The fight of the *Wiesbaden* helped also to strengthen my resolve to make an effort to render assistance to her and at least save the crew.

Accordingly, after we had been on the new course about a quarter of an hour, the line was again swung round to starboard on an easterly course at 8.55 P.M. The battle-cruisers were ordered to operate with full strength on the enemy's leading point, all the torpedo-boat flotillas had orders to attack, and the First Leader of the torpedo-boats, Commodore Michelsen, was instructed to send his boats to rescue the *Wiesbaden's* crew. The boats told off for this purpose were compelled to relinquish the attempt. The *Wiesbaden* and the boats making for her were in the midst of such heavy fire that the leader of the torpedo-boats thought it useless to sacrifice his boats. In turning to go back "V 73" and "G 88" together fired off four torpedoes at the "Queen Elizabeths."

The battle that developed after the second change of course and led to the intended result very soon brought a full resumption of the firing at the van which, as was inevitable, became the same running fight as the previous one, in order to bring the whole of the guns into action. This time, however, in spite of "crossing the T," the acknowledged purpose was to deal a blow at the centre of the enemy line. The fire directed on our line by the enemy concentrated chiefly on the battle-cruisers and the Fifth Division. The ships suffered all the more as they could see but little of the enemy beyond the flash of fire at each round, while they themselves apparently offered a good target for the enemy guns. The behaviour of the battle-cruisers is specially deserving of the highest praise; crippled in the use of their guns by their numerous casualties, some of them badly damaged, obeying the given signal, "At the enemy," they dashed recklessly to the attack.

The conduct of Squadron II (Rear-Admiral Behncke) and the action of the ships of the Fifth Division are equally worthy of recognition. They, together with the battle-cruisers, bore the brunt of the fight, and thus rendered it possible for the torpedo-boat flotillas to take so effective a share in the proceedings. The systematic procedure of our ships in the line was a great help to the flotillas on their starboard side in opening the attack. The first to attack were those ahead with the cruisers, the boats of Flotillas VI and IX. Next came Flotillas III and V from the Main Fleet. Flotilla II was kept back by the Second Leader of torpedo-boats, for fear it might be left unprotected behind VI and IX. This action was justified by the course of events. The 1st Torpedo Half-Flotilla and a few boats from Flotillas VI and IX were occupied in covering the damaged *Lützow*. There was no longer any opportunity for an attack by Flotilla VII which had been in the rear of our fighting line. As they advanced Flotillas VI and IX were met by the heavy enemy fire that until then had been directed against the battle-cruisers; they carried the attack to within 70 hm. against the centre of a line comprising more than twenty large battleships steering in a circle E.S.E. to S., and opened fire under favourable conditions. In the attack "S 35" was hit midships and sank at once. All the other boats returned, and in doing so sent out dense clouds of smoke between the enemy and our own Main Fleet. The enemy must have turned aside on the attack of Flotillas

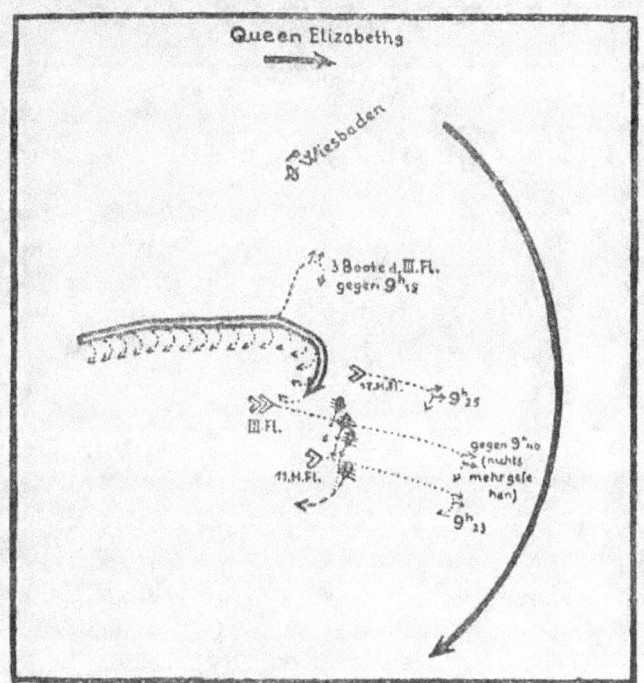

Position at 9.17 P.M.

VI and IX. Flotillas III and V that came after found nothing but light craft, and had no opportunity of attacking the battleships. The action of the torpedo-boat flotillas had achieved its purpose.

At 9.17 P.M., therefore, the line was again for the third time swung round on to a westerly course, and this was carried out at the moment when the flagship *Friedrich der Grosse* was taking a southerly course close by the turning point. Although the signal to swing round hung on the starboard side and was being carried out by the neighbouring ships, I made the Chief of the *Friedrich der Grosse* carry out the turn to larboard [port].

Germany's High Sea Fleet

This might have led the ships following behind to think that there was a mistake in the signalling. But my intention to get through and save the ships in front of the *Friedrich der Grosse* from a difficult situation in carrying out the manœuvre was rightly understood by Vice-Admiral Ehrhardt Schmidt in the *Ostfriesland*, the Leader of Squadron I. He did not wait, therefore, for the carrying out of the movement from the rear—which is the general rule to avoid all danger of collision—but himself gave the lead in the turning of his squadron by starting the turn to starboard

Swinging Round to Starboard

with the *Ostfriesland*—and thus forced his ships round. This action was a very satisfactory proof of the capable handling of the ships and the leaders' intelligent grasp of the situation.

After the change to a westerly course the Fleet was brought round to a south-westerly, southerly, and finally to a south-easterly course to meet the enemy's encircling movement and keep open a way for our return. The enemy fire ceased very soon after we had swung round and we lost sight of our adversary. The enemy's casualties at this stage of the fighting cannot be given.

Excepting the effects of direct hits which we were able to confirm from the flames of explosions, the enemy has only admitted the damage to the *Marlborough* by torpedoes.* On our side all the ships were in a condition to keep up the speed requisite for night work (16 knots) and thus keep their place in the line.

* Admiral Jellicoe admits that torpedoes reached his line, but claims to have escaped further damage by the clever handling of his ships. Our assumption that he had already turned back before the attack by the torpedo-boats is thus confirmed.

The Battle of the Skagerrak

NIGHT MOVEMENTS AND BATTLES

Twilight was now far advanced, and it was only by personal observation that I could assure myself of the presence and external condition of those ships that chiefly had been under fire, and especially that the *Lützow* was able to keep with the unit. At 9.30 the battle-cruiser was seen to larboard [port] of the flagship, and had reported that she could do 15 knots. The report made by the torpedo-boat flotilla as to the enemy's strength and the extension of his firing line made it quite certain that we had been in battle with the entire English Fleet. It might safely be expected that in the twilight the enemy would endeavour by attacking with strong forces, and during the night with destroyers, to force us over to the west in order to open battle with us when it was light. He was strong enough to do it. If we could succeed in warding off the enemy's encircling movement, and could be the first to reach Horns Reef, then the liberty of decision for the next morning was assured to us. In order to make this possible all flotillas were ordered to be ready to attack at night, even though there was a danger when day broke of their not being able to take part in the new battle that was expected. The Main Fleet in close formation was to make for Horns Reef by the shortest route, and, defying all enemy attacks, keep on that course. In accordance with this, preparations for the night were made.

The Leaders of the torpedo-boats were instructed to arrange night attacks for the flotillas. At 9.20 a southerly course was ordered. In changing to this course Squadron II had fallen out on the starboard side as the leading ship of Squadron I fell into the new course, not being able to fix the position of Squadron II. Owing to the latter's inferior speed it fell behind the ships of Squadrons III and I in the last part of the day's battle. Squadron II now attempted, at full speed and manœuvring to larboard [port], to resume its place in front of Squadron I, which was its rightful position, after the Fleet had been turned. It came, therefore, just in time to help our battle-cruisers that were engaged in a short but sharp encounter with the enemy shortly before it was quite dark. While Scouting Divisions I and II were trying to place themselves at the head of our line they were met at 10.20 by heavy fire coming from a south-easterly direction. Nothing could be seen of the enemy beyond the flash of the guns at each round. The ships, already heavily damaged, were hit again without being able to return the fire to any purpose. They turned back, therefore, and passed in between Squadrons II and I to leeward of the firing.

The head of Squadron I followed the movements of the cruisers, while Squadron II (Rear-Admiral Mauve) stood by and took the enemy's fire. When Squadron II became aware that the failing light made any return fire useless it withdrew, thinking to attract the enemy to closer quarters with Squadron I. The enemy did not follow, but ceased firing.

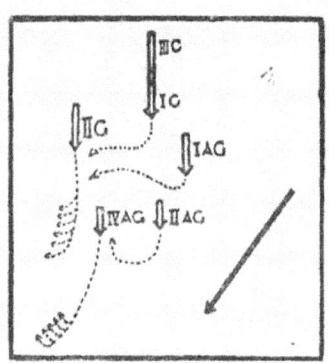

The Situation at 10.30 P.M.

Almost at the same time the Leader of Scouting Division IV. Commodore von Reuter, under similar conditions, had been engaged in a short encounter with four of five cruisers, some of them ships of the "Hampshire" class.

Following on this attack, we took a south-easterly course which was at once seen to be necessary and adopted by Squadron I, bringing Squadron II again on the starboard side of the Fleet. In view of the fact that the leading ships of the Main Fleet would chiefly have to ward off the attacks of the enemy, and in order that at daybreak there should be powerful vessels at the head, Squadron II was placed in the rear. At 11 P.M. the head of the line stood at 36″ 37′ North latitude, and 5″ 30′ East longitude. At 11.6 P.M. the order for the night was "Course S.S.E. ¼ E, speed 16 knots."

Out of consideration for their damaged condition, Scouting Division I was told off to cover the rear, Division II to the vanguard, and the IVth to cover the starboard side. The Leaders of the torpedo-boat forces placed the flotillas in an E.N.E. to S.S.W. direction, which was where the enemy Main Fleet could be expected. A great many of the boats had fired off all their torpedoes during the battle. Some were left behind for the protection of the badly damaged *Lützow;* others were retained by the flotilla leaders in case of emergency. The rescue of the crews of the *Elbing* and *Rostock* was due to that decision.

The Second, Fifth and Seventh, and part of the Sixth and Ninth were the only Flotillas that proceeded to the attack; the boats had various nocturnal fights with enemy light forces. They never

The Battle of the Skagerrak

sighted the Main Fleet. At 5 A.M. on June 1 "L 24" sighted a portion of the Main Fleet in Jammer Bay. It was as we surmised —after the battle the enemy had gone north. Flotilla II, which had been stationed at the most northerly part of the sector, was forced back by cruisers and destroyers, and went round by Skagen; at 4 o'clock when day broke the other flotillas collected near the Main Fleet.

The battleship squadrons proceeded during the night in the following order: Squadron I, Flagship of the Fleet, Squadron III and Squadron II. Squadrons I and II were now in reversed positions; that is to say, the ships previously in the rear were now at the van.

Other attempts to bring the admirals ahead were abandoned owing to the darkness and lack of time. The conduct of the line was entrusted to Captain Redlich on the *Westfalen*. The enemy attacked from the east with both light and heavy forces during the night almost without ceasing. Scouting Divisions I and II and the ships in Squadron I in particular were to ward off the attacks. The result was excellent. To meet these attacks in time, bring the enemy under fire and by suitable manœuvring evade his torpedoes, demanded the most careful observation on board the vessels. Consequently the line was in constant movement, and it required great skill on the part of the commanders to get into position again, and necessitated a perpetual look-out for those manœuvring just in front of them. Very little use was made of the searchlights. It had been proved that the fire from the attacking boats was aimed chiefly at these illuminated targets. As our light guns and the navigation control on the ships were close to the searchlights, and because of the better view to be obtained the officers and men on duty there would not take cover, several unfortunate casualities occurred. On board the *Oldenburg* the commander, Captain Höpfner, was severely wounded by a shell, and several officers and many of the crew were killed.

Utterly mistaking the situation, a large enemy cruiser with four funnels came up at 2 A.M. (apparently one of the "Cressy" class), and was soon within 1,500 metres of Squadron I's battleships, the *Thüringen* and *Ostfriesland*. In a few seconds she was on fire, and sank with a terrible explosion four minutes after opening fire. The destruction of this vessel, which was so near that the crew could be seen rushing backwards and forwards on the burning deck while the searchlights disclosed the flight of the heavy projectiles

L

till they fell and exploded, was a grand but terrible sight. Squadron I reported during the night that after carrying out an evading manœuvre the *Nassau* had not returned in her place, and as she did not answer a call it was feared she had been torpedoed. Towards morning, however, there was a faint wireless from her reporting that she was standing by the Vyl Lightship at Horns Reef, and during the night had rammed and cut through a destroyer. After this exploit the commander preferred not to return to our darkened line but made for the morning's rendezvous.

A careful estimation showed that during the night one battle-cruiser, one light cruiser and seven destroyers were sunk on the enemy's side, and several battle-cruisers and destroyers badly damaged. The 2nd Division of Squadron I at the head of the line were specially successful in the defence they put up against torpedo attacks, as they themselves accounted for six destroyers.

On our side the old light cruiser *Frauenlob*, the battleship *Pommern* and "V 4" were sunk; *Rostock* and *Elbing* were abandoned and blown up. At 12.45 A.M. the *Frauenlob* (Captain Georg Hoffmann), during a fight between Scouting Division IV and four cruisers of the "Town" class, was hit by a torpedo and, according to the accounts of the few survivors, went down fighting to the last.

The *Pommern* (Captain Bölken) was torpedoed at 4.20 A.M. and went down with a violent explosion. Unfortunately none of the crew could be saved, as the wreckage drifted away so quickly that nothing was seen on the water by a ship following at 500 m. distance.

At 4.50 A.M. "V 4" struck an enemy mine; the crew was not saved. At 1.30 A.M. the *Rostock* and *Elbing* to the larboard [port] of the head of Squadron I were engaged in a fight with destroyers, but had finally to withdraw from the enemy's torpedoes and break through Squadron I's line, so as not to impede the firing from the ships of the line. While doing this the *Rostock* was hit by a torpedo, and the *Elbing* and *Posen* collided. Both cruisers were put out of action. The *Rostock* kept afloat till 5.45 A.M., but as enemy cruisers were then sighted she was blown up, the entire crew and the wounded having previously been taken off by the boats of Flotilla III. The crew of the *Elbing* was also taken over by a boat belonging to Flotilla III. The Commander, Captain Madlung, the First Officer, the Torpedo Officer and a cutter's crew remained on board to keep the ship afloat as long as possible. When, however,

The Battle of the Skagerrak

enemy forces were sighted at 4 A.M. the *Elbing* was also blown up. The remainder of the crew got away in the cutter and were subsequently picked up by a Dutch fishing-smack and returned home *via* Holland.

The *Lützow* was kept above water until 3.45 A.M. The *König*, the rear ship of the Fleet, lost sight of her at 11.15 P.M. The vessel was at last steered from the stern. All efforts to stop the water pouring in were fruitless; the fore part of the ship had been too badly damaged, and she had at last 7,000 tons of water in her. The screws revolved out of the water, and she had to be given up. The crew with all the wounded were taken off by the torpedo-boats "G 40," "G 37," "G 38" and "V 45," and the *Lützow* was sunk by a torpedo. Altogether the four boats had 1,250 men from the *Lützow* on board. Twice they encountered enemy cruisers and destroyers, but on each occasion, led by the senior officer, Commander Beitzen (Richard), they attacked and successfully made their way into the German Bight. In the last engagement "G 40" had her engines hit and had to be towed.

When this report reached the Main Fleet the Second Leader of Torpedo-Boats on the *Regensburg* turned at once, regardless as to whether he might meet with superior English forces or not, and took over the towing party. "S 32," Leader of Flotilla I (Captain Fröhlich), was hit in her boiler at 1 A.M. and rendered temporarily useless. By feeding the boiler with sea water the captain succeeded, however, in taking the boat into Danish waters. From thence she was towed through the Nordmann Deep by torpedo-boats dispatched to her assistance.

These events prove that the English Naval forces made no effort to occupy the waters between the scene of battle and Horns Reef.

It was only during the night that there was opportunity for the ships to report on the number of prisoners they had on board and to gather from them some idea of the enemy's losses. Then I learned that the *Warspite*, which we had observed to be badly damaged in the battle, was sunk. Among other vessels reported sunk were the battle-cruisers *Queen Mary*, *Indefatigable*, and *Invincible*. This was all news to me, and convinced me that the English losses were far more considerable than our own.

On arriving at Horns Reef at 5 A.M. I decided to remain there and await the *Lützow*. I had not then heard of her fate. From 11.30 P.M. on, the vessel had been able to do 13 knots. The last report from her was at 1.55 A.M.—transmitted by convoy-boat

163

"G 40"—stating that she was making very slow way, that the means of navigation were limited, that the gun power was reduced to a fifth, course south, station E 16. At 5.30 A.M. came a message that the *Lützow* had been abandoned at 4 A.M.

After that I had no difficulty in drawing my own conclusions. As the enemy did not come down from the North, even with light forces, it was evident that he was retiring, especially as nothing more could be seen of him notwithstanding that his torpedo-boats were about until dawn.

6

THE SITUATION ON THE MORNING OF JUNE 1

"L" 11, 13, 17, 22 and 22 had gone up during the night for an early reconnaissance. At 5.10 A.M. "L 11" reported a squadron of twelve English battleships, numerous light craft and destroyers on a northerly course about the centre of the line Terschelling—Horns Reef, and immediately afterwards enemy battleships and battle-cruisers north of the first unit. The airship was heavily fired at but kept in touch until compelled to retire and lost sight of the enemy in the thick atmosphere. The airship's reports taken from its war diary are as follows:

Reconnaissance Trip of "L 11" on June 1, 1916

"On June 1 at 1.30, after midnight 'L 11' went up at Nordholz with the following orders: As fourth airship to cover flank of High Sea forces, course N.W. to W. by Heligoland. Full crew on board, fresh south-westerly wind, visibility limited owing to ground fog and later to a fog-like atmosphere high up extending over 2 or at most 4 nautical miles. Heligoland was not visible through the fog. At 5 A.M. clouds of smoke were seen north of the ship in Square O 33 B and were made for. At 5.10 it was possible to make out a strong enemy unit of twelve large warships with numerous lighter craft steering north-north-east full speed ahead. To keep in touch with them 'L 11' kept in the rear and sent a wireless report, circling round eastwards. At 5.40 A.M. east of the first unit the airship sighted a second squadron of six big English battleships with lighter forces on a northerly course; when sighted, they turned by divisions to the west, presumably to get into contact with the first unit. As this group was nearer to the Main Fleet than the first one, 'L 11' attached itself to it, but at 5.50 a group of three English battle-cruisers and four smaller craft were sighted to the north-east, and, cruising about south of the airship, put themselves between the enemy Main Fleet and 'L 11.' Visibility was so poor that it was extremely difficult to keep in contact. For the most part only one of the units was visible at a time, while, apparently, the airship at an altitude of 1,100—1,900 m. was plainly visible to the enemy against the rising sun.

"At 5.15, shortly after sighting the first group of battleships, the enemy opened fire on the airship from all the vessels with anti-aircraft guns and guns of every calibre. The great turrets fired broadsides; the rounds followed each other rapidly. The flash from the muzzles of the guns could be seen although the ships were hidden by the smoke. All the ships that came in view took up the firing with the greatest energy, so that 'L 11' was sometimes exposed to fire from 21 large and numbers of small ships. Although the firing did not take effect, that and the shrapnel bursting all around so shook the ship's frame that it seemed advisable to take steps to increase the range. The firing lasted till 6.20 A.M. At that time the battle-cruisers bearing down from S.W within close distance of 'L 11' forced her to retire to N.E. to avoid their fire. At the same time the visibility became worse and the enemy was lost to view.

"'L 11' again took a northerly course and went as low down as 500 metres, in the hope of better visibility. It was impossible to see beyond 1 to 2 nautical miles, and as under these conditions no systematic plan for keeping in contact could be made, N. and S. course was followed so as to keep between the enemy and our own Main Fleet. The enemy did not come in sight again.

"At 8 A.M. the Commander-in-Chief of the High Sea Fleet dismissed the airship, and 'L 11' returned. On the way back the ship came across a number of our own torpedo-boats exchanging bases, and messages were given for further transmission. The airship remained close to those boats as far as Sylt. Landed at Nordholz at 2 P.M."

At 4 A.M., 50 nautical miles west of Bovbjerg, "L 24" sighted a flotilla of enemy destroyers, was fired at and returned the fire with bombs, then got away further north, and at 5 A.M. discovered a unit of twelve ships in Jammer Bay, steaming rapidly to the south. It was impossible to keep in contact for further reconnaissance as there was a bank of cloud as low down as 800 m.

From the Main Fleet itself no signs of the enemy were visible at daybreak. The weather was so thick that the full length of a squadron could not be made out. In our opinion the ships in a south-westerly direction as reported by "L 11" could only just have come from the Channel to try, on hearing the news of the battle, to join up with their Main Fleet and advance against us. There was no occasion for us to shun an encounter with this group, but owing to the slight chance of meeting on account of visibility conditions, it would have been a mistake to have followed them. Added to this the reports received from the battle-cruisers showed that Scouting Division I would not be capable of sustaining a serious fight, besides which the leading ships of Squadron III could not have fought for any length of time, owing to the reduction in their supply of munitions by the long spell of firing. The *Frankfurt*, *Pillau* and *Regensburg* were the only fast light cruisers now available, and in such misty weather there was no depending on aerial reconnaissance. There was, therefore, no certain prospect of defeating the enemy reported in the south. An encounter and the consequences thereof had to be left to chance. I therefore abandoned the idea of further operations and ordered the return to port.

On the way back, west of List, the *Ostfriesland*, at 7.30 A.M., struck a mine, one that evidently belonged to a hitherto unknown and recently laid enemy minefield. The damage was slight; the vessel shipped 400 tons of water, but her means of navigation did not suffer, and she was able to run into harbour under her own steam. I signalled, "Keep on." The last ships passed through the area without coming across further mines.

Several submarine attacks on our returning Main Fleet failed entirely, thanks partly to the vigilance of the airmen who picked up the Main Fleet over List, and escorted them to the mouth of the river. During the course of the day all the ships and boats were safely in their haven, Special mention must be made of the bringing-in of the *Seydlitz* (Captain von Egidy) badly damaged at her bows. That the vessel ever reached the harbour is due to the remarkable seamanship of her commander and crew. Finally she was run astern into the dock at Wilhelmshaven.

The U-boats lying off English harbours were told to remain at their posts a day longer. At 6.20 P.M., 60 miles north of Terschelling, the "U 46" came across a damaged vessel of the "Iron Duke" class (the *Marlborough*). She was, however, so well protected that it would have been impossible to get within firing distance of her. A torpedo was fired, but failed to reach the

The Battle of the Skagerrak

objective. Among the U-boats lying off enemy harbours the "U 21" on May 31 and "U 22" on June 1 both succeeded in hitting a destroyer. In each case, however, the sinking could not be observed owing to enemy counter-action. Besides this, one of our mine-layers, occupied in laying mines west of the Orkney Islands, achieved an important success. The English armoured cruiser *Hampshire* (11,000 tons) struck one of these mines on June 5 and sank; with her perished Field-Marshal Lord Kitchener and all his Staff.

LOSSES ON BOTH SIDES

According to careful estimation made by us the enemy lost:

		Tons.
1	Dreadnought of "Queen Elizabeth" class...	28,500
3	Battle-cruisers (*Queen Mary, Indefatigable* and *Invincible*)	63,000
4	Armoured Cruisers (*Black Prince, Defence, Warrior* and one of the "Cressy" type)	53,700
2	Light Cruisers	9,000
13	Destroyers	15,000
	Total	169,200

We lost:

		Tons.
1	Battle-cruiser (*Lützow* *)	26,700
1	older Battleship (*Pommern*)	13,200
4	Light Cruisers (*Wiesbaden, Elbing, Rostock* and *Frauenlob*)	17,150
5	Torpedo-boats	3,680
	Total	60,730

The enemy's were almost complete losses, whereas we had rescued the crews of the *Lützow, Elbing, Rostock* and half of those of the torpedo-boats.

---

* In my first report of the battle sent to the Admiralty at Berlin the loss of the *Lützow* was mentioned. The announcement of this loss was suppressed by the Naval Staff, though not at my request. The enemy could not have seen the ship go down. In the interests of naval warfare it was right to suppress the news. Unfortunately the secrecy observed produced the impression that it was necessary to enlarge our success to that extent.

Germany's High Sea Fleet

Our losses in personnel amounted to : 2,400 killed; 400 wounded. The enemy's losses may be estimated at over 7,000 killed.

According to a list which he added to his report of June 18, 1916, Admiral Jellicoe endeavoured to exaggerate our losses in the following manner :

BATTLESHIPS OR BATTLE-CRUISERS

	Correct facts.
2 Battleships, "Dreadnought" type (certain)	none
1 Battleship, "Deutschland" type (certain)	one
1 Battleship or Battle-cruiser (probable)	one
1 Battleship, "Dreadnought" type (probable)	none

LIGHT CRUISERS

4 Light cruisers (certain)	four
1 Large ship or light cruiser (certain)	none

TORPEDO-BOAT DESTROYERS

6 Torpedo-boat destroyers (certain)	five
3 Torpedo-boat destroyers (probable)	none

SUBMARINES

1 Submarine (certain)	none
3 Submarines (probable)	none

With regard to the submarines he was totally mistaken, as none took part in the battle. I sent my final impressions of the battle in a written report of 4/7/16 to H.M. the Emperor as follows :

" The success achieved is due to the eagerness in attack, the efficient leadership through the subordinates, and the admirable deeds of the crews full of an eminently warlike spirit. It was only possible owing to the excellence of our ships and arms, the systematic peace-time training of the units, and the conscientious development on each individual ship. The rich experience gained will be carefully applied. The battle has proved that in the enlargement of our Fleet and the development of the different types of ships we have been guided by the right strategical and tactical ideas, and that we must continue to follow the same system. All arms can claim a share in the success. But, directly or indirectly, the far-

The Battle of the Skagerrak

reaching heavy artillery of the great battleships was the deciding factor, and caused the greater part of the enemy's losses that are so far known, as also it brought the torpedo-boat flotillas to their successful attack on the ships of the Main Fleet. This does not detract from the merits of the flotillas in enabling the battleships to slip away from the enemy by their attack. The big ship—battleship and battle-cruiser—is therefore, and will be, the main strength of naval power. It must be further developed by increasing the gun calibre, by raising the speed, and by perfecting the armour and the protection below the water-line.

"Finally, I beg respectfully to report to Your Majesty that by the middle of August the High Sea Fleet, with the exception of the *Derfflinger* and *Seydlitz*, will be ready for fresh action. With a favourable succession of operations the enemy may be made to suffer severely, although there can be no doubt that even the most successful result from a high sea battle will not compel England to make peace. The disadvantages of our geographical situation as compared with that of the Island Empire and the enemy's vast material superiority cannot be coped with to such a degree as to make us masters of the blockade inflicted on us, or even of the Island Empire itself, not even were all the U-boats to be available for military purposes. A victorious end to the war at not too distant a date can only be looked for by the crushing of English economic life through U-boat action against English commerce. Prompted by the convictions of duty, I earnestly advise Your Majesty to abstain from deciding on too lenient a form of procedure on the ground that it is opposed to military views, and that the risk of the boats would be out of all proportion to the expected gain, for, in spite of the greatest conscientiousness on the part of the Chiefs, it would not be possible in English waters, where American interests are so prevalent, to avoid occurrences which might force us to make humiliating concessions if we do not act with the greatest severity."

I followed up my report on the battle with a more detailed account on July 16, 1916, after Admiral Jellicoe's report had appeared in the English Press. I quote here from the above-mentioned account :

"Admiral Jellicoe's report, published in the English Press, confirms as follows the observations made by us:

Germany's High Sea Fleet

1

Grouping of the English Forces

Under Vice-Admiral Beatty :

1st and 2nd Battle-Cruiser Squadrons.
5th Battle Squadron (" Queen Elizabeths ").
1st, 2nd and 3rd Light Cruiser Squadrons.
1st, 9th, 10th and 13th Destroyer Flotillas.

Admiral Jellicoe led :

1st, 2nd and 4th Battle Squadrons (Fleet Flagship at the head of 4th Battle Squadron).
3rd Battle-Cruiser Squadron (" Invincibles ").
1st and 2nd Cruiser Squadrons.
4th Light Cruiser Squadron.
4th, 11th and 12th Destroyer Flotillas.

2

Intervention in the Battle by the English Main Fleet

" When he first had news that the enemy was sighted, Admiral Jellicoe was north-west of Admiral Beatty's forces. He thereupon advanced at full speed in column formation on a S.E. course, put the 1st and 2nd Cruiser Squadrons for reconnaissance at the head of his formation, and sent forward the 3rd Battle Cruiser Squadron (apparently reinforced by the *Agincourt* *), to support Admiral Beatty. The 3rd Battle-Cruiser Squadron passed east of Admiral Beatty's leader at 7.30 P.M.; they heard in the south-west the thunder of guns, and saw the flashes, sent out the light cruiser *Chester* to reconnoitre, and themselves took a N.W. course. Shortly before 8 o'clock the *Chester* encountered our Scouting Division II and was set on fire by them. After pursuing the *Chester*, Scouting Division II came across the 3rd Battle-Cruiser Squadron, which opened fire on them. The attacks at 8 P.M. by our Torpedo-Boat Flotilla IX and the 12th Half-Flotilla were launched against this 3rd Battle-Cruiser Squadron.

"Admiral Beatty sighted the 3rd Battle-Cruiser Squadron at 8.10 P.M., and at 8.21 P.M. had it ahead of the 1st and 2nd Battle-Cruiser Squadrons he was leading.

"At 7.55 P.M. Admiral Jellicoe sighted the fire from the guns.

* Observed by Scouting Division II.

170

The Battle of the Skagerrak

It was impossible for him to make out the position of our Fleet. The difference between his and Admiral Beatty's charts added to the uncertainty in judging of the situation. The report says it was difficult to distinguish between friend and foe. At 8.14 P.M. the battleship squadrons turned east into the line between the 1st and 2nd Battle-Cruiser Squadrons and the 5th Battle Squadron. At 8.17 P.M. the 1st Battle Squadron opened fire on the leaders of

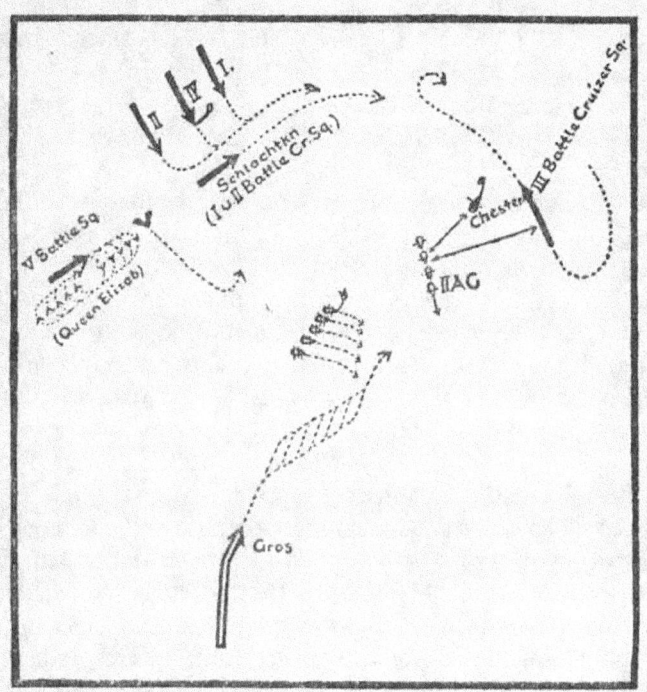

Position of English Forces at 8 p.m.

our ships of the line. Up to 10.20 P.M. those squadrons, with some few pauses, took part in the fighting.

"Shortly before the battleship squadrons arrived, the 1st Cruiser Squadron, together with light forces from the Main Fleet, joined in the fighting. At 8.50 P.M., therefore, between our first and second blows, Admiral Beatty put the 3rd Battle-Cruiser Squadron in the rear of the 2nd. At 9.6 P.M. the leaders of the battleships made for the south. The total impression received by us of the battle is made more complete by the statements in the English Press, and is not altered.

3

The Enemy's Action during the Night

"At 9.45 P.M. Admiral Beatty had lost sight of our forces. He sent the 1st and 3rd Light Cruiser Squadrons to reconnoitre in the west, and at 10.20 P.M. went to their support with the 1st and 2nd Battle-Cruiser Squadrons, also on a westerly course. Immediately after came the encounter described in my report with the leading ships of our Main Fleet, consisting of Scouting Divisions IV and I and Squadron II. The fact that our forces turned westward must have led the English Admiral to assume that our Main Fleet had taken a westerly course, and made him follow in that direction. The fact that we at the same time put Squadron II in the rear, and with the new leader, Squadron I, again took a S.E. course, resulted in Admiral Beatty's forces passing west in front of us and ultimately losing contact. It was obvious that after the battle the English Main Fleet was divided into two. Admiral Jellicoe's report makes no mention of this. The one portion, consisting of large battleships and light craft, took apparently northerly and easterly courses, as one group of ships was sighted by 'L 24' at 5 A.M. on June 1 in Jammer Bay, close under land. It may perhaps have been both those rear squadrons which made off on the attack by our Torpedo-Boat Flotillas VI and IX, and then apparently lost touch with the Main Fleet.* The other portion, under Admiral Jellicoe, consisting, according to observations by 'L 11,' of eighteen large battleships, three battle-cruisers (probably the 3rd Battle-Cruiser Squadron) and numerous light forces, had, up to 10.46 P.M., been steering south and then south-west. It would appear, from intercepted English wireless messages, that he covered 15 nautical miles. Based on these courses and the speed, he must have crossed our course at midnight, 10 to 15 nautical miles in front of us, and have taken later a course to the centre of the line Horns Reef—Terschelling, where he was seen at 5 A.M. by 'L 11' on a N.N.E. course.

4

The Consequence of the Enemy's Action during the Night

"Admiral Jellicoe must have intended to resume the battle with us at dawn. It is inexplicable, therefore, why a portion of the

* According to Admiral Jellicoe's book, one group of battleships did not rejoin him till 6 P.M. on June 1.

The Battle of the Skagerrak

Main Fleet made for Jammer Bay during the night. Nor can it be understood how it was that the enemy's light forces, which were engaged with our Main Fleet up to 4.36 A.M. and thus were in touch with us the whole night, could find a way to inform Admiral Jellicoe and Admiral Beatty of our course and navigation. But even apart from that, it must be assumed that the fire from our guns and the enemy's burning cruisers and destroyers would have pointed out the way to the English Main Fleet. In any case it is a fact that on the morning of June 1 the enemy's heavy forces were broken up into three detachments; one in the North, a second with Admiral Beatty in the North-west, and the third with Admiral Jellicoe South-west of Horns Reef. It is obvious that this scattering of the forces—which can only be explained by the fact that after the day-battle Admiral Jellicoe had lost the general command—induced the Commander to avoid a fresh battle."

CHAPTER XI

AFTER THE BATTLE

ON June 1 at 3 P.M. the *Friedrich der Grosse* anchored in the Wilhelmshaven Roads. Meanwhile the crews on all our ships had attained full consciousness of the greatness of our successes against the superior enemy forces, and loud and hearty cheers went up as they steamed past the flagship of their leader. Though they had been under such heavy fire, very little external damage on the ships was apparent; none keeled over or showed an increased draught. On a closer inspection, however, considerable damage was disclosed, but the armour-plating had so thoroughly served its purpose of protecting the vital parts of the ships that their navigating capabilities had not suffered. The *König* and *Grosser Kurfürst* went into dock as their anchor cables had been shot away. The battle-cruisers were also docked to find out to what extent repairs would be necessary. In their case the exterior damage was considerably greater. It was astonishing that the ships had remained navigable in the state they were in. This was chiefly attributable to the faulty exploding charge of the English heavy calibre shells, their explosive effect being out of all proportion to their size. A number of bits of shell picked up clearly showed that powder only had been used in the charge. Many shells of 34- and 38-cm. calibre had burst into such large pieces that, when picked up, they were easily fitted together again. On the other hand, the colour on the ships' sides, where they had been hit, showed that picric acid had been used in some of the explosive charges. A technical Commission from the Imperial Naval Department made a thorough investigation of the effects of the shots in order to utilise the experience gained. We arrived immediately at one conclusion—a final decision on the much-debated question of protective torpedo-nets for the Fleet to the effect that the nets must be done away with. On most of the ships they were so damaged as to make it impossible to remove them after the fighting; they hung, for the most part, in a dangerous fashion out of their cases and it was a wonder that they did not get entangled in the propellers, an occurrence which, during the battle— or at any time for that matter—might have greatly inconvenienced

After the Battle

the Fleet. The total impression produced by all the damage done was that by their splendid construction our ships had proved to be possessed of extraordinary powers of resistance.

The next step was to make arrangements for the repairing of the ships as the docks at Wilhelmshaven were not able to cope with all the work, and it was essential that the Fleet should be brought as quickly as possible into a state of preparedness for action. The Wilhelmshaven yard was entrusted with the repairs of the *Seydlitz*, and the ships of Squadron I, of which the *Ostfriesland*—owing to a mine explosion—and the *Helgoland*—hit above the water-line—had to be placed in dock. The *Grosser Kurfürst*, *Markgraf*, and *Moltke* were sent to Hamburg to be repaired by Blohm & Voss and the Vulcan Works. The *König* and the *Derfflinger*, after the latter had been temporarily repaired in the floating-dock at Wilhelmshaven, proceeded through the Kaiser Wilhelm Canal to the Imperial Yard and Howaldt's yard at Kiel.

The Imperial Dockyards at Kiel under the management of Vice-Admiral von Henkel-Gebhardi, and those at Wilhelmshaven under Rear-Admiral Engel, as well as the private yards occupied on repairs, deserve the greatest credit for the excellent work done in restoring the Fleet.

If the English Fleet had fared as well as the English Press accounts led us to believe we might count on their immediately seizing the opportunity for a great attack. But it never came off. Our efforts were centred on putting to sea again as soon as possible for a fresh advance. By the middle of August the Fleet was again in readiness, with the exception of the battle-cruisers *Seydlitz* and *Derfflinger*. But a new ship, the *Bayern*, had been added to the Fleet, the first to mount guns of 38 cm.

Immediately after the battle joyful messages and congratulations on the success of the Fleet poured in from all divisions of the army in the field, from every part of the country and from all classes of the people. I welcomed with special gratitude the many sums received towards the support of the families of the fallen and wounded, which showed in a touching manner the sympathy of the donors, and which, in a very short space of time, reached the sum of one million marks.*

The first honour paid to the Fleet was a visit from His Majesty the Emperor on June 5, who, on board the flagship, *Friedrich der*

* The fund, called " The Skagerrak Gift," is now administered by the Imperial Naval Institute, Berlin.

Germany's High Sea Fleet

Gross, made a hearty speech of welcome to divisions drawn from the crews of all the ships, thanking them in the name of the Fatherland for their gallant deeds. In the afternoon the Emperor visited all the hospitals where the wounded lay, as well as the hospital ship *Sierra Ventana,* where lay Rear-Admiral Behncke, Leader of Squadron III, who was wounded in the battle, and who was able to give the Emperor a detailed account of his impressions while at the head of the battleships. Several of the German princes also visited the Fleet, bringing greetings from their homes to the crews and expressing pride in the Fleet and the conduct of the men. The Grand Dukes of Mecklenburg-Schwerin and of Oldenburg came directly after the battle and were followed very soon after by the Kings of Saxony and Bavaria.

All this afforded clear proof that no other organisation in the Empire was so fitted to signify its unity as the Navy, which brought together in closest contact those belonging to all classes in the Fatherland and united them by common action in fortune and misfortune. Apart from the inspection of the ships, these visits also offered an opportunity of gaining information respecting the general duties of the Fleet and the plans for the impending battle that was expected, for, as those visits proved, the battle had greatly enhanced the interest in the Fleet throughout the whole country.

The development of the battle and its lessons were thus summarised by me at the time :

"The battle was brought about as a result of our systematic efforts to attract the enemy out of his retirement, especially of the more drastic operations which culminated in the bombardment of the English coast. England's purpose of strangling Germany economically without seriously exposing her own Fleet to the German guns had to be defeated. This offensive effort on our part was intensified by the fact that the prohibition of the U-boat trade-war made it impossible for us to aim a direct blow at England's vital nerve. We were therefore bound to try and prove by all possible means that Germany's High Sea Fleet was able and willing to wage war with England at sea and thus help to establish Germany's claim to independent overseas development.

"The German idea incorporated in the founding of the Fleet had to hold its own in battle in order not to perish. The readiness to face a battle rests on the fundamental idea that even the numerically inferior must not shirk an attack if the will to conquer is supported by a devoted staff, confidence in material, and a firm conviction of perfect training.

"A preliminary fight between cruisers lasting about two hours, which proved the superiority of our guns, was followed by the engagement with the vastly superior enemy Main Fleet. The clever attempts made by the English to surround and cut us off from home by their Main Fleet were turned into a defeat, as we twice succeeded in pushing into the enemy formation with all our strength, and in withdrawing from the intended encircling movement. In spite of various attacks during the night we forced a way for ourselves to Horns Reef, and thus secured an important strategical point for the following morning.

"The enemy suffered twice as much material loss and three times as many losses in personnel as we did. English superiority was thus wrecked, for the Fleet was unable to keep in touch with us at the close of the day-battle and its own formation was broken.

"After an encounter with our leading ships, as darkness came on the English battle-cruisers lost touch with us in a mysterious way. They advanced into an empty North Sea.

"At the end of the battle the English Main Fleet had lost touch with its other units and they only came together again the following day at 6 P.M.

"After a continuous, and for the English very disastrous, night's fighting, Jellicoe did not seek us out the following morning, although he possessed both the power and the requisite speed to do so.

"We have been able to prove to the world that the English Navy no longer possesses her boasted irresistibility. To us it has been granted to fight for the rights of the German Nation on the open seas, and the battle proved that the organisation of our Navy as a High Sea Fleet was a step in the right direction. The German national spirit can only be impressed on the world through a High Sea Fleet directed against England. If, however, as an outcome of our present condition, we are not finally to be bled to death, full use must be made of the U-boat as a means of war, so as to grip England's vital nerve."

I expressed these views to the Imperial Chancellor, who visited the Fleet on June 30 in company with the Under-Secretary of State, von Stumm, and laid great emphasis on them in my report of July 4, as I noticed from communications from the Chief of the Naval Staff and the Naval Cabinet that efforts were on foot for resuming the U-boat warfare in its inadequate form. The Imperial Chancellor gave me a detailed but gloomy picture of the situation which forced him for the time to ward off any further enemies from Germany, who, he was convinced, would soon show themselves on the proclamation of unrestricted U-boat warfare. I explained to him the military reasons which would render ineffectual the carrying on of the U-boat war on a cruiser basis.

Whether political circumstances would permit us to employ the most effective weapon against England was, however, a matter for the Cabinet to decide, and for my part as Chief of the Fleet I would not attempt to exert any pressure in that direction, as that was the business of the Naval Staff. But I could not approve of carrying on the U-boat campaign in a milder form, for that would be unsatisfactory from every point of view. The Imperial Chancellor agreed with me, but declared, for various reasons, that he could not embark on a course of unrestricted U-boat warfare, because it was impossible to avoid incidents which might lead to complications, and the result would be that the fate of the German nation might lie in the hands of one U-boat commander. Before leaving Wilhelmshaven he met at dinner all the admirals then stationed there, and on this occasion he expressed the hope that in this war we should succeed in making good use of all the weapons of the Navy.

After this visit, however, it became abundantly clear to me that for the time being we were hardly likely to resume the active U-boat campaign against English commerce. In a long interview with the Imperial Chancellor that afternoon, I gathered from his remarks that he was very anxious not to incense England further, or to provoke that country to "war to the death."

Very soon all sorts of rumours arose concerning this visit: the Chancellor had gone with the object of persuading the admirals to weaken their attacks upon the British; he had more especially objected to the airship raids. All these reports were absolutely unfounded, for these matters were never touched upon, and moreover, I could not have considered it within his province to give me advice as to the manner in which war was to be waged.

Until the active operations of the Fleet were resumed, the torpedo-boats continued their efforts to get in touch with the enemy. As the base in Flanders offered better opportunities for this, while the Fleet was restricted in its activities, a flotilla was despatched there. This arrangement was continued later. At first detachments of the various flotillas were sent in turn, in order as far as possible to afford all boats the opportunity of becoming familiar with the methods of attack from that point. Later on, it appeared more advantageous to place a single flotilla for this purpose under the control of the Naval Corps, so as to make full use of the knowledge they had acquired of the local conditions.

At the beginning of August it was possible to resume the air raids again, as the nights had by then grown darker. The first attack was made in the night of the 2nd and 3rd, and was directed upon the counties of Norfolk, Suffolk and Essex. London, too, was extensively bombed. In the night of August 8th—9th there was another attack, this time upon the Midlands; and at the end of the month, in the night of August 24th—25th, there was a third raid upon the City, and the south-west district of London, as well as upon Harwich, Folkestone and the roads at Dover. One army airship took part in this. In spite of active opposition the airships returned safely from all three expeditions.

We learnt that the English defences had been decidedly improved, which rendered our attacks more difficult. The greater the effort England made to maintain her army on the Continent and in the other theatres of war, in order to do her part in forcing the decision against us on land, the more embarrassing she must have found it to organise a strong defence against airships.

Between these two periods of attack the airships were placed at the service of the Intelligence Department in connection with an attack which was planned as soon as the ships had been made ready, and which was to be again directed towards Sunderland. No change in the strategic disposition of the English Fleet had been observed. The U-boat campaign against commerce in the war-zone round about England was still in abeyance, and the U-boats were ready to be used for military purposes. These two weapons, the airships and the U-boats, would, I thought, make up for the superiority of the English Fleet in other respects.

The disposition of U-boats outside British ports on May 31 in accordance with the plan we had adopted had resulted in no success worth speaking of; it was bound to fail if the English Fleet was already at sea when the flotilla put out. Nor was their method of attack satisfactory. Before the Firth of Forth each of the seven U-boats which had been dispatched thither had a certain sector assigned to it, and these sectors radiated from a central point at the mouth of the estuary. The nearer the boats came to the estuary, the nearer they approached each other in the neighbourhood of this central point, so that they were liable to get in each other's way, or mistake one another for hostile craft. If they stood farther out to sea, the distance between them was increased and they lost their formation, thereby making it easier for the enemy to get through.

The matter was, therefore, reconsidered, and new arrangements made which promised greater success. Trial was first to be made of the method of a movable base line in the direction of the probable approach of the enemy, on which line the U-boats were to take up positions. The boats in commission in the middle of August were divided into three groups, two of which consisted of boats belonging to the Fleet, and the other of boats attached to the Naval Corps in Flanders. The two former were first to take up positions indicated in the accompanying plan by "U-Line I" and "U-Line III." In this way they afforded protection to the Fleet on either flank when proceeding to attack. The U-boats of the Fleet took up a position of defence for flank and rear against possible attacks from the Channel. In addition to the Lines I and III, other positions had been provided, which the boats were to take up either after a certain interval of time, or upon a prearranged signal. In order that the boats should be directed in accordance with the aims and movements of the Fleet, the officer commanding the U-boats was on board one of the battleships for the duration of the Fleet's attack.

The following was the plan for this enterprise against Sunderland: The Fleet was to put out by night, to advance through the North Sea towards the English coast, so that the line of U-boats might come into action, if required. If no collision with the enemy occurred, and there were no indications that the English Fleet would cut off our retreat from the sea, the ships were to push on to the English coast and bombard Sunderland at sunset. After the bombardment, while the Fleet returned in the darkness to the German Bight, the U-boats were to take up their second positions in the direction of the probable approach of the enemy, if, as was expected, he should come up as a result of the bombardment.

On August 18, therefore, at 10 P.M., the Fleet put to sea from the Jade and set out upon the course indicated in the diagram.

Squadrons III and I took part in full force; to Squadron II had been assigned the duty of protecting the German Bight. The cruisers were stationed at a distance of 20 nautical miles in advance of

After the Battle

the Fleet and were to maintain this distance throughout the advance. They were reinforced by the following battleships : The *Bayern*, which had only newly joined the Fleet after the battle of the Skager-rak; the *Grosser Kurfürst* and the *Markgraf*, because Scouting Division I was short of two battle-cruisers still under repair. A further reason for the reinforcement was the possibility that the fast squadron of "Queen Elizabeth" ships might have joined the English battle-cruisers. The distance of 20 nautical miles between our cruisers and the main body of the Fleet was to ensure immediate tactical co-operation in the event of our meeting the enemy, and to prevent the Cruiser Division, together with the three valuable battle-ships which had been assigned to it, from possibly failing to join up with the two other squadrons.

Thanks to the clear weather which prevailed during the advance on the following day, August 19, the smoke of the cruisers was visible all the time. Eight airships, among them three of a new and improved type, had taken up their positions encircling the Fleet. I hoped by this means to be able to get early news of the approach of any considerable English force within the area covered by the airships. The advance of the Fleet took place according to plan along the course indicated in the sketch, up till 2.23 P.M.

At 5.30 A.M. our advance guard met a submarine, which induced me to manœuvre the Fleet so as to evade this danger. Nevertheless, the submarine succeeded in getting within striking distance of the last ship of our line. At 7.5 A.M. the *Westfalen* reported that she had been hit amidships on the starboard side. Though the ship was not seriously damaged, I nevertheless feared that if she were struck by another torpedo she might be sunk, so I gave up the idea of her going on with us. The *Westfalen* was able to return to the Jade under her own steam, and on her way was attacked a second time, but the torpedo missed. In the course of the morning various items of information as to enemy movements were received from the airships and U-boats. The positions of the various fighting units and groups of the enemy that were notified, are indicated in the diagram.

At 8.30 A.M. the "L 13" sighted two destroyer flotillas and behind them a cruiser squadron proceeding at full steam on a south-westerly course, and at 10.40 A.M. some small cruisers with three flotillas on a north-easterly course were seen. At our chief wireless station at Neuminster, owing to the many messages intercepted, they concluded that the English Fleet was at sea, and informed us to this effect. "U 53"—Lieutenant-Commander Rose—sighted three large battleships and four small cruisers at 10.10 A.M. on a northerly course, and towards noon "L 21" announced hostile craft on a north-easterly course. At 1.40 P.M. "U 52" reported that she had sighted four small hostile cruisers on a northerly course at 9 A.M., and had sunk one of them. Thus the line arrangement had already proved effective. But from all the information received no coherent idea of the counter-measures of the enemy could be formed. We could safely assume that he was aware of the fact that we had put to sea, for the submarine that had hit the *Westfalen* had had ample time since 7 A.M. to send messages to England. Up to this time the remaining airships had reported no movement of larger forces, and the visibility in the locality of the Fleet justified the assumption that our airships commanded a clear view over the whole sea area.

At 2.22 P.M. the "L 13" reported that at 1.30 P.M. it had sighted in the south strong enemy forces comprising about 30 units, on a northerly course in Square 156, and I determined to advance against these forces. The Cruiser Division was called up, and when they had joined us, they were pushed forward in a south-easterly direction in column formation. At 2.30 P.M. another report came from the "L 13" that the hostile forces were now in Square 144 on a course north by east, that they consisted of 16 destroyers, small and large cruisers and battleships. If we and they continued on our respective courses, we might expect to encounter them in two hours. The Scouting Division and Torpedo Flotilla II were sent ahead to reconnoitre. At 3.50 P.M. the "L 13" reported that it had lost touch with the enemy forces because it had been forced to turn aside from its course in order to avoid thunderstorms. Unfortunately the airship failed to get into touch with them again. I hoped, however, soon to get news of the enemy from our ships, since, according to our reckoning, it was now the hour when the encounter should take place; but I received no information from them. Either the enemy had changed his course, because he was disquieted by the presence of the airship which he assumed was scouting for the Fleet, or the airship, owing to its unreliable navigation, had incorrectly reported his position.

The bulk of the fleet continued to advance until stopped by the minefields in the south. It being then 4.35 P.M., our course was altered to E.S.E., and we began our return journey. There was no further prospect of coming up with the enemy in the south, and it had grown too late to bombard Sunderland. While the Fleet was moving in a south-easterly direction, reports came in from "U 53" and two airships, "L 11" and "L 31," which indicated that strong enemy forces had assembled at a spot about 60 nautical miles north of our course, and were steaming in such a direction that they would have met the main portion of our Fleet had it held on its course. "U 53" had followed the hostile fleet until 4.30 P.M., when she lost sight of it as it was on a southerly course. Later, at 9 P.M., by chance she again met the enemy ships, which were then on a north-westerly course. At 10.45 P.M. this enemy squadron passed within range of "U 65," so that this boat had a chance to attack, which it accepted and succeeded in damaging a large battle-ship with her torpedoes. The British Fleet then disappeared in a northerly direction under full steam.

Another of our U-boats, "U 66" (Lieutenant-Commander von Bothmer) encountered six battle-cruisers and a number of small, fast cruisers towards 6 P.M.; these were steaming, when first seen, south-east, but later on their course was north-west. She succeeded in hitting a destroyer with a torpedo which sank her, and badly damaged a small cruiser of the "Chatham" class with two torpedoes. This same group was also sighted by the "L 31."

From reports received at 6 P.M. from "U 53" and "L 31," it was apparent that the British Main Fleet discontinued its advance to the south about 6 o'clock and turned back in a north-westerly direction. As to the movements of the hostile craft reported by "L 13" at 2.23 P.M., nothing further was discovered, except that from 7.40 P.M. onwards six small cruisers and two destroyer flotillas accompanied the main German force on its easterly course until darkness fell. They were first reported by "L 11" and then sighted by our ships as their funnels and masts were just visible above the horizon. There was no doubt but that the English light craft must have recognised our big ships with their heavy smoke-clouds, and as they kept on the same course it was to be inferred that they would keep in touch with us until there was a chance of making a night attack. I had to decide whether or not I should send our light cruisers and torpedo-boats against them to drive them off, and I relinquished the idea of doing so, because I reckoned that the English would have the advantage of us in speed. Moreover, I thought that after our lucky experience on the night of June 1, I might run the risk of a combined night attack. But so as not to be surprised by torpedo-boat attacks a strong guard of torpedo-boats was placed in our van, for the return journey by night. The English torpedo-boats, however, did not take advantage of this favourable opportunity to make a night attack upon our whole fleet. To our great surprise, "L 11" reported at 8.10 P.M. that the enemy was sheering off in a south-easterly direction, and that at 10.10 P.M. he had turned completely and disappeared from view. Probably these light craft belonged to the group first reported by "L 13," and had separated from the battleships.

No further special incidents occurred during our return journey. The cruiser attacked by "U 66" was met by "U 63" the next day while she was being towed into port. "U 63" attacked the towing convoy, which had strong protection, and succeeded in sending two torpedoes into the cruiser, which then sank. The protecting destroyers immediately gave chase to "U 63"; one of them ran her down and rammed her slightly, without, however, doing any serious damage. "U 66" sent the following report of her encounter with the enemy: At 5 P.M. she sighted small cruisers, two destroyer flotillas, and in the rear six battle-cruisers, all on a south-easterly course, and she managed to attack a four-funnel destroyer, apparently of the "Mohawk" class. Shortly after being torpedoed the destroyer lay with her stern projecting from the water, while her deck was submerged as far as the third funnel. A little later the whole cruiser squadron returned. "U 66" then attempted an attack on the small cruisers, that were now in the rear, steaming 25 knots. She got within range of a cruiser of the "Chatham" class, and struck her first in the forecastle and then in the turbine room. The ship stopped at once and lay with a strong list. Kept under water by the hostile destroyers, it was two and a half hours before "U 66" found an opportunity to attack for the second time. Shortly before firing this torpedo, our U-boat sighted a destroyer 300 metres away bearing down upon her at full steam. The U-boat quickly submerged. Immediately after a loud explosion occurred above the boat, the lights went out, the gratings burst off two hatches, the hatch-covers sprang open so that the water poured in fore and aft, but luckily they were closed again by the pressure of the water. The boat was chased by destroyers until dark, and was then out of sight of the cruiser.

"U 65," which encountered the English Fleet towards evening, made the following report. In the twilight she saw the English Fleet approaching on a westerly course. Its formation was three divisions in single line abreast, of which two comprised seven or eight large battleships, and the other five ships of the "Iron Duke" and "Centurion" classes, and a group of three battle-cruisers, one of

After the Battle

which belonged to the "Indefatigable" class. The first squadron proceeded on a N.W. course, and the others followed; the battle-cruisers, bringing up the rear, were disposed about 500 metres to port. Pushing forward at full speed, "U 65," at an estimated range of 3,000 metres, fired four torpedoes at the leading battle-cruiser. The U-boat was half submerged, and the observers in the conning-tower. After a lapse of some three minutes, the time required by a torpedo to traverse a distance of 3,000—4,000 metres, a column of fire, 20 metres wide and 40 metres high, rose behind the stern funnel of the last battleship and was visible for about a minute, while the funnel itself, white hot, was clearly discernible through the flames. At the same time there was a violent escape of steam. The fire lasted one minute. When the ship became visible again only the hull, without funnels or masts, was to be seen, whereas the silhouettes of the ships near by, with their funnels and masts, were clearly visible. This attack was made at 10.45 P.M., Lat. N. 55° 25', Long. W. 0° 30'. The commander, the officer of the watch, and the U-boat pilot were all unanimous in their description of this phenomenon. After this the U-boat had to submerge very deeply, as the Main Fleet was surrounded by a considerable number of destroyers (about forty). The only difference of opinion among the observers was as to whether the ship that had been hit was the last battleship of the 3rd Squadron or the leading battle-cruiser.

The disposition of our U-boats in a movable line had met with the desired success, and certainly was more advantageous than stationing them outside the enemy ports of issue, a proceeding which must be worthless if the ships were at sea beforehand. The U-boats also accomplished good service in scouting on this occasion, and the perseverance with which "U 53" clung to the enemy was especially praiseworthy. Unhappily, her speed was not sufficient to enable her to follow the enemy all the time.

The reports from the airships were not entirely reliable, chiefly because they were only eight in number and were expected to keep such a large area in view. Scouting by airships is, in any case, somewhat negative in character, since the fleet is only informed by them that the main hostile fleet is *not* within their field of vision, whereas the important thing is to know where it actually is.

Although on this occasion the expected naval action with the enemy did not take place, and we had to content ourselves with the modest success of two small cruisers destroyed and one battle-ship damaged, while on our side the *Westfalen* received injuries, yet we had conclusively shown the enemy that he must be on the watch for attacks by our Fleet. From English reports received subsequently we know that the British Admiral, when he ran up against our line of U-boats, felt as if he were in a hotbed of submarines and consequently quickly retired to the north.

There was a possibility that we might have joined battle with the enemy fleet at 4 P.M., if the report of "L 13" had not induced me to turn south with a view to attacking the ships sighted in that direction. The main object of our enterprise was to defeat portions of the English Fleet; the bombardment of Sunderland was only a secondary object, merely a means to this end. Therefore, when an opportunity seemed to offer to attack hostile craft in the south, I had to seize it and not let it slip.

A similar enterprise was planned for the beginning of September. The disposition of the U-boats was again based on the idea of protecting our flanks. But this time there was to be a modification, because with the single base-lines there was no guarantee that the U-boats on the line would be sure of a chance to open fire if the enemy should run into the line. The enemy's protective craft were in a position to prevent that U-boat which first sighted the enemy from attacking, and the other U-boats of the line would be too far away to take a hand. A new disposition was consequently made, in which only the enemy's probable direction of approach was taken into account; the U-boats covered a larger area, altogether 100 nautical miles, and were placed in three rows, opposite the gaps between the leading craft. Unfortunately, we were prevented from carrying out this plan because unfavourable weather made scouting impossible.

When, at the beginning of October, orders were given to carry out the same scheme, a new obstacle arose, owing to the issue of instructions from the Supreme War Council for an immediate resumption of the U-boat campaign against commerce. Lacking U-boats, I was forced to adopt quite a different scheme; instead of making for the English coast and luring the enemy on to our line of U-boats before the actual battle took place, I had to make a widespread advance with torpedo-boats to take stock of the commercial traffic in the North Sea and capture prizes. The Fleet was to serve as a support to the light craft that were sent out. As I was not in a position to reinforce the fighting power of the Fleet with U-boats, I had to try and choose the battle-ground so that we might join battle under the most favourable conditions to ourselves as possible.

After the Battle

Judging by the experience gained in the Battle of the Skagerrak, the position with regard to wind and sun must play an important part in the outcome of the artillery battle; again, the interval of time before darkness fell after the commencement of the battle was important, since the enemy had strong reserves at his disposal which, as yet untouched, could enter the fight when our ships were already damaged.

The sinking of the *Pommern* had unfortunately proved that this class of ship could not be risked in heavy fighting, owing to their being insufficiently protected against the danger of being sunk. The tactics of the British made it unlikely that our Squadron II would be able to take part in another big battle, on account of its artillery and its old type of torpedo, which had a range of less than 6,000 metres. I did not, therefore, take these ships with me, but assigned to them the duty of guarding the German Bight in the absence of the Fleet. When the Fleet went out in this way, a torpedo flotilla was sent on ahead to the probable vicinity of the guard-line of English submarines, the object being to keep the latter under water and so prevent them from giving too early a warning of our approach.

On October 10 the Fleet advanced according to this plan to the centre of the North Sea, but the torpedo-boats were unable to go as far afield as had been arranged, owing to adverse weather conditions. There was no encounter with the enemy.

The resumption of the U-boat campaign against commerce, which was to begin early in October, had to be supported as far as possible, even though it was little to the taste of the Navy, and had also been adversely commented upon by Admiral von Schröder, the head of the Naval Corps in Flanders.

After our sortie on October 19, two torpedo-boat flotillas were sent to Flanders, and from that base they were to attack the guard-boats at the entrance to the Channel, so as to make it easier for our U-boats to get through. The First Leader of the Torpedo-Boats, Commodore Michelsen, was sent to Flanders for the same purpose, and to gather information about the local conditions there. On October 23, 1916, the Flotillas III and IV started for Zeebrugge, which they reached without incident before dawn on October 24. The carrying out of these voyages to and from Heligoland and Zeebrugge marks the change in the development of conditions between that time and October, 1914, when the seven half-flotillas were sent out from Ems and utterly destroyed. From now on, there was frequent traffic between these points, as the flotillas were changed and new boats were sent to Flanders. As a rule the movements took place without incident, so that they came to be looked upon more and more as ordinary trips and not as risky enterprises.

On the night of October 26-27, the two flotillas, reinforced by the half-flotilla attached to the Naval Corps, carried out an attack on the ships guarding the entrance to the Channel and on the transports west of this line. According to previous observations, the boats on guard consisted of a few destroyers, but chiefly of small craft and trawlers, some of which had nets. These were always a very great hindrance to our U-boats when they wanted to get through, for they were forced to go under water and thus run the risk of getting entangled in the nets. An advance farther west beyond this line was an enterprise in which strong opposition was always to be expected. Even if our boats succeeded in reaching the line of guard-ships unnoticed, from the moment the Admiral in command at Dover heard of our approach, we had to reckon that in a short time strong forces would be assembled in the Straits between Dover and Calais.

A glance at the map will show that vessels which penetrated farther west could be cut off from their base at Zeebrugge both from Dover and Dunkirk; so they could, if they went to the southern end of the Downs to attack the mouth of the Thames. For this reason the half-flotilla in Flanders was not strong enough to carry out such expeditions unaided.

The following orders were issued relative to any ships that might be met with:

Ships without lights crossing the Channel were to be regarded as military transports and torpedoed without warning; ships with prescribed lights were to be treated according to prize law, unless they were convoyed by warships or became involved in a fight by their own fault.

Torpedo-Boat Flotillas III and IX and the Flanders Half-Flotilla set out at 6.30 P.M. from Zeebrugge; Commodore Michelsen was on board the leading boat of the Fifth Half-Flotilla. It was a clear starlit night, with a new moon. The surprise of the enemy was complete. The results we achieved were: eleven hostile guard and outpost ships sunk, and some other guard-ships badly damaged, from one of which ten men were taken prisoners. Besides this two enemy destroyers were sent to the bottom, and an English steamer, the *Queen*, was sunk, eight miles south

After the Battle

of Folkestone. This steamer, according to English information, was a transport, but she declared herself to be a packet-boat. She could make 25 knots. On our side we sustained no loss. The only damage done was to a torpedo-boat with which a rudderless and burning guard-ship collided while her engines were still running.

As usual this surprise resulted in greater watchfulness on the part of the enemy. Commercial traffic eastward bound from the Channel was stopped, and aeroplane reconnaisance to observe movements in Zeebrugge harbour were considerably increased. When therefore in the afternoon of November 1 our boats intended to repeat the enterprise, everything pointed to the fact that the enemy was informed of our intentions, so that it was probable the blow would either miscarry or be turned into a reverse. Consequently when the flotillas had been at sea for some hours, and flash signals had shown that the enemy was on the watch, they were recalled. In these circumstances it was not desirable to keep the two flotillas any longer at Zeebrugge, especially as the nights were getting lighter and on that account unsuitable for such enterprises. Flotilla III was, therefore, sent back to Wilhelmshaven. Nevertheless, we decided to keep similar raids in mind, since the sudden appearance at considerable intervals of torpedo-boat flotillas in the Channel and near the south-eastern coast of England might bring about favourable results.

One difficulty connected with the sending out of large numbers of torpedo-boats from Zeebrugge was, that in order not to expose them to aerial bombardment, they were not allowed to lie by the Mole, but sent up to Bruges. This entailed very considerable delay, on account of the lock, for it took 2½ hours to get four torpedo-boats through.

As soon as they left Bruges harbour it was not possible, as a rule, to conceal the movements of the boats from the enemy.

The behaviour of the enemy after the battle of the Skagerrak showed clearly that he intended to rely entirely on economic pressure to secure our defeat and would continue to keep his fleet in the northern waters of the British Isles. Nothing but serious damage to his own economic life could force this opponent to yield, and it was from him that the chief power of resistance of the hostile coalition emanated. As English economic life depended on sea trade, the only means of getting at it was to overcome the Fleet, or get past it. The former meant the destruction of the Fleet, which, in view of our relative strength, was not possible.

Germany's High Sea Fleet

But so long as the Fleet was not destroyed, we could not wage cruiser warfare—which alone could have badly damaged British trade—on a large scale. The U-boats, however, could get past the Fleet. Free passage to the open sea had been gained for these in the naval action on May 31, for the English Fleet stayed far North and did not dare to attack our coast and stamp out the U-boat danger at its source.

The recognition of this necessity to attack British trade as the only means of overcoming England, made it very clear how intimate was the connection between the conduct of the war by land and by sea.

The belief that we could defeat England by land had proved erroneous. We had to make up our minds to U-boat warfare as the only means we could employ that promised a measure of success. The ultimate decision was left to the Supreme Army Command, which was taken over on August 30, 1916, by Field-Marshal von Hindenburg. After the discussions with Roumania, however, it did not seem advisable to the Supreme Command to begin an unrestricted U-boat campaign at once, in view of the fact that no additional troops were available in the event of neutral nations, such as Holland and Denmark, joining the enemy.

On October 7 the Fleet Commanders received the order to resume cruiser warfare with U-boats in British waters, and also to send four U-boats to the Mediterranean, where submarine warfare had been carried on during the summer months with quite good results. In September the Chief of the Naval Staff had been of opinion that the general situation would permit of the full development of the U-boat campaign at latest by the middle of October. I had counted on the co-operation of the U-boats with the Fleet up to that date.

When, however, orders came through that the economic war against England was to be resumed in a modified form, although it was known that I considered the scheme to be useless, there was no chance of my opposition having the least effect in the face of this definite order, and in view of the fact that the Supreme Army Command considered it as a matter of principle. Unfortunately, I could adduce no great successes achieved by the Fleet in conjunction with the U-boats, and I could hardly take the responsibility of prolonging the immunity which British trade had enjoyed since the end of April.

The support to be given by the Fleet to this form of warfare

After the Battle

became a question of increasing importance, as the enemy recognised the danger of the U-boats, and strained every nerve to get the better of it. A curious incident early in November emphasised the necessity for the co-operation of the Fleet in this phase of the war. On November 3 at 8 A.M. "U 30," then on her way home and about 25 miles north-west of Udsire (an island off the south-west coast of Norway), reported that both her oil engines were not working. The question for the Fleet was, how to get help to this boat so as to enable it to reach the Norwegian coast? A few hours later there was a report from "U 20," commanded by Lieutenant-Commander Schwieger, who was returning from a three weeks' long distance trip off the west coast of England, that he had hastened to the assistance of "U 30." The two boats then continued their journey in company, first to the latitude of Lindesnaes and then on November 3, at 10 P.M., they made for Bovsbjerg, on the coast of Jutland, where "U 30" could be met by tugs. The charts of both boats, compared at frequent intervals, indicated that the next day at 10 P.M., they should be about 15 miles from Bovsbjerg. Towards 7 P.M. on November 4 a fog came up, and at 8.20 P.M. both boats ran aground. As appeared later, they lay 5 sea miles north of Bovsbjerg; they had steered considerably more to the east than, according to their observations, they thought they were doing, and in the fog they had not been able to see the land properly. After two hours "U 30" succeeded in getting clear by reducing her load by about 30 tons, but she was no longer able to submerge freely, and could not be steered under water. Her commander remained in the neighbourhood of "U 20." This boat, owing to the prevailing swell, had got on the farther side of a sandbank, and in spite of efforts continued throughout the night, was unable to get off. The Fleet received the news of their stranding soon after 10 P.M. Hostile patrols of cruisers and destroyers had repeatedly been reported in the neighbourhood of Bovsbjerg, so it seemed desirable to send a considerable protecting force with the light craft which were despatched thither. The Danes would certainly notice the stranded boats at dawn, and we might assume that the news would quickly find its way to England, and that in consequence enemy ships which happened to be near by would hasten to the spot. It was not to be supposed that the whole English Fleet would just happen to be at sea, but single groups might well be cruising in the neighbourhood. Assistance must, therefore, be as swift and as well protected as possible. The officer in command of the scouting craft received the order to send the Fourth Half-Flotilla of Torpedo-boats ahead immediately, and to cover them with the *Moltke* and Squadron III. If we did not succeed in getting "U 20" off quickly, it was to be feared that the Danish Government would intervene and intern her. At 7.20 A.M. on November 4 Commander Dithmar arrived on the scene of the accident with the Fourth Half-Flotilla. The leading vessel anchored 500 metres from "U 20." A strong swell was running from the south-west, which increased greatly in the course of the morning, and caused a ground swell on the sandbank. Three times attempts were made to tug the U-boat off, and each time the ropes and chains broke. "U 20," in spite of all efforts and favourable conditions—it was high tide at 11 A.M.—did not budge. She lay too high on the shore. As further efforts seemed hopeless she was blown up, her crew taken on board and the return journey was begun.

The cruisers and Squadron III, in the meantime, had followed to the spot and patrolled near by until the attempts at rescue were abandoned. At 1.5 P.M. the *Grosser Kurfürst,* and immediately afterwards the *Kronprinz,* were each hit by a torpedo just as the squadron was executing a turning movement. Both torpedoes must have been fired by a submarine. The submarine itself was not sighted, owing to the waves ; the course of the torpedoes was not observed until it was too late to avoid them. The *Grosser Kurfürst* was hit in the steering gear and the helm on the port side rendered useless. The *Kronprinz* was hit under the bridge and sustained only slight damage in her bunkers and gangway. The *Grosser Kurfürst,* which at first had to fall out because of her difficulty in steering, was able to follow the squadron later at 19 knots, and the *Kronprinz* was able to keep her place in the line, steaming at 17 knots.

Upon receipt of the news of this incident, His Majesty the Emperor expressed the opinion that to risk a squadron for the sake of one U-boat, and in so doing almost lose two battleships, showed a lack of sense of proportion and must not occur again. Now this dictum might easily have imposed too great a restraint upon the Fleet merely for fear of submarines. We should have lost the confidence in our power to defend the Bight which we had gained as a result of the sea fight, and which became manifest when we sent these scouts 120 nautical miles from Heligoland, a distance which had hitherto been regarded as the ultimate limit to which our Fleet could advance.

After the Battle

On November 22 I received a summons to General Headquarters at Pless, and had the opportunity to submit my view of the case to His Majesty, to which he gave his concurrence. It was as follows:

"In view of the uncertainty of naval warfare, it is not possible to determine beforehand whether the stakes risked are out of proportion or not. England, threatened anew by the U-boat campaign, as the increase in shipping losses in October clearly proved, is very anxious to allay popular anxiety on the score of this new danger. No better means to achieve this can be imagined than the news that they had succeeded in destroying a German U-boat close to the German coast. If, in addition to this, the number of the U-boat were ascertained—in this instance 'U 20,' which had sunk the *Lusitania*—this would indeed be glad tidings for the British Government. On the other hand, the dangers that threaten our U-boats on these expeditions are so great that they are justified in demanding the utmost possible support that our Fleet can give them in case of need. On no account must the feeling be engendered amongst the crews that they will be left to their fate if they get into difficulties. That would diminish their ardour for these enterprises on which alone the success of the U-boat campaign depends. Moreover, English torpedoes have never yet proved fatal to our big ships, a statement which was again confirmed in this case.

"The temporary loss of the services of two ships while under repair is certainly a hindrance, since, for the time being, the Fleet cannot undertake any considerable expedition. But, on the other hand, incidents such as occurred on the occasion of the stranding of these boats afford the junior officers an opportunity to develop their independence. There is no doubt that in this case a few torpedo-boats would have sufficed to drag the stranded U-boats free and tow them home. But if they had been surprised by a larger force of English boats that happened to be passing, or had been notified of their whereabouts, then further losses were possible, and the expedition would have failed in its aim. You can only make each expedition as strong as the means at your disposal at the moment permit. Fear of loss or damage must not lead us to curb the initiative in naval warfare, which so far has lain mainly in our hands.

"The bombardment of the enemy coast, airship attacks, the U-boat campaign, as well as the sea-fight itself, have shown that our Fleet has hitherto taken the offensive to a far greater extent than the English Fleet, which has had to content itself entirely with defensive action. Apart from a few unsuccessful aeroplane raids—the last was on October 21 of this year and made no impression—the English Fleet cannot boast of its achievements. The whole organisation for holding the Fleet in readiness is directed towards affording every enterprise the greatest possible security, and towards leaving out of account those ships which have come to port for necessary rest. It is of great value to uphold this principle, because in the course of the U-boat campaign, upon which, in my opinion, our entire naval strategy will sooner or later have to be concentrated, the Fleet will have to devote itself to one task—to get the U-boats safely out to sea and bring them safely home again. Such activities would be on precisely the same lines as the expedition to salve ' U 20.' To us every U-boat is of such importance, that it is worth risking the whole available Fleet to afford it assistance and support."

While at Pless I took the opportunity of making myself known to Field-Marshal von Hindenburg, and also to have an interview with General Ludendorff. I discussed the U-boat campaign with both officers, and it was agreed that if the war should drag on for so long, February 1, 1917, was the latest date at which to start the unrestricted U-boat campaign, that is to say before England could revictual.

The Field-Marshal, however, added that now that matters had taken such a favourable turn in Roumania, he could not for the moment face the possible complications that the declaration of unrestricted U-boat warfare might entail, although, at the same time, he was convinced that it was the right step to take. He went on to say that he had charged our ambassador at the Hague, Herr von Kühlmann, on his honour, to give his candid opinion as to Holland's attitude, and had received a definite assurance that an aggravated form of the U-boat campaign would force Holland to come in against us.

It was of great importance to me, to have found complete understanding of the circumstances and conditions of our naval warfare at Headquarters, and to be assured that those in authority were determined not to let the suitable moment slip for the employment of all means that would lead to a speedy termination of the war.

In order to resume cruiser raids on the open sea, the auxiliary cruisers *Moewe* and *Wolf* were sent out at the end of November, the former under the officer who had commanded her on her first cruise, the latter under Captain Nerger, and both reached the high seas without hindrance from the enemy.

After the Battle

The peace proposals of Germany and her Allies, made on December 12, had little prospect of finding acceptance with our enemies; but the fact that they had been made would tend to simplify the situation and, in case of refusal, to rouse the will of the people to strain themselves to the uttermost for the final conflict. There was no hope of yielding on the part of those who had recently come into power in England—Lloyd George, with Carson as First Lord of the Admiralty. Thus the die was cast in our country for the employment of the most extreme measures, which it had been Bethmann's policy hitherto to avoid.

Towards the end of the year I regrouped the High Sea Fleet and took Squadron II out of the tactical group. One ship, the *Lothringen*, had already been put out of commission, and another ship of this same squadron was permanently needed to guard the Sound in the Baltic and had to be relieved from time to time. Thus for one reason and another (e.g. repair work) the squadron only consisted of five, or even fewer, ships. The fighting value of the ships had decreased with age, and to take them into battle could have meant nothing but the useless sacrifice of human life, as the loss of the *Pommern* had already proved. The creation of a new U-boat fleet demanded numerous, efficient young men, with special technical knowledge, and these could only be drawn from the Fleet. As Squadrons IV, V, and VI had already been disbanded for similar reasons, the reduction of Squadron II was only a question of a short time, as we were bound to have recourse to their crews. The U-boat flotilla had by this time a greater number of officers than all the large battleships of the Fleet.

When the two new battleships, the *Baden* and *Bayern* (with 38-cm. guns), joined up, it was possible to dispose the battleships in the High Sea Fleet in the following manner :

Baden, Flagship of the Fleet.

SQUADRON I—Vice-Admiral Ehrhardt Schmidtt.

Ostfriesland.   Thüringen.   Heligoland.   Oldenburg.
Posen.   Rheinland.   Nassau.   Westfalen.

SQUADRON II—Vice-Admiral Behncke.

König.   Bayern.   Grosser Kurfürst.   Kronprinz.
Markgraf.

195

Germany's High Sea Fleet

SQUADRON IV—Rear-Admiral Mauve.

Friedrich der Grosse.   König Albert.   Kaiserin.
Prinzregent Luitpold.   Kaiser.

When in column formation, Squadrons III and IV formed a division and Squadron I was divided into two divisions. These three squadrons had their headquarters in the Jade. Squadron II lay in the Elbe when, as was often the case, it was not sent to the Baltic to provide target-ships for the torpedo-boat flotillas and U-boats which practised there, and to undertake manœuvres in common with them.

The chief duty at this time was to protect the Bight when the Fleet put to sea. During the winter the number of large battleships in the English Fleet had been materially increased, and by the spring of 1917 we should have to reckon with 38 large battleships (of which 14 had 38 cm. guns) and 10 battle-cruisers (of which 3 had 38 cm. guns). On our side we had 19 battleships (two with 38 cm. guns) and five battle-cruisers whose biggest guns were 30·5 cm. In place of the *Lützow*, which had been lost, we had the *Hindenburg*.

This relative strength indicated, from a tactical point of view, the desirability of our making as much use as possible of the advantages to be derived from the short days and long nights of winter. The long nights afforded our torpedo-boats good chances of success and prolonged the time during which our Fleet could approach unperceived. On the other hand, the short days had this advantage: that we could time a battle so that our munitions did not give out and so that the enemy could not bring up fresh reserves against our damaged ships.

At the close of 1916 the idea prevailed among the commanders of our Fleet, that England, anxious about her future, and pressed by her Allies, intended to develop greater activity at sea. The fall of the old Ministers, and the change in the command of the Grand Fleet might be looked upon as steps to prepare the way for this.

It was decided that the U-boats were to carry on the campaign against commerce in accordance with Prize Law during the winter, and a number of these were detailed for special duty off the east coast of England. It was possible to connect these up with an advance of the Fleet, whenever a fair number of U-boats was ready to put to sea or had been at sea a short time. By the middle of January we had ten ready for this purpose, and they received orders,

196

in addition to their campaign against trade, to take up two lines south-west of the Dogger Bank on a certain date, when the Fleet was to undertake an advance to the west, south of the Dogger Bank. Support by the U-boats of the Naval Corps was arranged for in the usual way. The bad weather which prevailed in January prevented the realisaton of this scheme, which was again to depend on airship scouting.

As we had to reckon on the possible failure of airship scouting within the time available for such an enterprise—boiler-cleaning in the flotillas, repairs on the wharves and the preparedness of the U-boats also influenced our arrangements—another plan was drawn up in which the weakness of airship scouting was not of such importance as to necessitate the abandonment of the enterprise on that account. This was not to be carried out until March, and was to take place during the light nights at the period of full moon—which would last until March 12.

The idea was to make a raid into the Hoofden to interfere with the convoyed traffic between England and Holland—from Rotterdam to the Thames. In the meantime unrestricted U-boat warfare had commenced on February 1, but our U-boats could not get at this traffic very well. At night it was difficult for them to get an opportunity to open fire, especially when the vessels were protected, and by day the shallowness of the water made submarine attacks impracticable, especially if the accompanying ships used depth charges. As the crossing took so short a time, and moreover could be carried out by night, this traffic was exposed to no risks worth speaking of and there was a noticeable increase on this route.

Our boats were to advance to a line Schouwenbank—Galloper, make a night raid through the Hoofden, and then at 6 A.M. steam in a northerly direction to meet the Main Fleet which would follow them. The Main Fleet itself, consisting of Squadrons I, III and IV, was to lie off Braune Bank at 6 A.M., and for that purpose would have to leave the Jade at 2.30 P.M. on the previous day. It was not expected that the enemy would notice our putting to sea in the afternoon before dusk. Success in this case depended entirely on surprise, for otherwise the steamers would simply postpone their journey. The raid by night through the Hoofden was designed to cover the whole area.

The officer commanding the scouting craft had at his disposal Scouting Divisions I, II, and IV—with the exception of two small cruisers which had to remain as vanguard with the Main Fleet

—and 22 torpedo-boats. In view of the large number of boats taking part, it was necessary to choose a light night for the enterprise, so that the ships should not foul each other, and should be free to act so as to hold up the steamers.

Though the heads of the Fleet enjoyed complete independence in organising and arranging their operations, they nevertheless had to inform the Naval Staff of them. This was imperative, if only because all information was collected and we might consequently receive valuable hints in good time. In this case it was especially important for us to know whether there was any news of the Anglo-Dutch traffic, and if so, what. The remark in my orders to the effect that the enterprise was to be carried out even if air-scouting were lacking gave rise to direct intervention on the part of the All Highest, who declared we were on no account to do without air-scouting.

The stormy spring weather made it extremely doubtful whether we could carry out the plan under these conditions, and in fact, when the time fixed for the enterprise arrived, the weather was quite unsuitable for airships and thus the scheme collapsed.

In a petition to the Kaiser, I clearly showed that from a military point of view my plan was practicable, and urgently requested him to withdraw his restriction, pointing out that such a restriction even in one direction only, would paralyse the power of a leader to carry out an enterprise which he had carefully planned, and which was well within the scope of the Fleet. The only reply I received was that the order had been issued after due consideration and must stand. I did not carry away the impression that when this decision was arrived at the Chief of the Naval Staff had presented the point of view of the Fleet with sufficient emphasis to dissipate the Supreme War Lord's fears—which was a pity. These fears were probably due to the idea that now that our ultimate success was entirely dependent on the results of the U-boat campaign, there must be no deviation from the course on which we had embarked, or any risk incurred which might force the Fleet to give up its support of the U-boats, before the goal had been reached.

It must be admitted that in principle these considerations were sound, for events might occur—e.g. the loss of the U-boat base in Flanders—which would confront the Fleet with tasks for which it would require all its strength. On the other hand, there was the consideration that every successful fight stimulates the confidence of those who take part in it. In a Fleet there are numbers of men

After the Battle

who, in a certain sense, are merely onlookers in a fight, who are unable to join in as individuals, as soldiers do on land, and thus develop each man's pride in having "done his bit." On the other hand, in a sea fight, perhaps to a greater extent than anywhere else, the intervention of an individual may have a decisive influence, if he has the presence of mind to ward off some great danger by resolute and skilful action, and thereby save the whole ship and her crew and ensure victory for his side.

So long as there is no actual fighting, these men, who take no immediate and active part, are very apt to criticise the initiative and resolution of the leaders of the Fleet and of individual ships. They want no cravens at their head, for they know that cowardice in their leaders may prove fatal to themselves and because each man feels in a measure responsible for the ship to which he belongs. When battle is once joined, ship against ship, each man's strength must be strained to the utmost, whether he be a member of a gun-crew or a stoker, a munition man or a man on look-out duty who gives timely warning of the course of a torpedo. Co-operation on the part of all these, of whom no single one can be dispensed with, is absolutely essential in an action if success is to be achieved.

The Fleet had little rest in 1917, even though the success of its activities was barely apparent. It found expression in the effect of the U-boat campaign, for the work of the Fleet was from that time onward chiefly directed to the support of the campaign.

The U-boat could only prove effective against British trade if the boats succeeded in going to and fro unharmed between their base and their areas of activity. To achieve this, strong opposition on the part of the enemy in the North Sea had to be overcome. This opposition was planned on a large scale. We know from Lord Jellicoe's own lips that at the beginning of 1917 he had ordered 100,000 mines to be placed round the Heligoland Bight, and we were very soon to feel the effect of this. The belt of mines which curved round from Terschelling to Horns Reef grew closer and closer. At the same time our mine-sweeping operations were subjected to closer scrutiny on the enemy's part, so that very often, by the sowing of fresh mines in the path we had cleared, the work of many days was undone in a single night. As the enemy laid his mines in concentric circles west of the line originally laid, the area over which our mine-sweepers had to work was constantly enlarged. Unhappily we never had the luck to catch the enemy mine-layers at their work, which they probably mostly undertook when darkness shielded them, at any rate when the mines were not laid by submarines.

To explain what might appear to be crass incompetence on our part, we may remark that, so far as we know, the enemy's efforts in this direction met with little more success. I remember that on the return of one of our submarine mine-layers I was told that this boat had laid her two thousandth mine on this journey. How many difficulties she must have overcome before that work was achieved! The cruiser *Hampshire,* on which Lord Kitchener went down, was sent to sea in a heavy storm in the belief that in such weather little danger was to be apprehended west of the Orkneys from mines or U-boats; and yet one of our boats (Lieutenant-Commander Curt Beitzen) had been at work, and had made use of the opportunity provided by the bad weather to lay the mines to which this ship was to fall a victim. We, too, often noticed that after stormy days, when the mine-sweepers' work had to be interrupted, new mines had been laid in places which had been cleared just before the storm began.

Another difficulty that our mine-layers had to contend with was that they had to lay their mines quite near the British coast or the entrance to ports, where closer watch was kept and defence was more effective than in the open North Sea. There, at a distance of 100 sea miles from Heligoland, we had to keep watch on what was being done at night on the extreme edge of the wide curve which stretched from the East Frisian coast right up to Jutland. The great distance at which the mine-sweepers had to work made it necessary for us to send a strong protective force with them, for fear they should be surprised by a squadron of destroyers, which were greatly superior to them in armament and speed, and would make short work of them. A few attempts at catching them unawares had been made by the English, but these had been so half-hearted that our boats had got away with very slight damage and loss. After we had opened fire, the enemy ships soon gave up the pursuit.

We could, however, not rely on the mine-sweepers getting off so lightly on every occasion. The more the English felt the unpleasant effects of the U-boat campaign, the more they would be likely to make great efforts to combat the U-boat danger in all its manifestations, wherever they had a chance. Only our light cruisers could afford effective protection to the mine-sweepers, because their guns were superior to those on the English destroyers. Just so many torpedo-boats were assigned to them as appeared necessary for their protection from submarines. If we had had to protect the mine-sweepers by torpedo-boats alone, we should have had to employ the latter in greater numbers than was compatible with their other duties. From the very beginning of the war the importance of the work of mine-sweepers was recognised, and much time and care were devoted to developing these flotillas and equipping them with increasingly better material.

Instead of the old boats that had been turned out of the torpedo service, which found a place in our first mine-sweeper flotillas, and the trawlers which were used provisionally to assist in this work, new craft were built specially designed for mine-sweeping; moreover, they were built in such numbers that in the course of 1917 almost all the mine-sweeping divisions were provided with them. We also had large demands for similar craft from the Baltic, where they were needed to enable us to maintain commercial traffic, and the more so as the offensive activities of the Russians were entirely devoted to mine-laying operations.

The development of seaplane activity in the North Sea afforded good support to the mine-sweepers. At the beginning of the war the number of really efficient seaplanes was so small as to be of no value, for the only seaplane station we then had at Heligoland boasted but five machines, to which after a time three more were added. Both pilots and observers had to be trained. Thanks to the energy of those at the head of the Air Service (Rear Admiral Philipp as Chief of the Naval Air Service and Captain Brehmer as officer commanding the North Sea Seaplane Division), this arm of the service developed tremendously and rendered us invaluable services as scouts, thereby relieving the fighting forces on the water of a great burden.

Bases for seaplanes were constructed on the North Sea at List (Sylt), Heligoland, Norderney, Borkum and, in addition to these, at Zeebrugge and Ostend. Further, the small cruiser *Stuttgart* was arranged as a seaplane carrier, after the necessary experiments had been made on the auxiliary cruiser *Santa Elena,* and when, as the flying machines were perfected, it seemed desirable not to confine their activities to the coast of the North Sea, but to make use of them at sea as well. This development of flying became necessary, and was encouraged by the urgent need of the mine-sweeping service.

Thus the requirements of the U-boat campaign demanded many-sided service from the Fleet; this applied more particularly to officers and men, for in addition to the existing Navy a second one had to be created for submarine warfare, one which had to be developed out of the old Navy sailing *on* the water, and which was dependent on it in every respect, although it represented an absolutely new creation.

CHAPTER XII

AIRSHIP ATTACKS

ON the outbreak of war the Navy had at its disposal only three airships, "L 3," "L 4 " and "L 5," of 15,000 cb.m. capacity. The last Zeppelin built during the war bore the number "L 71," and its capacity was 62,000 cb.m. These figures express the gigantic development which the airships underwent. The airships placed at the service of the Fleet were almost all of the Zeppelin type. The firm of Schütt-Lanz built a few ships as well, which at first were only used experimentally, but subsequently were put to practical use.

Probably no arm of any service has suffered such severe losses as our airships, with the exception of the U-boats. Out of 61 Zeppelins which were assigned to the Fleet in the course of the war, 17, with their whole crews, were destroyed by the enemy, namely "L " 7, 10, 19, 21, 22, 23, 31, 32, 34, 39, 43, 44, 48, 53, 59, 62 and 70.

Twenty-eight airships were lost through stranding and other accidents, such as the burning of sheds in consequence of explosion. The crews of these were all saved, though in six instances they were made prisoner. Six ships had to be placed out of service as being useless; at the end, ten were still left in a condition fit for use.

Owing to the ever-increasing defensive measures of the enemy the airships at the front were built in two sizes, the types being "L 50 " and "L 70."

The chief distinctive features of the former were five motors, each of 260 h.p., and such as could develop sufficient speed even in the highly rarefied atmosphere of the upper strata of the air; four propellers, all coupled directly to the shafts (the two rear motors are coupled to one propeller); a central gangway, 196.5 m. long; a breadth of 23·9 m.; a gas capacity of 55,000 cb.m.; a speed of 30 metres per second (about 110 km. per hour); a load of 38 tons.

Type "L 70 ": Seven motors, each of 260 h.p.; six propellers; central gangway, 211.5 m.; greatest diameter, 23·9 m.; volume of gas, 62,000 cb.m.; speed, 35 metres per second (equal to 130 km. per hour); load, 43 tons.

The "L 50 " carried a crew of 21, and "L 70 " one of 25, among whom were 1 Commander, 1 Officer of the Watch, 1 Quarter-Master, 1 Chief Artificer-Engineer, 2 men for lifting gear (Yeomen of Signals,) 2 men for the balancing gear (Boatswains), 2 Motormen (stoker petty officers) for each motor, 1 Sailmaker, 1 Petty Officer Telegraphist, and 1 Ordinary Telegraphist for the wireless installation.

They carried two machine-guns, and later on a 2-cm. gun as well. The supply of bombs consisted of incendiary bombs of 11.4 kilo weight, and explosive bombs of 50, 100, and 300 kilos.

In order to gain some idea of the difficulties encountered by airships, it may not be out of place to make a few general remarks on the navigation of these ships.

Their main task was scouting. That is why they were retained during the war as a weapon by the Navy; the Army had no use for them. The development of the aeroplane produced a keen competitor and a dangerous opponent. The Flying Service could not, at first, overcome the difficulty of covering the great distances which scouting at sea entailed. It was a question of flying over large sea areas, such as the North Sea, and providing the Fleet with trustworthy information and reports. Flights of twenty-four hours and longer had to be reckoned with, and no flying-man could hold out for so long.

The great load that the airship could carry, combined with its high speed, made it especially suitable for purposes of attack. The dangers to which the airship itself was exposed were best overcome by assigning to the crews some definite task for furthering the war, for which they gladly risked their lives. Any activity which did not bring them into contact with the enemy would not have satisfied them for so long a period as the duration of the war, and this would have hindered the development of this arm of the service.

When navigating on the water you steer for a goal which lies in a horizontal plane, on the surface of the water; the airship has to negotiate a second dimension due to differences in height. And this it is which presents such great difficulties in aerial navigation. In contradistinction to the aeroplane, the load, including its own weight, carried by an airship is not borne by motor power, dynamic lifting power, but by gas-tight cells filled with a gas lighter than air.

Hydrogen of specific gravity 0.07, which was used for filling, gave a ship of 55,000 cub.m. capacity a lifting power of 64,000 kilos. Of these, 26,000 kilos, in round numbers, were taken up by the weight of the unloaded airship itself, so that the load she could carry would be 38,000 kilos, i.e. 38,000 kilos can be packed into the ship before she will float in the air, being neither heavier nor lighter than air. The weight of the crew, stores of benzine and oil, spare parts, supplies of oxygen for the passage through high altitudes, and bombs amounted to about 10,000 kilos; the remainder, about 26,000 kilos, is available for water ballast. This is most essential in order to neutralise such influences as affect unfavourably the carrying power of the ship. At first, when ascending, the upward pressure of the atmosphere ceases; the gas pressure in the cells becomes proportionately greater. In order to equalise the pressure, every gas-cell is provided with a safety-valve through which the superfluous gas escapes; in this event, carrying power is consequently diminished; the ship becomes too heavy. To remedy this, a proportionate weight of water must be thrown out, so that the ship recovers her equilibrium. As a standard of measurement, we may state that for every 100 m. that she rises the ship loses 1 per cent. of her carrying power, that is 640 kilos.

Temperature, both of the air and the gas, also exerts an influence. Cold air is heavier than warm, whereas the lifting power of the gas, on the contrary, is increased by warmth and diminished by cold. In this respect the following law obtains: a change of one degree in the temperature increases or diminishes the carrying power by about 240 kilos. The commander must, therefore, constantly keep an eye on the temperature and judge by the changes what the behaviour of his ship is likely to be.

The weight is also affected by the amount of moisture that collects on the ship's envelope when passing through clouds; ice is also easily formed if the temperature becomes sufficiently low. The additional weight on the ship due to rain may amount to 3,000 kilos, and owing to ice as much as 5—6,000 kilos. The heat of the sun's rays and the strong draught caused by rapid progress soon make the deposit due to rain disappear. Ice has the further disadvantage that pieces of it may be hurled through the envelope by the revolving propellers and may possibly pierce the cells so that gas escapes.

Further, since hydrogen is highly explosive, and when mixed in certain proportions with air becomes very dangerous, care must be taken to prevent fire or electric sparks from coming into contact with escaping gas. But the gas escapes of its own accord when the cells are deflated—when in rising the limit of elasticity of the cells is reached, and when the cells are injured, either by pieces of ice or by hostile projectiles. If the hostile missiles generate flames, as incendiary weapons do, then the ship is inevitably destroyed.

Care is needed, too, in thunderstorms. It is best to avoid clouds charged with electricity. If you cannot go round them you should go under or over them. When rising to get over them, in no circumstances must you rise to such a height that the gas completely fills the cells, for owing to the diminished pressure of the air the superfluous gas is bound to escape and a flash of lightning striking this gas mixture will immediately destroy the ship by fire. There is no danger if the ship is struck by lightning if no gas is escaping, so long as the framework is intact. The framework of aluminium is connected throughout the ship, and acts as a conductor for lightning, which passes out by the stern. Instances in which this has happened have been quite frequent; but such experiences, of course, are better avoided if it is possible to do so.

As regards the height at which the gas completely fills the cells, the following should be noted: When a ship has risen above this height and has let off gas and then descended again, the remaining gas is insufficient to fill the cells completely; it only fills the upper part of them, while the lower part remains empty. The ship is then no longer buoyant. The higher the ship has risen, the less gas the cells contain upon descent; it is forced to come down to earth. In this case more water-ballast must be discharged. The water is distributed in ballast-bags along the entire length of the ship. The valves of these bags are connected by wires with the steering car. Every bag holds 1,000 kilos. The commander is in a position to release any quantity of water desired, from whatever part of the ship he thinks proper. Fore and aft of the ship there are another four bags each containing 250 kilos, which can also be opened from the steering car. These differ from the others in that when opened they discharge their contents instantaneously, whereas the others let the water run out slowly. When the ship needs to be lightened suddenly, the containers fore and aft are used, e.g. for a sharp rise when attacked by aeroplanes, or when a cell has been emptied owing to damage by a missile or some other accident, and one end of the ship suddenly becomes heavy; and also when landing with a "heavy" ship.

Below the ship and at the sides the cars are hung. The foremost and largest of these contains the steering-gear in front, next to that the wireless installation, and at the rear a motor. The last car, which hangs on the central line of the ship, carries two motors

Airship Attacks

which both connect with one propeller. The cars at the side each carry one motor.

As soon as there are no guiding objects in sight, such as land, lightships, or one's own warships, by which the ship's position may be ascertained, navigation becomes very difficult, because of the leeway when the wind blows at an angle to the course of the ship. That is where wireless telegraphy comes in; the installation is such that the ships can be called up by directing stations, their whereabouts calculated, and their position reported to them by wireless.

The airship stations were so placed that they lay as near as possible to the coast, and had a sufficient extent of level ground for ascending and landing; but they had to be placed sufficiently inland to obviate the danger of an unexpected attack from the sea. The Navy possessed the following airship stations on the coast of the North Sea: Nordholz near Cuxhaven, Ahlhorn near Oldenburg, Wittmundshaven (East Friesland), Tondern (Schleswig-Holstein). Hage, south of Norderney, was abandoned.

The ideal airship hangar is a revolving shed which can be turned according to the direction of the wind. Unfortunately, we possessed only one such shed, that of Nordholz, as it involves a great deal of time and uncommonly large expense to build them. The problem of building material had also to be faced in the course of the war. Most of the sheds were placed in a position suitable to the prevailing wind in the neighbourhood. It is not possible to take an airship into or out of a shed if the wind blows across its path at a speed of more than 8 metres per second.

This consideration, and the fact that airship attacks had to be made during the time of the new moon, occasioned the long pauses between the raids, pauses which often gave rise to the impression that other influences had led to the abandonment of these activities. This was not so. From the date of the first airship raid on England, on January 15, 1915, no regulations were made limiting the offensive action of the airships. So far as London was concerned, we had orders at first only to attack such establishments as were immediately connected with military work, such as arsenals, docks, batteries and so forth. But this limitation could not be adhered to in the long run, partly because of the difficulty of discovering these particular places, partly because just round London the defences were especially intense. But it never was the object of an airship raid to attack defenceless dwelling-places. Their aim always was to destroy those establishments which, either directly or indirectly, served some military purpose: munition factories, arsenals, stores, docks, wharves, etc. Airships frequently returned from their expeditions with their full complement of bombs, because they had not been able to make out such targets with sufficient certainty. It would have been easy enough for them before returning to get rid of their bombs and drop them on any place over which they happened to fly, if they had wanted to kill harmless citizens.

Once the airship was in the air there were no further difficulties except such as arose from thunderstorms or very high winds; just as at sea, very bad weather hinders and circumscribes the activities of ships. The revolving shed is a factor of the utmost importance for the future of the airship.

While the U-boats were at full swing at their work of destroying English commerce, the airships with dogged perseverance did their best to contrive their attacks on the island. In March, 1917, a raid was made by five airships. Two of them reached London. In consequence of a considerable freshening of the wind, the return journey became very difficult. "L 2" was forced to descend in Juteborg, "L 35" in Dresden, "L 40" and "L 41" came down to their shed at Ahlhorn. "L 39" (Commander, Lieutenant-Commander Rob. Koch) was driven by the storm to the south-west, passed over the enemy lines in France, and, according to a wireless message from the Eiffel Tower, was shot down at Compiègne. Her crew perished in the flames.

A raid which was started in April had to be abandoned because on the outward journey the weather became unfavourable.

May again presented an opportunity for a successful raid which took place in the night of May 23—24. The following took part in it: "L 40," "L 42," "L 43," "L 44" and "L 45." Captain Strasser, the Head of the Airship Service, was on board the "L 44." The officer commanding the airships made the following report:

"Towards 1.45 A.M. we crossed the coast near Harwich; cloudy sky with breaks in it. A number of searchlights tried in vain to pick up the ship. Very little gunfire; no aeroplanes. In consequence of three motors missing simultaneously, did not carry out attack on London, as ship lost height rapidly, but dropped bombs to amount of 2,000 kilos on Harwich. Shortly after attack all engines began to miss, ship travelled like a balloon for three-quarters of an hour over enemy country and fell from 5,700 m. to 3,900 m. After this till 10 A.M. travelled with one motor only going, from 10 A.M. with 2-3 motors; landed at Nordholz 7.20 P.M. 'L 43' had to pass through severe thunderstorms with extraordinarily heavy hail-showers on return journey. Lightning struck ship in forepart and ran along the framework without doing any damage."

The next attack was on June 17. "L 42," "L 43," "L 44" and "L 45" took part in it. The raid on London was again prevented owing to the shortness of the night, because detours had to be made to avoid several thunderstorms. "L 42" could not reach London, and at 3 A.M. expended her entire ammunition on Dover. There was severe gun-fire during the attack, but the searchlights could not hold the ship for long owing to heavy mist. The bombs fell on their targets. Violent explosions at intervals of 10 minutes followed one detonation; whole districts of houses seemed to be hurled into the air; fires could be observed for a long time afterwards. Shortly after the raid the ship was pursued and violently bombarded by light craft, apparently torpedo-boats or small cruisers.

From "L 42" it was observed that one of our ships was being attacked by an airman. The airship was at a height of 4,500—5,000 metres, with the airman 300—500 metres above her. As "L 43" (Lieutenant-Commander Kraushaar) did not return from this journey, we were forced to assume that the airman had destroyed her. Later reports from England confirmed this.

On August 30 the *Ordre pour le Mérite* was conferred on Captain Strasser. I took the opportunity of handing him this distinction personally, and for that purpose went to the airship station at Ahlhorn, 20 km. south of Oldenburg, which had been erected during the war and was now the chief base of the airships.

It is to Captain Strasser's credit that he developed Count Zeppelin's invention to military perfection, and made the airship a weapon of great efficiency, besides rousing the enthusiasm of the crews of the airships by his example. He was the life and soul of the whole and made everyone under his command share his conviction that airships had a great future before them. He was particularly gifted in estimating meteorological conditions. He had an almost prophetic instinct for the weather. How often we have had to apologise mentally to him, when in apparently favourable weather the airships did not go out; for he was always right, and shortly afterwards there was invariably a change in the weather which would have endangered the ships and made their return impossible. But he recognised no such thing as insurmountable difficulties; the stronger the enemy's defence grew, the more energetically did he concentrate on counter-measures. Thus he sent his ships up higher and higher, and ultimately they worked at an altitude of 6,000 metres—a height which was considered impossible at the beginning of the war. For this he needed elbow-room and sympathetic co-operation in his technical suggestions. His organisation of the airship service did not immediately place him in such a position as to get his demands satisfied quickly enough. Moreover, he was hampered by all sorts of difficulties connected with technical administration. Those in command of the Fleet, however, did not rest until the organisation had been changed so that he could have free play in the conduct and development of the airship service.

Captain Strasser took part in most of the airship raids, although permission was often given him very grudgingly. The loss of airships was so considerable that I was always afraid that one of these days he would not come back, and he was too valuable to the airship service for such a risk. But just because the difficulties always grew, I had to admit that he was right in considering it necessary to see for himself what conditions were like on the other side, so as to judge what he could demand of his crews and how he could improve the efficiency of the ships.

An attack in October, 1917, brought about the loss of five airships out of the eleven which set out. This was due to such a strong head wind setting in that four ships were blown far over into France, and one, though it reached Middle Germany, was lost in landing. The six others, thanks to their timely recognition of the change in the weather conditions, came home safely. The individual ships had more than the usual difficulty in determining their positions, because the angles from the directing stations became very steep when the ships were over the South of England, and consequently the calculation of their positions was less accurate.

Another very painful set-back for navigation by airships occurred in January, 1918, when owing to the spontaneous combustion of one of the airships in Ahlhorn, the fire spread by explosion to the remaining sheds, so that four Zeppelins and one Schütt-Lanz machine were destroyed. All the sheds, too, with one exception, were rendered useless.

After this the Fleet had, for the time being, only 9 airships at its disposal. From the autumn of 1917 onwards, the building of

Airship Attacks

airships had been restricted, because the material necessary for the building of aircraft was needed for aeroplanes for the army. From that date only one ship was placed on order every month. But even this did not prevent us from repeating our attacks on England from time to time, though we had to be careful to incur no further losses, so as not to be without airships for scouting which was so important for the other activities of the Fleet.

On August 5, 1918, the airships of the Navy attacked England for the last time. Captain Strasser was on board the "L 70," the latest ship, commanded by Lieutenant-Commander von Lotznitzer. He did not return from this journey; his ship was the only one of those that took part in the raid that was shot down over England. Thus this leader followed his comrades who had preceded him and to whom he had always given a glowing example.

The value of airships as a weapon has been much called into question.

In the beginning of the war, when seaplane-flying was quite undeveloped, they were indispensable to us. Their wide field of vision, their high speed, and their great reliability when compared with the possibilities of scouting by war-ships, enabled the airships to lend us the greatest assistance. But only in fine weather. So the Fleet had to make its activities dependent on those of the airships, or do without them.

A weakish fleet needs scouts to push as far in advance as possible; scouts, too, which can make observations without being driven off. The airships could do this. The danger from aeroplanes only arose later, and was never very serious at sea; but over the land it was extremely unpleasant. Although as a rule an airship can mount more quickly than an aeroplane, yet it has far less chance of hitting its opponent. Ultimately the airships were forced up so high that it was beyond the power of human endurance (altitudes of more than 6,000 m.). That meant the end of their activities as an attacking force. But for far-reaching scouting they retained their importance and their superiority to other aircraft, for they can remain in the air much longer and are independent of assistance from other ships. But the bigger they grew the rarer grew the opportunities for them to go on expeditions, because of the difficulties of getting them out of the sheds, and later on of their landing.

We have no reliable information as to the results of their attacks. They were the first war-engines which scared the English out of their feeling of security on their island, and they forced them to organise a strong defence. To judge by this, their visits must have been regarded as a very considerable menace.

It was our business to make as much use as possible of our superiority in airships, and to increase their efficiency so that fear of them might be a contributory cause in inducing England to make peace possible.

Such an ideal of perfection can only be attained if it is perseveringly sought in spite of the set-backs we endured, and although the opposition we had to overcome was increasingly great.

That is the right warlike spirit—not to give in, but to redouble one's efforts as our airship men did in an exemplary way.

We may probably look upon the military career of the airship as over and done with. But the technical side of airship navigation has been developed in such a high degree by our experience in war, that airship traffic in peace times will derive great advantages from it, and the invention of Count Zeppelin will be preserved as a step in the progress of civilisation.

PART III
The U-Boat Campaign

CHAPTER XIII

THE MILITARY AND POLITICAL SIGNIFICANCE
OF THE U-BOAT CAMPAIGN.

OUR fleet was built for the protection of German interests at sea; its object was quite definitely a defensive one. This was proved in its construction; its main strength consisted in battleships and torpedo-boats which were meant exclusively for naval battles. There were so few cruisers that they barely sufficed to scout for a fleet on the move. Both in numbers and in construction they were unfitted to threaten the trade of the enemy; they could not touch English world-trade, because the British Isles formed a barrier in the North Sea. We had no naval bases abroad. Thanks to English policy, in this war a hostile fleet ran little risk in attacking ours, though it was built as a defence against such attacks. England had secured the co-operation of the next strongest land and sea Powers, and could count on the benevolent neutrality of the United States of America, until they, too, sided with our enemies. Nevertheless, England forbore to risk her superior fleet in battle, and her naval policy in the war was confined to this: to cut Germany off from all supplies by sea, and to starve her out by withholding food and raw materials.

On October 2, 1914, the British Admiralty published a warning that it had become necessary to lay a large minefield at the entrance of the Channel into the North Sea; this was 1,365 square sea-miles in extent. It left free a narrow channel near the English coast, which was only passable within British territorial waters. On November 2, 1914, the whole of the North Sea was declared to be in the War Zone. Any ships which crossed it other than by routes prescribed by the British Admiralty would do so at their own peril, and would be exposed to great danger from the mines laid in these parts and from warships which would search for suspicious craft with the greatest vigilance. This was the declaration made by the British Government. The provisions of the Declaration of London of 1909 had not been ratified by England at the time, and she therefore did not consider herself bound by any international laws which would have made it possible to get articles of trade through neutral countries into blockaded Germany. The result of the measures adopted by the British Government were as follows:

1. All import trade into Germany both by land and sea was strangled, and in particular the importation of food was made impossible, because the distinction between absolute and relative contraband was done away with. Even the importation of goods that were not contraband was prevented, by taking them off the ships on the plea that contraband might be hidden in them; then when they were landed, either they were requisitioned or detained on the strength of some prohibition of export so that they had to be sold.

2. The neutral states in order to obtain any oversea imports for themselves were forced by England's demands to forbid almost all export of goods to Germany. The British Government even demanded the cessation of trade in free goods and their own produce between these countries and Germany, threatening to treat the neutral country as an enemy if these demands were not complied with.

3. In neutral countries, especially in the United States of America, whole industries were forced to stop all trade with Germany. In addition to this, the neutral countries of Europe were compelled to set up organisations which controlled all the trade of the country, and thereby placed it under the control of England. Persons and firms who did not comply with the regulations were cut off from sea trade, because all cargoes addressed to them were detained under suspicion of being destined for the enemy.

4. Free trading of neutral merchant vessels on the North Sea was made impossible when that was declared to be in the War Zone, because every ship that did not follow the instructions of this declaration was exposed to the risk of destruction. In this way all shipping was forced to pass through English waters and so to submit to English control. Winston Churchill, at that time First Lord of the Admiralty, openly expressed the aim of the British Government in his speech at the Guildhall on November 9, 1914. He said the British people had taken as their motto, "Business as usual during alterations in the map of Europe," and they expected the Fleet, on which they had spent so much care and money, to make it possible for them to adhere to this motto, and the Fleet was at the moment about to do so. It was very difficult at the beginning of war to estimate the full effect of the pressure exerted by sea power. The

The U-Boat Campaign

loss suffered was obvious and easily computed; the loss they inflicted was often invisible, or if it was visible its extent could not be determined. *The economic stringency of the blockade required time to attain its full effect.* They saw it then only in the third month. They must have patience and consider it in the sixth, the ninth, the twelfth months; then they would see the success which would be achieved gradually and silently, which meant the ruin of Germany as surely as the approach of winter meant the fall of the leaves from the trees.

The attitude of the English Fleet was absolutely in keeping with this declaration. They avoided battle or any attempt to destroy the German Fleet. They thought they could force Germany's submission without any fear that the English Fleet might forfeit its superiority to the other fleets of the world. Their strategy also gave their fleet certain tactical advantages if we should seek to join battle in those waters which it had selected for its stand. From this position the English Fleet was enabled to carry out the system they had planned of watching the approaches to the North Sea and the routes which lead to Scandinavia, and at the same time most effectively to protect this system from German attacks emanating from the Bight.

The English plan, however, was based on the further assumption that the Fleet would be able effectually to protect English trade. They probably counted upon the life of our cruisers in foreign waters being a short one, and reckoned that only in exceptional cases auxiliary cruisers would evade the watch in the North Sea and get out. These might temporarily disturb trade, but could never have any decisive effect. The English were not mistaken in this assumption; and in their certainty of controlling the seas, without any regard to the rights of neutral countries from whom they were not likely to meet with serious opposition, they took such measures as were best adapted to cut off Germany. When they declared the War Zone they dropped the old idea of a blockade, because mines and submarines made it impossible to carry out a regular blockade effectively. So far as the Englishman was concerned, that was the end of the blockade, and he proceeded to introduce an innovation which, to his idea, was suited to the times, and therefore justified; nor did he trouble in the least about the protest of neutrals.

To English ideas it is self-understood that naval warfare is directed towards the destruction of enemy trade, and equally so that all means that can promote this end are right. Their practicability was founded on the might of the English Fleet, from which neutral protests rebounded unheeded. This war has made it clear that the neutrals were mistaken when they thought that they could demand of that great Sea Power, England, the same rights that she had secured by treaties when she herself had been neutral. These rights of neutrals are nothing but pretensions which a mighty Sea Power would like to turn to its own advantage if on some occasion it should not be one of the belligerents and should wish to carry on its trade regardless of whether one of the parties at war should suffer thereby or not.

This was typical of the relations between us and America. Of course, the semblance of right must be maintained, and for that purpose any catchword which happens to appeal most to the people is made use of.

In this war it was the "dictates of humanity" which had to bolster up American trade interests. No State, not even America, thought it against the dictates of humanity to build submarines for war purposes, whose task it should be unexpectedly to attack warships and sink them with all on board. Does it really make any difference, purely from the humane point of view, whether those thousands of men who drown wear naval uniforms or belong to a merchant ship bringing food and munitions to the enemy, thus prolonging the war and augmenting the number of women and children who suffer during the war?

What England considered to be maritime law is most clearly seen by the layman in her attitude towards the Declaration of London. On the invitation of the British Government there was a conference in connection with the Second Hague Peace Conference in 1909 by which a number of rules were drawn up, the signatory Powers—amongst them England, France, Russia, the United States, Germany and others—"*had agreed in the statement that the rules drawn up in the Declaration were in all essentials in conformity with the generally accepted principles of International Law.*" England had not ratified this treaty owing to the veto of the House of Lords because it did not take British interests sufficiently into consideration. She therefore had the formal right not to abide by these rules, but at the same time she ran counter to the principles of International Law recognised by every State. On August 20, 1914, the British Government announced that it had decided to accept the Declaration of London in general, but with certain changes and additions that it considered absolutely imperative *in order to be*

The U-Boat Campaign

able to carry out operations at sea effectively. Here with touching ingenuousness it is stated that the Englishman considers himself bound by law only in so far as it does not hinder his operations, and that he will allow himself such deviations as will ensure the effective execution of his plans. That meant that he contravened the right of neutrals to send any goods to Germany and put obstacles in the way of such trade by every means in his power. The Neutral States even had to give an undertaking to consume all food received from overseas in their own countries and not to make use of foreign imports to set free a like quantity of home-grown food for transport into Germany. Anyone who wished to defend himself by means of remonstrances or protests in law was foredoomed to defeat owing to this *brutal policy of might;* but unfortunately this was the form our own policy had taken.

Moreover, we looked in vain for sympathy from the neutrals. America declared that if England ignored International Law that did not give us the right to pursue a course contrary to International Law to which America would be expected to submit. On the contrary, she demanded for her citizens the right to travel anywhere by sea unmolested. If we did not refrain from the counter-measures we had announced, which she considered contrary to the dictates of humanity, she would hold us responsible. Such a peremptory tone was not employed towards England. And why should it have been? The Englishman was only too glad of the visits of American ships, for they brought him everything that he badly needed. No disturbance of trade was to be expected from him, for he would have thereby injured his own interests and could, therefore, never be in the awkward position of running counter to the dictates of humanity as understood by America. How the efforts of Americans to tighten the screw of hunger on our people could be reconciled with humanity is a question that can only be explained by the peculiar maxim of the Anglo-Saxons that "business" has nothing to do with it.

When the starvation of Germany was recognised as the goal the British Government were striving to reach, we had to realise what means we had at our disposal to defend ourselves against this danger. England was in a position to exert enormous pressure. We could not count on any help from the neutrals. Without exception they had submitted to England's will, though they had not all sought their advantage in it as Norway and America had done. As we have explained in the preceding chapters, in view of the attitude of the English Fleet, our Fleet with its smaller numbers, and as it was constituted at the outbreak of war, could not hope to score a decisive success by means of which German trade might revive and British trade be at the mercy of our cruisers. The assumption that we might have done this is Utopian and does not take into account the subsidiary means of controlling sea traffic which would still be left to England, even if her war Fleet proper were badly damaged.

The help of such neutrals as were left in this war would not have afforded us sufficient security to enable us to maintain our economic life, so long as imports from overseas were lacking, even if they had been in a position to treat us in a more friendly manner after their spines had been stiffened by a severe English defeat at sea. We could only escape from this tight corner if we could find the means of exerting a still more stringent pressure upon English trade and so force England to yield. The U-boat might rescue us, because the protection which the English afforded trade was powerless against this weapon.

A military and political problem of the utmost importance thus arose : Germany was in possession of a weapon which would render the English Fleet ineffective and was capable of upsetting England's whole plan of starving us out. It was only when the effectiveness of these boats under the pressure of war had proved to be far beyond all expectations that it became clear that the U-boat could attain such importance as a weapon in naval warfare. The closest understanding between the political leaders and the Naval Command was requisite for the use of this weapon. The first considerations were of course those concerning Maritime Law.

It would take too long to reproduce here all the legal discussions that took place on this question. The novelty of the weapon demanded new methods which the opposition considered unjustifiable and which they, of course, opposed with the greatest vigour, since they were contrary to their interests. But there was no doubt that the English conduct of the war had given us the right to use retaliatory measures, especially since they had shown by example that it was a simple law of necessity imposed by war, to make use of the means at one's disposal "in order to carry out operations at sea effectively."

The submarine was a weapon of war adopted by every state. This gave us the right to make use of it in the manner to which, owing to its peculiar nature, it was best adapted. Any use of it which did not take this peculiar nature into account would be nonsensical and unmilitary. The U-boat's capacity for diving made it

The U-Boat Campaign

specially suitable for war on commerce, because it could appear unexpectedly and thereby cause fear and panic and scare away trade, while at the same time it could escape the pursuit of the enemy. The fact that it could travel under water made the new weapon particularly promising. If it sank merchant vessels, including their crews and any passengers, the blame would attach to those who despised our warnings and, open-eyed, ran the risk of being torpedoed, in exactly the same way as the crews of those steamers that would not submit to English dictation, and in spite of the English warnings, took the risk of crossing the areas where mines were laid.

Was the audacity of the merchant seamen to prevent us from seizing a weapon on the use of which our fate depended? Certainly no legal considerations could stop us from pursuing this course, but only political considerations as to whether we were strong enough to disregard unjustified protests. It was imperative to make the most of the advantages arising from the submersibility of the boat, otherwise the weapon would be blunted at the start and bound to be ineffectual. The U-boat must constitute a danger from which there was no escape. Neither watchfulness nor speed could afford ships sufficient protection. That was the consideration on which the conclusion was based, that, as the loss of ships increased, trade with the British Isles must ultimately cease. The submersibility of the boats would also leave the enemy in doubt as to the number of boats with which he had to wrestle; for he had no means of gaining a clear idea of the whereabouts of his opponent. One single successful U-boat that had made a route dangerous might produce the impression that two or more had been at work. For it is human nature to exaggerate unknown dangers. The target of attack presented to the U-boats by English trade, spread all around the British Isles, was vulnerable at every point of the coast. Therein lay a great advantage as compared with the conduct of war against trade as carried on by cruisers. They had to seek the open sea where there was little traffic in order to escape pursuit; the U-boat on the contrary could frequent the neighbourhood of the coast where all traffic met, and could escape pursuit merely by diving.

All these considerations had led to the same suggestion being made at one and the same time from the most varied sections of the navy—that our conduct of naval warfare must follow the example given by England, and be directed towards the destruction of com-

Germany's High Sea Fleet

mercial traffic, because in that way we can hit England in a vital spot. The U-boat will serve as a suitable weapon for this purpose.

In November, 1914, the Leaders of the Fleet laid this suggestion before the Chief of the Naval Staff, Admiral von Pohl, advancing the following arguments:

"As our coast is not blockaded, our trade with neutrals, in so far as it does not involve contraband, might continue in the usual way. Nevertheless all trade on the North Sea coast has ceased. England exerts strong pressure on our neighbours to put a stop to all trade between them and us in goods which we need for the conduct of the war. Their most vigorous efforts are directed towards preventing the import of food from neutral countries. This does not apply merely to food imports destined for the troops; England wants to starve our whole nation. In this she overrides all rules of International Law, as food is only conditional contraband and only liable to stoppage, therefore, when intended to assist in the conduct of the war. According to the provisions of the London Conference, conditional contraband can only be stopped when it is shipped direct to the enemy country. If it be sent *via* a neutral country, e.g. Holland, it is not permissible to stop it. In spite of this a large number of steamers carrying food, oil, metals, etc., to neutral countries have been held up on the way, although it had not been ascertained with certainty that their further destination was Germany.

"As England is trying to destroy our trade it is only fair if we retaliate by carrying on the campaign against her trade by all possible means. Further, as England completely disregards International Law in her actions, there is not the least reason why we should exercise any restraint in our conduct of the war. We can wound England most seriously by injuring her trade. By means of the U-boat we should be able to inflict the greatest injury. We must therefore make use of this weapon, and do so, moreover, in the way most suited to its peculiarities. The more vigorously the war is prosecuted the sooner will it come to an end, and countless human beings and treasure will be saved if the duration of the war is curtailed. Consequently a U-boat cannot spare the crews of steamers, but must send them to the bottom with their ships. The shipping world can be warned of these consequences, and it can be pointed out that ships which attempt to make British ports run the risk of being destroyed with their crews. This warning that the lives of steamers' crews will be endangered will be one good

The U-Boat Campaign

reason why all shipping trade with England should cease within a short space of time. The whole British coast, or anyway a part of it, must be declared to be blockaded, and at the same time the aforesaid warning must be published.

"The declaration of the blockade is desirable in order to warn neutrals of the consequences. The gravity of the situation demands that we should free ourselves from all scruples which certainly no longer have justification. It is of importance too, with a view to the future, that we should make the enemy realise at once what a powerful weapon we possess in the U-boat, with which to injure their trade, and that the most unsparing use is to be made of it."

Such action was suggested on military grounds. As was only natural, the political leaders were filled with grave doubts on account of its probable effect upon neutrals. The Imperial Chancellor sent a reply to the Admiralty on December 27, 1914; in this he summed up his reflections on the subject and declared that there was nothing from the legal point of view to be urged against the U-boat campaign, but that the decision must depend upon military and political considerations as to its advisability. The question was not whether it *should* be done, but *when it could be done without ruining our position*. Such a measure as the U-boat blockade would react detrimentally upon the attitude of neutrals and our imports; it could only be employed without dangerous consequences when our military position on the Continent was so secure that there could be no doubt as to the ultimate outcome there, and the danger that the neutrals would join our opponents might be regarded as out of the question. At the moment these conditions did not exist.

This answer shows that the importance of this matter was not fully recognised or appreciated.

It was not a question of whether the Navy might make use of a new and peculiar weapon in order to make the conduct of war at sea more effective and many-sided; the question was whether the gravity of the situation had been truly appreciated. The Imperial Chancellor's answer culminated in the remark: *First the war on land must be successful; then we can think of attacking England.*

Enemies on all sides! That was the situation. Could the war on land alone rescue us from the position, or war at sea as carried on heretofore? How could we increase our efforts so as not to be defeated? Simple and straightforward reflection on this question pointed to the U-boat campaign against commerce as the way out. Of course it was our duty thoroughly to weigh its political consequences, its practicability from the military point of view, and its chances of success on a careful estimate of English economic conditions. But the study of these points ought to have preceded the war. It was neglected then because no one foresaw that a fight with England would mean a fight against her sea traffic with all the consequences it would entail. For who anticipated that we could possibly be in a position to inflict as severe an injury on English trade as that which we must expect to receive from the effects of the English blockade? It is no reproach to anybody not to have foreseen this. On the contrary, such aggressive ideas were quite foreign to our naval policy. In the course of the world-war, under the necessity of defending ourselves against the nations opposed to us, when we recognised the magnitude of the disaster which England had planned for us then, and then only, we descried a prospective possibility of winning freedom. It was lucky for us that our naval policy made it possible for us to carry out this plan; that we could pass from the state of defence, in which the enemy would cheerfully have allowed us to go on stewing, to an offensive; that we not only possessed this weapon in our naval armament but that we also had the men to use it, men with sufficient technical knowledge and the necessary courage; and lastly that the U-boats could rely on the security of their bases which the Fleet was called upon and ready to maintain.

The prospect was one of overwhelming magnitude, for it meant neither more nor less than the realisation of Germany's demand for the freedom of the seas. If we compare the importance of this undertaking with the manner of its execution we are filled with bitter disappointment over the lack of farsightedness and resolution amongst those with whom the ultimate decision lay; and with deep regret for the great and heroic sacrifices that were made in vain.

Thus the U-boat campaign became almost entirely a question of politics. It was originally suggested by the Navy for military reasons; for it was the Fleet that had to bear the brunt of English pressure at sea, and it was the Fleet's duty to neutralise the effect of that pressure, which was very definitely directed against our economic life. Considering the strength of the English Fleet and its strategy, it was impossible to remove this pressure directly, but all the same the U-boat had proved to be a weapon with which we could inflict direct injury on English economic life, notwithstanding the protection which the Fleet afforded it. Economic life in

The U-boat Campaign

England was almost entirely dependent on shipping, and so there was a prospect of our inflicting such material injury upon that island State that it would be unable to continue the war; four-fifths of the food of the country and all raw materials it needed, excepting coal and half of the iron ore, had to be imported by sea. Neutral shipping also took part in supplying these imports. That is why the U-boat war against English trade became a political question, because it might do very considerable injury to the interests of countries which so far were not involved in the war.

There is such an enormous literature on the subject of the economic as well as the legal conditions, that I shall content myself with an account of the political developments of the U-boat campaign and of its military realisation as it affected us in the Fleet.

The suggestion made by those in command of the Fleet to inaugurate a U-boat campaign against commerce was adopted by the Chief of the Admiralty Staff, von Pohl, in the form of a declaration of a War Zone which was published on February 4, 1915, of which the wording was as follows:

NOTICE IN THE IMPERIAL GAZETTE (*Reichsanzeiger*)

1. The waters around Great Britain and Ireland, including the whole of the English Channel, are herewith declared to be in the War Zone. From February 18, 1915, onward, every merchant ship met with in this War Zone will be destroyed, nor will it always be possible to obviate the danger with which the crews and passengers are thereby threatened.

2. Neutral ships, too, will run a risk in the War Zone, for in view of the misuse of neutral flags ordained by the British Government on January 31, and owing to the hazards of naval warfare, it may not always be possible to prevent the attacks meant for hostile ships from being directed against neutral ships.

3. Shipping north of the Shetland Islands, in the eastern part of the North Sea, and on a strip at least 30 nautical miles wide along the Dutch coast is not threatened with danger.

Chief of the Naval Staff,

(Signed) v. POHL.

This declaration was made with the consent of the Government, which sent a memorandum to the Powers affected, in which it was clearly indicated that the declaration referred to the use of U-boats. The idea of declaring a blockade of the whole British coast, or individual ports, had been dropped. In declaring a War Zone we were following the English example. The characteristic of a blockade had always been that it must be rendered effective. But the number of boats at our disposal at that date could not be considered sufficient for such a purpose. The blockade of individual ports would not have fulfilled the object of spreading consternation amongst the whole English shipping community, and would make it easy for the English to take defensive measures if these could be confined to certain known areas.

Unfortunately, when they declared the War Zone, those in authority could not bring themselves to state in so many words that all shipping there was forbidden. Such a prohibition would not have been in accordance with the Chancellor's ideas as expressed at the end of December in the memorandum stating his doubts of the political wisdom of the move. This new declaration represented a compromise. We know from Grand-Admiral von Tirpitz, Secretary of State to the Imperial Admiralty, that he was given no opportunity to influence this decision. This is all the more incomprehensible, because he had to furnish the necessary material, and therefore should have had the casting vote as to whether the scheme were practicable or no. There seems to be no particularly valid reason why the announcement should have been hurried on in this way, except that perhaps Admiral von Pohl wanted to close the discussions with the Foreign Office by publishing this declaration before he took up his new post as head of the Fleet, to which he had already been appointed. This undue haste proved very awkward for him in his new position when he realised that the U-boats could not act in the way he had planned, on account of the remonstrances of the neutral States. He found himself obliged to protest against the orders issued for these reasons, orders which endangered the vital interests of the U-boats.

The success of this declaration of a War Zone depended upon whether the neutrals heeded our warning and refrained, for fear of the consequences, from passing through the War Zone. If they did not wish to lose the advantages accruing to them from their sea trade with England they had to take the risks.

The memorandum issued by the Government had characterised our action as a retaliatory measure against Great Britain, because the latter conducted the war against German trade in a manner

The U-boat Campaign

which ignored all principles of International Law. It then proceeded:

"As England has declared the waters between Scotland and Norway to be part of the War Zone, so Germany declares all the waters round Great Britain and Ireland, including the whole English Channel, to be in the War Zone, and she will combat hostile shipping in those parts *with every weapon at her disposal*. For this purpose, from February 18 and onward, she will seek to destroy every hostile merchant ship which enters the War Zone, and it will not always be possible to obviate the danger with which the persons and goods on board will be threatened. Neutrals are therefore warned in future not to risk crews, passengers and goods on such ships. Further, their attention is drawn to the fact that it is highly desirable that their own ships should avoid entering this zone. For although the German Navy has orders to avoid acts of violence against neutral ships, so far as they are recognisable, yet, in view of the misuse of neutral flags ordained by the British Government, and owing to the hazards of warfare, it may not always be possible to prevent them from falling a victim to an attack directed against an enemy ship."

Our U-boats received orders to adhere to the following rules while conducting their campaign against commerce:

"The first consideration is the safety of the U-boat. Consequently, rising to the surface in order to examine a ship must be avoided for the sake of the boat's safety, because, apart from the danger of a possible surprise attack by enemy ships, there is no guarantee that one is not dealing with an enemy ship even if it bears the distinguishing marks of a neutral. The fact that a steamer flies a neutral flag, and even carries the distinguishing marks of a neutral, is no guarantee that it is actually a neutral vessel. Its destruction will therefore be justifiable unless other attendant circumstances indicate its neutrality."

This attitude was all the more justified because the object of the whole enterprise was to make use of the U-boats to compensate us, since, owing to our geographical position, it was impossible for our surface ships to touch English world commerce. A perceptible effect of the campaign against commerce could only be achieved if the peculiarities of the U-boat were taken into consideration, as they were in the instructions issued to them. The U-boat, as a special weapon in the war upon sea-borne trade, was to carry out the blockade in the War Zone. Its strength lay in the difficulty of perceiving an under-water attack, and it had to make use of this in the interests of self-preservation. You do not demand of an aeroplane that it should attack the enemy on its wheels.

The danger which the neutrals ran arose from the difference in their attitude towards the two declarations of a War Zone made by England and by Germany. *Never* did a single ship, not even an American, defy the British order, and thereby test whether, in an extreme case, England would have carried out her declaration of a War Zone by the exercise of violence. On the contrary, the neutral ships voluntarily followed the routes prescribed by the English Admiralty, and ran into British ports. In our case the neutrals, despite all warnings, tried to break through again and again, so that we were forced to carry out our declaration in such a way that the threatened danger became a reality.

The assumption that the neutrals would accept our attitude without protest was not fulfilled. The United States especially raised very decided objections, accompanied by threats. In view of the attitude they observed towards England they could not contradict the statement that the new conditions of naval warfare formed a reason for new laws; but they made use of the maxim that the dictates of humanity set limits to the creation of new laws. That was equivalent to saying that human life must be spared under any circumstances, a demand which the U-boat is not always able to fulfil, owing to its very nature. This is an extraordinary example of the Anglo-Saxon line of thought. You may let old men, women and children starve, and at the same time you insist that they must not be actually killed, because the English blockade of the North Sea could be carried out in such a manner that the ships only needed to be taken into port and not sunk.

It appears very curious to-day that the possibility of such objections was not foreseen and their consequences carefully examined. Owing to such objections our Government was faced with the following alternatives: Either it must retract its declaration of a War Zone, or, in carrying out activities in the War Zone, should consider the neutrals, and in so doing gravely diminish the chances of success, if not destroy them altogether. Once we had shelved

The U-boat Campaign

the question of our moral right to carry on the U-boat campaign, because of the American demands made in the name of humanity, it became increasingly difficult to take it up again later in an intensified form, if this should prove necessary; for if there were need of an amelioration of the military situation, which the U-boat campaign could have brought about, then we must expect that the politicians would object on the grounds that the employment of this weapon would only make the general situation worse.

That is the key to the continued opposition of the Imperial Chancellor to the initiation of a mode of warfare which could have dealt an effective blow at England. He had made it impossible from the very start. For in their answer to the American protest our Government said that they had announced the impending destruction only of enemy merchant vessels found in the War Zone, but not the destruction of all merchant shipping, as the American Government appeared erroneously to believe; and they declared that they were furthermore ready to give serious consideration to any measure which seemed likely to ensure the safety of legitimate neutral shipping in the War Zone.

This recognition of legitimate shipping was in direct contradiction to the intentions of the Naval Staff. It is not clear why the declaration of the U-boat campaign should have been made so hastily, if the political leaders had not the will to carry it through. But there had to be a clear understanding on this point, if we intended to institute a U-boat campaign at all. One almost is tempted to think that this was a feeler to see if the neutrals would tamely submit to our action. But the consequences which a refusal must entail were far too serious. *The form of the announcement of February 4 made it possible for our diplomats to maintain their declaration, and at the same time, in the conduct of the campaign, to grant the neutrals the immunity which they demanded. This restriction was forced upon the U-boats, and thus the U-boat campaign was in fact ruined.*

The Note could not have been worded with greater diplomatic skill if we had wished not to carry out the will of our leaders responsible for the conduct of the war, but rather to protect the interests of our enemies, which in this case were identical with those of the neutrals.

Before the date fixed for the opening of hostilities had arrived, two telegrams were received by the Fleet on February 14 and 15. They ran as follows:

Germany's High Sea Fleet

1. "For urgent political reasons send orders by wireless to U-boats already dispatched for the present not to attack ships flying a neutral flag, unless recognised with certainty to be enemies."

2. "As indicated in the announcement on February 2, H.M. the Emperor has commanded that the U-boat campaign against neutrals to destroy commerce, as indicated in the announcement of February 4, is not to be begun on February 18, but only when orders to do so are received from the 'All Highest.'"

Thereupon the head of the Fleet telegraphed to the Naval Staff:

"'U 30' already in the neighbourhood of the Irish Sea. The order only to destroy ships recognised with certainty as hostile will hardly reach her. This order makes success impossible, as the U-boats cannot determine the nationality of ships without exposing themselves to great danger. The reputation of the Navy will, in my opinion, suffer tremendously if this undertaking, publicly announced and most hopefully regarded by the people, achieves no results. Please submit my views to H.M."

This telegram reflects the impression made upon Admiral von Pohl, as head of the Fleet, by the receipt of the two orders, which so utterly contradicted the hopes he had placed on his declaration of a War Zone. And it also proved how unwilling the Admiral himself was to demand such action from the U-boats. But the doubts which had arisen among our political leaders as to the wisdom of risking America's threatened displeasure continued to hold sway. I do not intend to question that their estimate of the general situation, combined with our capacity to carry on energetic U-boat warfare, justified their doubts; but then it was a grievous mistake to allow such a situation to arise, for it blocked the way for an unrestricted U-boat campaign in the future.

On February 18 instructions in conformity with the new conditions were issued to the U-boats with regard to their course of action. They ran as follows:

"1. The U-boat campaign against commerce is to be prosecuted with all possible vigour.

"2. Hostile merchant ships are to be destroyed.

The U-boat Campaign

"3. Neutral ships are to be spared. A neutral flag or funnel marks of neutral steamship lines are not to be regarded, however, as sufficient guarantee in themselves of neutral nationality. Nor does the possession of further distinguishing neutral marks furnish absolute certainty. The commander must take into account all accompanying circumstances that may enable him to recognise the nationality of the ship, e.g. structure, place of registration, course, general behaviour.

"4. Merchant ships with a neutral flag travelling with a convoy are thereby proved to be neutral.

"5. Hospital ships are to be spared. They may only be attacked when they are obviously used for the transport of troops from England to France.

"6. Ships belonging to the Belgian Relief Commission are likewise to be spared.

"7. If in spite of the exercise of great care mistakes should be made, the commander will not be made responsible."

On February 22 the U-boats were to begin their activities on these lines. In these instructions the Naval Staff had been obliged to conform to the declaration which the Imperial Government had made to America, explaining its conception of the conduct of the campaign against trade in the War Zone, although they had had no opportunity of expressing their doubts of the possibility of carrying out these instructions in practice.

The activities of the U-boats were made much more difficult because, for the time being, all goods conveyed to the enemy in neutral bottoms reached him without obstruction, and their successes were thereby reduced to a third of what they would otherwise have been; for that was the extent to which neutral shipping was engaged in the commercial traffic with England. Further, neutrals could not be scared out of trading with England, because they knew by the declaration made to America that activities in the War Zone would be attended with less danger than had been threatened. Our intention of pursuing a milder form of activity was confirmed to Holland when, after the sinking of the steamer *Katwyk*, popular opinion in Holland grew very excited, and our Foreign Office assured the Dutch Government in the following Note that an attack on a Dutch merchant vessel was utterly foreign to our desires:

231

Germany's High Sea Fleet

"If the torpedoing of the *Katwyk* was actually the work of a German U-boat the German Government will not hesitate to assure the Dutch Government of its profound regret and to pay full compensation for the damage."

Besides the neutral ships, many enemy ships by disguising themselves with neutral distinguishing marks could get through with their cargoes in safety if the U-boat was not able to set its doubts on the subject at rest. This became very noticeable when the arming of steamers, which had meanwhile been carried out, had been added to the misuse of flags, and the U-boats were exposed to great danger in determining the nationality of ships.

All these circumstances contributed to lessen the results. Our enemies acted in an increasingly unscrupulous manner, especially when bonuses were offered for merchant vessels which should sink U-boats. A particularly crude case was that of the British auxiliary cruiser, the *Baralong*, whose crew shot down the whole crew of " U 27 " (Commander, Lieut.-Commander Neigener) when they were swimming defenceless in the water and some of whom had taken refuge on board an American steamer.

Regardless of all added difficulties, our U-boat crews devoted themselves to their task. Trying to achieve the greatest possible results, they nevertheless avoided incidents which might be followed by complaints, until on May 7 the sinking of the *Lusitania,* the English liner of 31,000 tons, aroused tremendous excitement.

The danger which England ran, thanks to our U-boats, was shown in a lurid light; the English Press expressed consternation and indignation. It was particularly striking how the English Press persisted in representing the loss of the *Lusitania* not so much as a British, but as an American misfortune. One must read the article in *The Times* which appeared immediately after the sinking of the *Lusitania* (8/5/1915) to realise the degree of hypocrisy of which the English are capable when their commercial interests are at stake. Not a word of sympathy or sorrow for the loss of human life, but only the undisguised desire (with a certain satisfaction) to make capital out of the incident in order to rouse the Americans and make them take sides against Germany.

They were not to be disappointed in their expectations. In an exchange of Notes, which lasted until well into July, the Americans demanded the abandonment of the U-boat campaign because the manner in which we used this weapon to destroy trade was in prac-

232

The U-boat Campaign

tice irreconcilable with America's demand that her citizens should have the right in the pursuit of their lawful business to travel by sea to any spot without risk to their lives in so doing. We expressed our willingness to abandon this use of the U-boat if America could succeed in inducing England to observe International Law. But this suggestion met with no success. The U-boat campaign was, however, further hampered by an order not to sink any big passenger steamers, not even those of the enemy.

On August 19, 1915, a further incident occurred when the steamer *Arabic* was sunk by "U 24"; although the boat acted in justifiable self-defence against a threatened attack by the steamer, yet the prohibition with regard to passenger boats was made more stringent, for the order was given that not only large liners, but *all* passenger steamers must be warned and the passengers rescued before the ship was sunk. On this occasion, too, when the answer to the objections raised by America were discussed, the Chief of the Naval Staff, Admiral Bachmann, was not allowed to express his views. Consequently he tendered his resignation to His Majesty, which was duly accepted. Admiral von Holtzendorff was appointed in his place.

In consideration of the small chances of success, the U-boat campaign off the west coast of the British Isles was abandoned. The Chief of the Fleet, Admiral von Pohl, also asked to be released from his office if this last order concerning the passenger ships were insisted on, because he could not take the responsibility of issuing such instructions, which could only be carried out at great risk to the U-boats, in view of the fact that so many losses had occurred since the first limiting order had been published; further, he held it to be impossible to give up the U-boat campaign, which was the only effective weapon against England that the Navy possessed. His objections to the limitation of the U-boat campaign were dismissed by the remark that he lacked full knowledge of the political situation.

Though the U-boat campaign west of England was given up, it was not stopped entirely, for subsequent to March, 1915, a U-boat base had been established at Zeebrugge, and another in the Mediterranean. "U 21" had been sent under Hersing's command in April, 1915, to assist our warships which were engaged in the defence of the Dardanelles, and this had given proof of the great capacity of our U-boats. Consequently the newest boats, "U 33" and "U 34," were sent to Pola, the Austrian Naval Base, in order to carry on the U-boat campaign in the Mediterranean. The secession of Italy (May 27, 1915) to our enemies gave our boats there a new field of activity, because practically all steamer traffic in these waters was carried on under enemy flags, and complications with neutrals were hardly to be feared.

Thus the U-boat campaign dragged on, though with but moderate success, to the end of the year. Yet it managed to deal wounds to English sea trade which exceeded in gravity anything that the island State had ever thought possible. The total sinkings from February to August amounted to 120,000 tons. Further results were:

> September, 136,000 tons.
> October, 108,000 tons.
> November, 158,000 tons.
> December, 121,000 tons.*

Before the U-boat campaign oversea traffic to and from England had hardly been seriously reduced. Although the cruiser campaign carried on by the *Emden*, the *Karlsruhe* and the *Kronprinz Friedrich Wilhelm* and the *Prinz Eitel-Friedrich* had had a disturbing effect, yet no decisive results could be achieved owing to the lack of oversea bases. The rise in freights was still moderate, and on the whole the Englishman hardly suffered at all. There was no question of want anywhere, and the rise in prices was slight. The U-boat campaign, however, changed British economic conditions fundamentally. Freights rose considerably. In May, 1915, they were double what they had been in January; in January, 1916, they had risen on an average to ten times the amount they had been before the war (January, 1914). Wholesale prices, of course, followed this movement, and though imports had not decreased so much that there was any talk of want, yet the U-boat campaign had led to a scarcity, because the demand, so much increased by the needs of the army, was greater than the supply.

Towards the end of the year the lack of tonnage began to be felt acutely, and it became clear that this lack was the chief difficulty that England had to face as a result of U-boat warfare. In January, 1916, the new Chief of the Naval Staff handed in a memorandum in which he subjected British economic conditions to a thorough examination, and drew the following conclusions from his investigations:

* These figures indicate gross tonnage.

The U-boat Campaign

1. The U-boat campaign of last year, gradually increasing its weapons but hampered by growing restrictions of a non-military nature, dealt a blow to a new economic entity hitherto little affected by the war and capable of strong resistance. By means of a scarcity which was mostly felt in a considerable rise in the price of important foodstuffs as well as of manufactured goods and raw materials, it reduced England's commerce to such an extent that serious economic and financial injury is apparent in all directions. This injury has aroused a feeling of considerable anxiety in England, where it was felt that a vulnerable spot was threatened; moreover, it was calculated gradually to make England inclined for peace. The effect wore off as soon as England was certain that for reasons due to considerations of a non-military nature the U-boat campaign would not be continued.

2. The economic changes set up by the U-boat campaign have persisted, though for the most part in a milder form. Towards the end of 1915 lack of transport reduced British sea-traffic to such an extent that the difficulties due to the interruption in British foreign trade were rendered more acute by the steady rise in the price of imports. Market prices followed suit. The financial situation, too, became disquieting owing to the drain on the country caused by the military and political situation.

3. A new U-boat campaign would be undertaken under much more favourable circumstances than that of February, 1915, because the amount of tonnage still available for British imports and exports cannot stand much further diminution, as in that case the transport of essential goods will suffer, and because England has been robbed of the better part of her power of resistance by shortage, rise in prices and financial overstrain. Moreover, a new U-boat campaign has such weapons at its disposal that it is in a position to achieve considerably more from a military point of view than last year's campaign, for though the enemy has increased his defensive power the U-boats are equipped with a number of new technical improvements.

4. If on this basis the U-boat campaign has to be carried on with the same restrictions of a non-military nature as last year no doubt England's economic, and consequently also her financial, position will be further damaged. But it cannot be assumed with any certainty that in this way England will be forced to make peace, partly because of the many difficulties of carrying out a U-boat

campaign with such a limitation of its specific activities, and the consequent greatly increased possibilities of defence, but especially because, judging by last year's experience, the effect of terrorising shipping is to all intents and purposes lost.

5. But if a new unlimited U-boat campaign is inaugurated on the principle that all shipping in the War Zone may be destroyed, then there is a definite prospect that within a short time, at most six months, England will be forced to make peace, for the shortage of transport and the consequent reduction of exports and imports will become intolerable, since prices will rise still more, and in addition to this England's financial position will be seriously threatened. Any other end to the war would mean grave danger for Germany's future economic life when we consider the war on German trade that England has planned and from which she could be deterred only by such a defeat as the U-boats could inflict.

6. The United States are not in a position to lend England effective aid against a new U-boat campaign by providing her with tonnage. In view of the ever-increasing burdens imposed by the war, it is not to be supposed that the United States will afford England financial support for an indefinite period. Such support would, moreover, be of no avail in an unrestricted U-boat campaign against English trade, as it could not prevent a scarcity of essential goods or make it possible for the English to carry on their export trade.

The proposal made by the Chief of the Naval Staff in January, 1916, to start an unrestricted U-boat campaign was based on the following estimates of success:

(a). From the beginning of the U-boat war in 1915 till the end of October of that year in the War Zone round England one or two steamers, averaging 4,085 tons, were sunk daily by each U-boat; this does not include steamers of less than 1,000 tons. It could, therefore, be assumed that in the future each U-boat would sink ships amounting to at least 4,000 tons daily. If it is reckoned that in a month only four stations are continuously occupied—a very low estimate in view of the increase in the number of U-boats during 1915—then you get a total of 16,000 tons a day, or 480,000 tons a month, in the War Zone round England.

The U-boat Campaign

(b). In the Mediterranean in the second half of the year 1915 an average of 125,000 tons of shipping was sunk every month. Assuming that traffic did not materially fall off, as a result of the U-boat campaign, and that in the course of the summer of 1916 the number of stations in the Mediterranean would be further increased, the same result might be counted on; that is, 125,000 tons per month.

(c). The amount of tonnage destroyed by mines had averaged 26,640 tons a month. The same number could be assumed for the future. This would bring the total result per month up to 631,640 tons, which would mean a complete loss of 3,789,840 tons in six months. But the effect of this loss upon English trade and economic conditions must be measured by a multiple of this figure, because every lost ship would affect imports and exports, and would, moreover, have made several journeys in six months. The total tonnage of the English Mercantile Fleet at the outbreak of war amounted to 20 million tons in round numbers. Judging by the rise in prices which became manifest a few weeks after the opening of the U-boat campaign, an idea can be formed of what the effect would be if more than a third of England's total tonnage were completely lost, when it is considered that England is dependent on it to supply her manifold wants and keep up her widely extended business connections. There could then be no question of "business as usual."

But the Imperial Government rejected the admiral's suggestion. So the Chief of the Naval Staff resolved to content himself with a kind of payment on account, which consisted in treating all armed enemy merchantmen as warships. But he did not give up all hope of soon being able to take up the U-boat campaign in its intensest form.

When in January, 1916, I took over the command of the Fleet I considered it my first task to ascertain what weapons against England lay at my disposal, and especially to make sure whether, and in what way, the U-boat campaign against English trade was intended to be carried out. On February 1 the Chief of the Naval Staff assured me that the unrestricted U-boat campaign would be inaugurated on March 1. All preparatory work for the operations of the Fleet were based on this assumption. As early as February 11 the officers in command of the Fleet received the order as to the treatment of armed merchant vessels. According to this order enemy merchantmen armed with guns were to be looked upon as warships, and to be destroyed by all possible means. The commanders were to keep in mind that mistakes would lead to a break with neutral Powers, and therefore the sinking of a merchant vessel on account of its being armed might only be proceeded with when the fact that it carried a gun had been positively ascertained. In view of the warning to neutrals, which was to be conveyed through diplomatic channels, this order was not to come into force until February 29.

The Government again issued a memorandum about the treatment of armed merchantmen. In this they explained at length that in view of the instructions issued by the British Government, and of the consequent conduct of English merchantmen, enemy merchant ships that were armed no longer had the right to be regarded as peaceful trading vessels. The German Government notified neutral Powers of this state of affairs, so that they might warn their people in future not to entrust their persons or their fortunes to armed merchantmen belonging to any of the Powers at war with the German Empire. After this explanation no neutral State could demand that its citizens should be entitled to protection if they travelled on armed enemy steamers into the War Zone.

We expected that in these circumstances there would be fewer difficulties in carrying out the U-boat campaign, while paying due consideration to neutral shipping. But if, as the Chief of the Naval Staff had told me, it had been decided to open the unrestricted U-boat campaign on March 1, it was not clear why this declaration relative to the treatment of armed steamers should have preceded it. My suspicion that the date of March 1 would not be adhered to was confirmed on the occasion of H.M. the Emperor's visit on February 23, of which I have given an account in an earlier chapter. The Emperor shared the political doubts which the Government had advanced, and wished to avoid a break with America. This announcement of the Government had received the assent of the Naval Staff, which was responsible for the war at sea, and so of course those in command of the Fleet had to submit to the order to resume the campaign against English trade with a few U-boats.

We would try this first and await the result. Judging by the assurance given me, I took it for granted that the Government had

The U-boat Campaign

learnt a lesson from the events of 1915, and that it would not again give way if objections were raised, but would on the contrary then proceed with the intensified form of U-boat warfare. We had far greater means at our disposal now to give emphasis to our threats.

I should like to point out here that those in command of the Fleet had no right to exercise a decisive influence on the conduct of the war, but the Chief of the Fleet, being responsible for the execution of orders, could make representations if he found the conditions imposed on him too disadvantageous. Added to this, the Fleet had only some—about half—of the U-boats at its disposal; the rest were in part attached to the Naval Corps, and in part under the orders of the Commander-in-Chief of the Baltic; those in the Mediterranean took their orders direct from the Naval Staff. But the problem of the U-boat campaign was so closely connected with the combating of the English Fleet—our own Fleet's main task—that it became a matter of the greatest importance in its effect on the decisions of the Navy. I therefore thought it my duty to point out the difficulties which would arise in our conduct of the war in every sphere, if the U-boat campaign were prosecuted on principles that were militarily unsound; all the more so as I was accountable to the U-boats under my orders, if they were assigned to tasks which would in the long run entail their destruction without their having achieved the success which they promised to do if rightly wielded as a weapon.

From this point of view I endeavoured to combat the tendency to give way, which the Chief of the Naval Staff betrayed when dealing with political objections, although in a long and well-thought-out memorandum he, as the proper representative of the naval fighting forces, had shown that unrestricted U-boat warfare was the best and safest means we possessed to subdue England and generally to bring the war to a successful close.

On March 4 the decisive session at General Headquarters took place, and the Chief of the Naval Staff informed me of the result as follows:

"For military reasons, the unrestricted U-boat campaign against England, which alone promises full success, must begin without fail on April 1. Till then the Imperial Chancellor must set in motion all political and diplomatic machinery to make America clearly understand our position, with the aim and object of securing our freedom of action. Up to that date the U-boat campaign shall be carried on against England as effectively as possible in conformity with the orders issued on March 1."

The following considerations were the means of bringing about this decision at the discussion on March 4:

"The general military situation is good. East and west we hold the territory that we have victoriously won. No serious danger is to be apprehended from America so long as our U-boats and Fleet remain afloat. Austria is effectively repulsing Italy's attempts at attack; Bulgaria has a firm hold on Serbian territory; the Salonika campaign is doomed to come to a standstill; the Russian offensive against Turkey has come to a stand on the Erzerum—Trebizond line; the English expedition in Mesopotamia has ended in a heavy defeat; Egypt is threatened from the direction of Syria and by the Senussi, which means that a considerable British army of defence must be kept there. Latterly, too, military forces have had to be sent to Ireland. No essential change in the favourable general military situation is to be expected, nor on the other hand is there any prospect of a decisive victory of all our forces.

"From the economic point of view the fact that we are cut off from all imports from overseas and neutral countries becomes increasingly apparent; even a good harvest cannot bring security for the future, as long as England's policy of violence, whose object is to starve us out, is not stopped. Thus the economic conditions are very different from the military. Our opponents can hold out longer than we can. We must, therefore, aim at bringing the war to an end. We shall not be mistaken in assuming that an injury inflicted on England, which induces her to regard the conclusion of peace as better business, can force the others to peace as well. England can only be injured by war on her trade. The only means to inflict this injury is a ruthless U-boat campaign, the effects of which England will not be able to withstand for more than six or eight months if she cannot get assistance from others than her present Allies. Ruthless U-boat warfare will not only inflict damage on England; neutral shipping will also feel the full brunt of it, and cargoes and lives will be imperilled. The small neutral States must give in and are willing to do so: that is, to stop trade with England. America opposes this manner of waging the U-boat campaign, and threatens us with war. From a military point of view, and especially from the standpoint of the Fleet, we might well risk this war. But economically it would fatally aggravate our situation. Such a rich and distant country could stand the war for ten years or more. But it would afford our flagging opponents very considerable moral and material support which would enable them, including England, to hold out for a longer period. Our aim, which is to bring the war to an end within a short time, would be farther than ever from realisation, and Germany would be exposed to exhaustion.

"As the present military situation is not such as to force us to stake everything on one throw of the dice, our superiority in the field must be maintained, and at the same time our diplomatists must do all in their power, first to prevent us from making new, dangerous enemies, and then to find ways and means of sowing discord among our present enemies and thereby open a prospect for a separate peace. If we succeed in keeping friends with America, and at the same time, by concessions in our manner of conducting the U-boat campaign, can induce her to exert strong and effective pressure on England, so that the legitimate trade of neutrals with the belligerents is re-established, then we shall obtain the economic aid which will enable us to maintain our favourable military situation permanently, and so to win the war. A break with America certainly affords us the tactical advantage of ruthless U-boat warfare against England, but only under conditions that will prolong the war, and will certainly bring neither relief nor amelioration to the economic situation. Should the attempt to keep America out of the war fail, it will still be our lot to face these conditions. We cannot take the responsibility of neglecting to make this attempt, for the sake of a few hundred thousand tons of enemy shipping that we might sink during the time the attempt is being made."

These attempts met with no success whatever; certainly not within the period set aside up to April 1. Neither was the assumption fulfilled that we might exert pressure upon England through the agency of America, so as to re-establish legitimate trade with neutrals, and thereby obtain the economic aid which would enable us to maintain our favourable military situation permanently. As soon as this was recognised we were confronted with the necessity of drawing the inevitable conclusions, and of beginning the economic war against England in its intensest form. Otherwise the dreaded state of affairs spoken of at the session of March 4 would become a reality, and our opponents would be able to hold out longer than we could if no change occurred in the economic situation. The

Germany's High Sea Fleet

dullest must have been forced into some recognition of this, when on April 20, in connection with the *Sussex* incident, America presented her threatening Note.

The date of April 1 had passed, and still the unrestricted U-boat campaign had not been started. But the leaders of the Fleet had no special reason for urging an early start, as the U-boats then at sea had not gathered sufficient experience on the basis of which we might make counter proposals.

On March 24, 1916, the steamer *Sussex,* with 300 passengers on board, among them being a number of American citizens, was torpedoed in the Channel while crossing from Folkestone to Dieppe. So far as German observation went, it was not made clear at first whether the steamer had been hit by a U-boat, or had struck a mine. Certainly a ship had been torpedoed on that day and in that neighbourhood, but the German commander, judging by the circumstances and the appearance of the ship, took it for a mine-layer of the new "Arabis" class. The American Government took occasion, in consequence of this incident, to send a very sharp Note to the German Government, protesting against the wrongfulness of the submarine campaign against commerce. It threatened to break off diplomatic relations with Germany if the German Government did not declare the abandonment of its present methods of submarine warfare against passenger and merchant vessels, and see that it was carried out.

As a result of this Note, presented on April 20, 1916, our Government decided to give in and sent orders to the Naval Staff to the effect that submarine warfare was henceforward to be carried on in accordance with Prize Law. This order reached the Fleet by wireless telegraphy when it was on the way to bombard Lowestoft. As war waged according to Prize Law by U-boats in the waters around England could not possibly have any success, but, on the contrary, must expose the boats to the greatest dangers, I recalled all the U-boats by wireless, and announced that the U-boat campaign against British commerce had ceased.

On April 30 I was informed by the Naval Staff that His Majesty approved of the interruption of the U-boat campaign against commerce ordered by the Commander of the Fleet, and he directed that the U-boat weapon should meanwhile be vigorously used for military purposes. The order to resume the U-boat campaign against trade would be given when the political and military situation should demand it.

The U-boat Campaign

Having U-boats at my disposal for military purposes gave me the desired opportunity of extending the operations of the Fleet, and it was owing to this circumstance that the Fleet had occasion on May 31 to meet the English Fleet in battle near the Skagerrak. To my idea the moral impression which this battle left on the neutral nations created a most favourable atmosphere for us to carry on the war against England by all possible means, and to resume the U-boat campaign in all its intensity. I took the opportunity of submitting this view to H.M. the Emperor, when he visited the Fleet at Wilhelmshaven on June 5.

In May the Naval Staff had again begun to try to persuade the leaders of the Fleet to change their mind and resume the U-boat campaign in accordance with Prize Law, so as to be able to inflict at least some injury on England. But as even the regulations as to the treatment of armed steamers had been rescinded, I refused to contemplate a resumption.

In June, soon after the battle, the Naval Staff again returned to this subject, and on June 20 invited me to state my point of view in order to incorporate it in a memorandum to be presented to the Emperor. I replied that in view of the situation I was in favour of the unrestricted U-boat campaign against commerce, in the form of a blockade of the British coast, that I objected to any milder form, and I suggested that, if owing to the political situation we could not make use of this, our sharpest weapon, there was nothing for it but to use the U-boats for military purposes. A few days later the Chief of the Naval Cabinet thought to persuade me to change my attitude. He wrote me the following letter on the subject, dated June 23, from General Headquarters:

"The Chief of the Naval Staff has given me your letter to read on this subject; its conclusions may be summed up in the words, 'Either everything or nothing.' I can fully sympathise with you in your point of view, but unfortunately the matter is not so simple. We were forced, though with rage in our hearts, to make concessions to America, and in so doing to the neutrals in general, but, on the other hand, we cannot wholly renounce the small interruptions of trade that it is still possible for us to carry out, which are proving of considerable value, too, in the Mediterranean. It is the thankless task of the Chief of the Naval Staff to try and find some way of making this possible in British waters as well.

Germany's High Sea Fleet

And it is my opinion that the Chief of the Fleet should assist him in this as far as in him lies, by bringing about a compromise between the harsh professional conception of the U-boat weapon and the general, political and military demands which the Chief of the Naval Staff has to satisfy. Of course, to that end it is necessary that the Chief of the Fleet should unreservedly acknowledge the decisions of the All Highest with regard to the limitation of the U-boat campaign, as the result of the most serious deliberation upon the military, political and economic situation. This is, of course, merely what is to be expected of him as a soldier. And further, that he should pledge himself to make use of the U-boat as a weapon, despite the limitations imposed, in order in the first place to injure, or at least continually to threaten, the import trade of England. I do not take it upon myself to offer any suggestions on the way in which such use can be made of the U-boats, especially as I know it is a far more difficult matter near the English coast than it is in the Mediterranean.

"What I ask of you is merely this: that you should personally try to arrive at some understanding with the Chief of the Naval Staff which will lead to some positive result, and by so doing put an end to a situation in which His Majesty might be forced to issue commands instead of merely approving; as, for instance, if he should order so many more U-boats to be given up for use in the Mediterranean, as offering a more fruitful field for the U-boat campaign against commerce.

"In conclusion, I should like to remark that for my part I still believe in the possibility of a ruthless U-boat campaign. The conflict between America and Mexico, the growing bitterness of the neutrals on account of England's blockade, increasingly good prospects for the harvest, and last but not least our successes on both fronts—all these are matters which tell in favour of such use of our U-boats, without involving us in an uncertain political adventure.

"(Signed) v. MULLER."

I replied that nothing more could be expected of me than that I should express my honest conviction, especially as it was in connection with new and far-reaching decisions to be taken by the Emperor that my opinion on the subject was asked.

On his visit on June 30 the Imperial Chancellor gave me the impression that he had not the slightest intention of employing

The U-boat Campaign

against England all the weapons at our disposal, but also that he would not give his consent to an unrestricted U-boat campaign, so as not to be faced with fresh troublesome incidents. The course of events hitherto had shown that America interfered on England's behalf as soon as the U-boat campaign began to have perceptible results. For ever so long America had systematically prevented us from using our most effective weapon. Our attitude gave our people the false impression that, despite America's objections, we were still going to use our U-boat weapon with all our might. The people did not know that we, pledged to the nation by our big talking, were only pretending to carry on the U-boat campaign, and America laughed because she knew that it lay with her to determine how far we might go. She would not let us win the war by it. So we did not wield our U-boat weapon as a sword which was certain to bring us victory, but, as my Chief of the Staff, Rear-Admiral von Trotha, put it, we used it as a soporific for the feelings of the nation, and presented the blunt edge to the enemy. Gerard was right; he never wanted a war between America and Germany —but he wanted our defeat. That suited his book ever so much better.

If we review the course of development of our policy from January, 1916, we find that it had zigzagged in the following manner:

1. On January 13, 1916, the Naval Staff declares: If the U-boat campaign is to achieve the necessary success it must be carried on ruthlessly.

2. On March 7, 1916: Decision of His Majesty's, passed on by the Naval Staff: For military reasons the inauguration of the unrestricted U-boat campaign against England, which alone promises full success, is indispensable from April 1 onward.

3. On April 25, 1916: We are to carry on the war against trade absolutely according to Prize Law, consequently we are to rise to the surface and stop ships, examine papers, and all passengers and crew to leave the ship before sinking her.

4. On June 30, 1916: The Imperial Chancellor informs the Commander of the Fleet that he personally is against any unrestricted form of U-boat campaign, " which would place the fate of the German Empire in the hands of a U-boat commander."

5. At the same time a proposal from the Chief of the Naval Staff: The war against merchant ships to be carried on in the following manner: They are to be approached under water to see

245

whether they are armed; if they are not armed, the boat is to rise to the surface at a safe distance, examine papers, and sink the ship when the crew is in safety.

All these impressions induced me, when I wrote my report of the Battle of the Skagerrak for the Emperor, to conclude by again pointing out most emphatically the necessity of taking up the unrestricted U-boat campaign at once, unless we wanted to give up all hope of defeating England. Now Admiral von Muller's letter seemed to imply that the Emperor disapproved of my urging this, whereas I was able to ascertain later that His Majesty, far from appending any disparaging remark to the conclusion of my report, had actually appended a note of approval to it, and had acquiesced in my report as a whole.

We should have begun the U-boat campaign in January, 1916, as the Chief of the Naval Staff proposed, or at latest immediately after the Battle of the Skagerrak, when, to my idea, the circumstances were particularly favourable. That we failed to do so fatally affected the outcome of the war. Thanks to the number constructed in 1915, we had a sufficiency of U-boats. We lost valuable time that year, when our nation's power of resistance was much greater than in 1917, when we were almost at our last gasp, and we were forced, after all, to seize the weapon which promised to prove our salvation. And in the course of this year England was able systematically to develop her defence.

The remainder of 1916 was taken up with similar discussions between the Naval Staff, Fleet and Government. The Chief of the Naval Staff endeavoured to persuade the Ministry to sanction the unrestricted U-boat campaign, and, on the other hand, urged the Fleet to agree to the boats resuming the war against commerce in a milder form. I was convinced that, if the leaders of the Fleet had given way in this matter, the worst would have happened—just what we most had to try and avoid, viz. that we should really have carried on a sort of pretence campaign to act as a soporific to the feelings of the people, and we should have presented the blunt edge of our weapon to the enemy.

At the beginning of the year 1916 the Chief of the General Staff of the Army, von Falkenhayn, had also strongly advocated our embarking on an unrestricted U-boat campaign, because he had realised that our only hope of future salvation lay in overcoming English resistance. In the autumn of 1916 Field-Marshal von Hindenburg took over the Supreme Command of the Army, to save

The U-boat Campaign

the serious situation which had arisen in the war on land. At that time there was under discussion a new demand on the part of the Chief of the Naval Staff to resume the U-boat campaign with full intensity. At the meeting of September 3 at General Headquarters in Pless, at which the matter was considered, the following were present: the Imperial Chancellor, the Field-Marshal, General Ludendorff, Admiral von Holtzendorff, Admiral von Capelle, as Secretary of State of the Imperial Ministry of Marine, the Secretary of State for Foreign Affairs, von Jagow, the Secretary of State, Helfferich, and the War Minister, Wild von Hohenborn. The outcome of the proceedings was that, after consulting all who were concerned in the question of the U-boat campaign, they unanimously declared that the decision must for the time being be postponed, because the general situation, and especially the military situation, was by no means clear, and they resolved that the final decision should lie with General Field-Marshal Hindenburg.

I took occasion after that to send the Chief of the Staff of the High Sea Fleet to General Headquarters, to consult with General Ludendorff, and they agreed upon the following:

1. There is no possibility of bringing the war to a satisfactory end without ruthless U-boat warfare.

2. On no account must a half-and-half campaign be started, which could not achieve anything of importance, but involved the same military dangers, and would probably result in a new limitation for the nation.

3. The U-boat campaign should be begun as soon as possible. The Navy is ready.

4. The separate treaties with the Northern States, who had received considerable concessions in the matter of exports to England, must be cancelled with all speed, so that we can act without interference.

5. In no circumstances must there be any yielding.

The Chief of the Staff returned from this conference under the impression that the question of the U-boat campaign could not be in better hands than in those of the Chief of the General Staff of the Army. I was able to confirm this view later, when on November 22 I had occasion myself to discuss the question at General Headquarters with the Field-Marshal and with General Ludendorff.

Germany's High Sea Fleet

The military situation in the autumn had led to a postponement of U-boat activity, so as to avoid complications in the War Zone 'round England; the only injury to commerce at the moment was that inflicted on ships in the Mediterranean. That is why the U-boat campaign was extended to Northern waters—to sink supplies which were sent via Archangel to the Russian seat of war.

The refusal of our peace proposals in December brought about a new situation in the U-boat war. Our enemies had given us clearly to understand that they would accept no peace of understanding. This led to the decision to open the unrestricted U-boat campaign on February 1, 1917. The Chief of the Naval Staff, with the approval of the General Field-Marshal, succeeded in bringing about this decision, in which the Imperial Chancellor acquiesced. So on that date the most effective period of our war against England actually began. On December 22, 1916, the Chief of the Naval Staff had again, in a detailed memorandum, given explicit reasons for adopting this form of campaign. He summed up his arguments as follows:

" 1. A decision must be reached in the war before the autumn of 1917, if it is not to end in the exhaustion of all parties, and consequently disastrously for us. Of our enemies, Italy and France are economically so hard hit that they are only upheld by England's energy and activity. If we can break England's back the war will at once be decided in our favour. Now England's mainstay is her shipping, which brings to the British Isles the necessary supplies of food and materials for war industries, and ensures their solvency abroad.

" 2. The present state of the tonnage question, which has already been described in detail, may be summed up as follows: Freights in the case of a large number of important articles have risen tremendously, some of them to ten times and more what they were before. From many other indications we can conclude with certainty that everywhere there is a shortage of tonnage. We may with safety assume that English shipping still amounts at the moment to 20 million tons, gross tonnage. Of this at least 3.6 million tons are requisitioned for military purposes, and half a million tons are occupied in coast traffic; about 1 million tons are under repair or temporarily unfit for use; about 2 million tons are taken up in supplying the needs of England's Allies; so that for her own supplies at most 8 million tons are available. Computations based on statistics

The U-boat Campaign

of traffic in English ports gives an even smaller result. According to that, from July to September, 1916, English shipping amounting to only 6¾ million tons, gross tonnage, was engaged in traffic to England. Other shipping going to England may be estimated at 900,000 tons of enemy—'non-English'—ships, and a good 3 million tons of neutrals. Taking it all round, the shipping which supplies England amounts to only 10¾ million tons, gross registered tonnage, in round figures.

"3. The results achieved hitherto in the war on shipping justify us in assuming that further activities in this direction promise success. But in addition to this, the bad harvests in wheat and produce all over the world offer us a quite unique opportunity of which it would be sinful not to take advantage. North America and Canada will, in all probability, be able to send no more grain to England after February, so the latter will have to draw her grain supplies from the distant Argentine; and as the Argentine can spare very little, owing to a bad harvest, it will have to come all the way from India, and to an even greater extent from Australia. The fact that the grain has to come from such a much greater distance involves the use of 720,000 more tons of shipping for grain carrying purposes. It practically comes to this, that until August, 1917, of the 10¾ million tons at their disposal ¾ million are required for a purpose for which they were never needed before.

"4. Such favourable conditions promise certain success to an energetic blow, dealt with our full force against English shipping. I can only repeat and emphasise what I said on August 27: 'Clearly what we must do is to bring about a decision in our favour by continuing to destroy shipping,' and, further, 'It is absolutely unjustifiable from the military point of view not to make use of the weapon of the U-boat.' I do not hesitate to assert that, as matters now stand, we can force England to make peace in five months by means of the unrestricted U-boat campaign. But this holds good only for a really unrestricted U-boat campaign, not for the cruiser warfare formerly carried on by the U-boats, even if all armed steamers are allowed to be torpedoed.

"5. Basing our calculations on the former monthly results of 600,000 tons of shipping sunk by unrestricted U-boat warfare, and the expectation that at least two-fifths of neutral sea traffic will at once be terrorised into ceasing their journeys to England, we may reckon that in five months shipping to and from England will be reduced by about 39 per cent. England would not be able to stand

249

that, neither in view of post-war conditions, nor with regard to the possibility of carrying on the war. She is already confronted with a shortage of food which forces her into attempting the same rationing measures that we, as a blockaded country, have had to adopt in the course of the war. The existing conditions with which such an organisation will have to reckon are very different and incomparably less favourable in England than here. The necessary authorities do not exist, and the people in England have not been educated to submit to such coercion.

"For another reason it would not now be possible to institute a uniform reduced bread ration for the large population of England. It was possible in Germany at a moment when the sudden reduction in the bread ration was counterbalanced for the time being by supplies of other food. They have missed that opportunity in England, and nothing can recall it. But with about three-fifths of her former shipping she cannot continue her food supply without a steady and vigorous reduction in the consumption of wheat, while at the same time she has to keep up her war industries. In the accompanying memorandum I have refuted in detail the objection that England might have enough grain and raw materials in the country to be able to carry on through this period of danger until the next harvest. Added to this, the unlimited U-boat campaign would mean an immediate shortage of fats, since she would be cut off from imports from Holland and Denmark; and one-third of her total imports of butter come from the latter country, while all the margarine comes from the former. Further, it will mean that the lack of wood and iron ore will be intensified, because the import of wood from Scandinavia will be threatened, while at the same time the imports of iron from Spain will be jeopardised. That will mean an immediate reduction in coal production, because the necessary wood will not be forthcoming; the same is true of iron and steel, and consequently of munitions, which are dependent on both. Finally, it will at length give us the desired opportunity of attacking the supply of munitions from neutral countries, and by so doing relieve our army.

"As opposed to this, cruiser warfare waged by U-boats, even if armed steamers were not exempt from sinking, would result in reducing shipping to England by one-fifth of 400,000 tons, or about 18 per cent. of the present monthly traffic, that is less than half of what would result from the unrestricted U-boat campaign. Judging by our experience up to date, we cannot assume that if the armed steamers were not exempt there would be a perceptible increase in the sinking of tonnage, which in the last two months amounted to about 400,000 tons a month. So far as one can see, any such increase would only serve to counterbalance the losses which must be expected to grow in number as the arming of the ships proceeds.

"I am quite clear on the point that the loss of one-fifth of British shipping would have a very serious effect on their supplies. But I think it out of the question that, under the leadership of Lloyd George, who is prepared to go to all lengths, England could thereby be forced to make peace, especially as the above-mentioned effects of the shortage of fats, wood and iron ore, and the continued influence on the supply of munitions would not come into play at all. Further, the psychological effects of panic and fear would be lacking. These, which can only result from unrestricted U-boat warfare, I hold to be indispensable conditions of success. Our experiences at the beginning of the U-boat war in 1915, when the English still believed we were in earnest about continuing it, and even in the short U-boat campaign of March and April, 1916, proved how potent these effects were.

"A further condition is that the declaration and commencement of the unrestricted U-boat war should be simultaneous, so that there is no time for negotiations, especially between England and the neutrals. Only on these conditions will the enemy and the neutrals be inspired with 'holy' terror.

"6. The declaration of unrestricted U-boat warfare will confront the Government of the United States with the question whether they are prepared to draw the logical conclusions from the attitude they have hitherto adopted towards the use of U-boats or not. I am most emphatically of opinion that war with the United States of America is such a serious matter that everything must be done to avoid it. But, in my opinion, fear of a break must not hinder us from using this weapon which promises success. In any case, it is desirable to envisage the consequences least favourable to us and to realise what the effect on the course of the war will be if America joins our enemies. So far as tonnage is concerned this effect can only be very small. It is not probable that more than a small fraction of the tonnage belonging to the Central Powers which is lying in America, and perhaps also in neutral ports, will be quickly available for voyages to England. By far the greater part of it can be damaged to such an extent that it would be useless during the first months, which will be the decisive period. Preparations for this have been made.

"Nor would crews be immediately available for them. Decisive effects need not be anticipated from the co-operation of American troops, who cannot be brought over in considerable numbers owing to the lack of shipping; similarly, American money cannot make up for the shortage of supplies and tonnage.

"The question is, what attitude America would adopt if England were forced to make peace. It is improbable that she would decide to carry on the war singlehanded, as she lacks the means to make a vigorous attack on us, and her shipping would meanwhile be damaged by our U-boats. On the contrary, it is probable that she would associate herself with the peace concluded by England so as to return to healthy economic conditions as soon as possible.

"I have therefore come to the conclusion that we must have recourse to unrestricted U-boat warfare, even at the risk of war with America, so long as the U-boat campaign is begun early enough to ensure peace before the next harvest, that is, before August 1; we have no alternative. In spite of the danger of a break with America, an unrestricted U-boat campaign, begun soon, is the right means to bring the war to a victorious end for us. Moreover, it is the only means to that end.

"7. The situation has improved materially for us, since in the autumn of 1916 I declared that the time had come to strike a decisive blow against England. The failure of the harvests all over the world, together with the effect of the war on England up to the present time, once more give us a chance of ending the war in our favour before the new harvests are reaped. If we do not make the best of this, the last opportunity as far as man can tell, I see no other possibility than exhaustion on both sides without our being able to end the war so that our future as a World Power is secured. In order to achieve the necessary effect in time the unrestricted U-boat campaign must begin on February 1 at the latest.

"I beg your Excellency to inform me whether the military situation on the Continent, particularly as regards the States which are neutral, will permit of this date being fixed. I require a period of three weeks to make the necessary arrangements.

"(Signed) v. HOLTZENDORFF."

There is no doubt that the Chief of the Naval Staff, although we in the Fleet had no special knowledge to that effect, must have

The U-boat Campaign

made known to the Cabinet the same views which he described in so much detail in his memorandum to General Field-Marshal von Hindenburg, viz., that it was high time to start the unrestricted U-boat campaign. In this quarter, though, he seems to have met with greater difficulties, so that he once more appears to have been inclined to compromise. When the orders regarding the date of the opening of the campaign failed to reach the Fleet in the middle of December, the time for which the admiral had announced them, and when, in reply to my inquiries, I received evasive answers, I feared that a new obstruction had arisen. I therefore sent Captain von Levetzow to Berlin to make inquiries. He was given to understand in an interview with Admiral von Holtzendorff on January 4 that for the moment he could only obtain permission to sink armed liners. A Note on this subject was ready and about to be dispatched to America. Again there was the danger that we should pursue exactly the same course as a year ago, a course which had led to such miserable results. I had commissioned my representative to warn them emphatically against this. He had occasion on January 8 to be received by the Imperial Chancellor and to point out to him the inadequacy of such a middle course which was bound to give offence and would be wrecked if America offered objections. The difficulty of determining whether a steamer were armed or not would seriously compromise the success of the undertaking. The Chancellor went the very same evening to Pless, where the decisive session took place the following day when the Chief of the Naval Staff insisted on the necessity of the step, as explained in his memorandum to the Field-Marshal, and convinced His Majesty as well.

On January 9 the officer commanding the Fleet received two communications at short intervals. The first stated that from February 1 onward all merchant ships as soon as it had been positively ascertained that they were armed were to be attacked forthwith. Up to that date only armed cargo boats were to be sunk without warning. This meant that after February 1 passenger ships also would be subject to submarine attack. The second telegram contained an order sent by the All Highest to the Chief of the Naval Staff to the following effect:

"I command that the unrestricted U-boat campaign shall begin on February 1 in full force. You are to make all necessary preparations without delay, but in such a way that neither the enemy nor neutrals can obtain early information of this intention. The fundamental plans of operation are to be submitted to me."

It struck me as odd that an order to proceed against armed steamers should be issued on February 1 while the unrestricted U-boat campaign was to start on the same day. The only explanation I could think of was that the aforementioned Note concerning the treatment of armed steamers from February 1 onwards had already been sent to the American Government, and that it was too late to stop it being delivered. The American Government would certainly be surprised if, after receiving such an announcement in the first half of January, it were informed a few weeks later (February 1) of the intensification of U-boat warfare. But it would make a vast difference to America whether the fundamental right of neutrals to send ships to the blockaded area was conceded, or as unrestricted U-boat war demanded, all shipping in those parts was exposed to destruction. It seems, judging by later communications, there was some idea of asking the American Government to mediate; if this was so the adoption of two such different attitudes on the U-boat question, one following the other in such quick succession, must have an awkward effect. Nothing was made known to the Naval Staff, nor to the officers commanding the Fleet (who certainly were not directly concerned in such matters), of any negotiations which were in progress at that time and which might have been unfavourably affected by the declaration of the unrestricted U-boat campaign. When later on I took over the duties of Chief of the Naval Staff I found no record that any letter from the Imperial Chancellor had been received before the actual commencement of the unrestricted U-boat campaign on February 1 asking for a postponement so as to make a last attempt to avoid this extreme measure. I am convinced, too, that if Admiral von Holtzendorff had had any knowledge of the matter he would have told me of it when he handed over affairs to me on our change of office, if not before. Owing to his severe illness in the summer of 1918 he never had an opportunity of making any statement on this question.

With the unrestricted U-boat campaign we had probably embarked on the most tremendous undertaking that the world-war brought in its long train. Our aim was to break the power of mighty England vested in her sea trade in spite of the protection which her powerful Fleet could afford her. Two and a half years of the world-war had passed before we addressed ourselves to this task, and they had taxed the strength of the Central Powers to the uttermost. But if we did not succeed in overcoming England's will to destroy us then the war of exhaustion must end in Germany's certain defeat. There was no prospect of avoiding such a conclusion by the war on land; nor could we assume that America's definitely unneutral attitude towards us would change, or that by her mediation any peace could be obtained with satisfactory results for us, since Wilson's proposal to act as mediator in a peace in which there should be neither victors nor vanquished had been so brusquely refused by our enemies.

In such a situation it was not permissible to sit with folded hands and leave the fate of the German Empire to be decided by chance circumstances. All in a position of responsibility felt it incumbent upon them to suggest any means that offered a prospect of warding off the impending disaster. An opinion from the military point of view as to the chances of success in war upon enemy sea trade had been expressed; it was based on the statistics of tonnage sunk in previous years. In this respect expectations were far surpassed in the coming year. But the effects of this blow dealt to English commerce could not be foretold in the same way. It was immediately obvious that a reduction of the English mercantile fleet by a third, or even a half, must have a catastrophic effect on English economic conditions, and make England incline towards peace.

The Naval Staff had made a point of carefully examining the economic conditions with the help of experts and had recorded the results of their researches in a number of detailed memoranda, which they had submitted to the responsible Imperial officials. These researches had included the complicated problems of traffic for military purposes as well as for the needs of civilians by land and sea; supplies for the whole country as well as for the troops in the various theatres of war; the food supply of the nation; distribution of goods; home production; stores controlled by the State and rationing. Moreover, all these inquiries and the considerations they gave rise to had to be carried out in unfamiliar circumstances due to the war. Further, an estimate had to be made of the probable direct and indirect effects of all these conditions on the psychological state of the people. The conclusions based on these researches were drawn up in outline so as to give some idea of the probable effects, and they confirmed the general impressions gathered from the beginning of our war on trade that success was certain to crown our efforts if we pursued this course. We had no alternative but to attack our enemy by trying to destroy his economic strength, since all his efforts were directed towards crushing ours. Now, as never before, it depended on which of us could hold out the longer.

In every great effort, if you want to develop it to its fullest strength, you must have the conviction that you can defeat your opponent. That is why the U-boat campaign required the support of all classes that expected the victory of our Fatherland. Every doubt of its success must strengthen the enemy's view that we would soon tire.

But the political leaders had already done all in their power to undermine confidence; and their fear that this kind of warfare might assume forms which would burden us with new enemies had affected timid souls, and it was bound to have a depressing effect if doubts of the final outcome were allowed to appear; should expectations not be fulfilled exactly within the periods mentioned by the Naval Staff. The enemy took full advantage of the discouragement thus aroused, when these people despaired of attaining the desired end; his courage and resolution to hold out were strengthened by it. It is a great pity that the calculations of the Naval Staff were published throughout the country; they had assumed the success of the U-boat campaign within a fixed period of time, and were meant for a narrow circle only. Many who would have held out but for this disappointment lost courage, realising that we had no choice and must bear the privations until success, which could not fail to come ultimately, was achieved. If the calculations of the Naval Staff had fixed too early a date for the effects, and it had taken a much longer time until England could not stand any further destruction of her merchant ships, even then no other choice would have been left us but to make use of these means. The refusal of our peace proposal had so clearly demonstrated the enemy's desire to destroy us that no one would have been prepared, in view of the general situation at the end of the year 1916, to accept a humiliating peace.

The strategic offensive passed definitely to the Navy on February 1, 1917. U-boats and the Fleet supplemented one another to form one weapon, which was to be used in an energetic attack on England's might. Our Fleet became the hilt of the weapon whose sharp blade was the U-boat. The Fleet thus commenced its main activities during the war to maintain and defend the new form of warfare against the English Fleet.

The English defence consisted in combating the U-boats in home

The U-boat Campaign

waters, and to this we could oppose nothing but the skill of the U-boats in evading the enemy. This skill never failed to the very end, although our losses grew heavy.

Our enemies had to go farther to defend themselves against the danger, and had to try to crush it at its source. Only our Fleet could make such efforts fruitless. It had to be in a constant state of readiness to meet the English Fleet in battle; there was no other way. It expected this battle, and had to maintain its strength as much as possible, so as to be fit to cope with the enemy. That is why our Fleet might not weaken itself in view of this last demand that would in all probability be made upon its strength. It found plenty of continued and exacting occupation in combating the means that England had devised to prevent the U-boats from getting out.

The conduct of the U-boat campaign was less a question of the number of boats than one of their peculiar qualities—their invisibility and their submersibility. The former enables them to attack unexpectedly, the latter to escape the pursuit of the enemy. It goes without saying that more can be achieved with 100 boats than with 20. But when the Naval Staff was considering the prospects of a U-boat campaign the first question was to determine the minimum number of boats that would suffice. Moreover, the U-boat campaign's effect was not confined to that of actual sinkings; it did much by disturbing and scaring away trade. Its results were soon perceptible, as it became necessary to regulate traffic according to the ports and districts threatened by U-boats for the time being. Very considerable disturbances in supply and delivery must have been caused, if it suddenly became necessary to alter all the arrangements for traffic, e.g. if railway transport had to be shifted, when ports on the south or west coast of England received no supplies from abroad, because all the ships had to be taken to ports in the north and east.

The number of boats we were able to use in the war against commerce at the beginning of 1915 was about 24; for the first months the new boats built just about covered the losses. It had also become necessary to provide several boats for the U-boat school, so that crews could be trained for the many new boats that were being built. With these 24 boats it was only possible to occupy permanently three or four stations on the main traffic route of English commerce. The tonnage sunk during the whole year of 1915 equalled the tonnage sunk in only six weeks when the unrestricted U-boat campaign was opened. In view of the attitude of conciliation

R

adopted towards the complaints of neutrals, it was premature to begin the U-boat campaign in 1915. It would have been better to wait until the larger number of boats, resulting from the intensive building of 1915, guaranteed a favourable outcome—and then to have persisted in the face of all objections. *Had there been no giving way in 1915, the right moment to start the campaign— the beginning of 1916—would not have been missed.*

CHAPTER XIV

OUR U-BOATS AND THEIR METHOD OF WARFARE

IN the year 1916, up to the time of the Battle of the Skagerrak, the following additions had been made to the U-boat fleet: 38 large U-boats, 7 large submarine minelayers, 34 U-B-boats. Two large submarine minelayers, 3 U-B-boats and 3 U-C-boats had still to undergo steam trials; 53 large U-boats, 10 large submarine minelayers, 27 U-B-boats and 66 U-C-boats were under construction. Since the outbreak of war we had lost 21 large U-boats, 1 large submarine minelayer, 6 U-B-boats, 7 U-C-boats, and 2 U-B-boats had been handed over to the Bulgarian Government. The distribution of all the U-boats was so arranged that half were under the orders of the Admirals of the Fleet, and of the rest one half were stationed in the Mediterranean, while the other half, the last quarter, were assigned to the Naval Corps in Flanders. For the sake of quick construction the new types of so-called "U-B"-boats and "U-C"-boats had been introduced, in addition to the main type of large U-boats similar to "U 19," the first one fitted with Diesel engines.

The chief characteristics of the different types were as follows:

"U 19," surface displacement, 650 tons; highest speed on the surface, 12 knots; under water, 9 knots; number of torpedoes, 9, of 50 cm. calibre. Improvements were made in the type. From "U 40" onwards the displacement was raised first to 700 tons, and from "U 80" onwards to 800 tons, the speed was raised to 17 knots on the surface, the number of torpedoes increased to 12, and from "U 90" onwards to 16. The torpedo of 50 cm. calibre had an explosive charge of 200 kilos. The first large submarine minelayers were not armed with torpedoes. They had a displacement of 760 tons, a surface speed of 9.5 knots, and under water 7.5 knots; they carried 34 to 36 mines.

Of the U-B-boats, at first a small number with a displacement of 125 tons was built for use in Flanders, with four torpedoes, speed of 8.5 knots on the surface and 5.5 knots under water. The U-B-boat was then enlarged to 500 tons, with a speed on the surface of 12.5 knots and of 7 under water.

Germany's High Sea Fleet

The U-C-boats were of a type designed both for minelaying and firing torpedoes. At first only a small number of these was built, with a displacement of 150 tons; ultimately the boats had a displacement of 400 tons, speed of 11 knots on the surface and 6.5 under water. They carried 18 mines and could take four torpedoes.

At the beginning the U-boats were armed with one 5 cm. gun as a defence against enemy submarines. But as their use was developed in the war, such various demands were made upon them that their armament had to be increased. One or two guns of 8.8 cm. were placed on the U-boats, U-B-boats, and the U-C-boats; the submarine cruisers were in part armed with a gun of 15 cm. calibre.

The majority of the large U-boats was assigned to the Fleet for use in the blockaded area west of England. The length of their trips was 21 to 28 days, but this was also dependent on the amount of ammunition used when the boats had found a favourable opportunity to fire their torpedoes soon after leaving port. The big minelayers were also under the command of the Fleet, and could be sent on distant expeditions—to the White Sea or to the Mediterranean.

The U-B-boats, being rather smaller, had proved to be very handy and quickly submersible, although they could not remain so long at sea. They were, therefore, mainly assigned to the base in Flanders, as were the U-C-boats, of which a small number, however, was at the disposal of the Fleet and used for laying mines on the east coast of England. The distribution of the boats among the various bases was carried out according to the facilities the latter had for repairing the boats on their return from expeditions. The large amount of technical apparatus in a U-boat required very careful overhauling and repair on her return from an expedition; also the damage due to the voyage or to enemy attacks had to be repaired. Generally speaking, after four weeks at sea a boat would need to lie in the dockyard for the same length of time for repairs. The Imperial dockyard at Wilhelmshaven had been enlarged and was the chief place to which the U-boats of the Fleet were sent for repair. The docks at Kiel and Danzig were needed for other purposes; the bases at Zeebrugge and Pola were used at first mainly for overhauling the boats. Until these dockyards had been altered so as to be able to undertake more extensive work the boats which belonged there had to return home for important repairs.

When the U-boat campaign was opened on February 1, 1917,

Our U-boats and their Method of Warfare

there were 57 boats already in the North Sea. The officer commanding the Baltic district had eight assigned to him, the Naval Corps
in Flanders had at its disposal 38, and the stations in the Mediterranean 31 U-boats of different types. The favourable experiences
of the commercial U-boat U-*Deutschland* had led to the construction
of U-cruisers, of which the first series had a displacement of 1,200
tons, which was later on raised to 2,000 and more. When they could
no longer be used for trade purposes the commercial U-boats were
taken over by the Navy and altered for use as warships. They were
fitted with two guns of 15 cm. calibre and two torpedo tubes, and
could carry about 30 torpedoes in accordance with the extended
period during which they could be used on cruises, cruises which
reached as far as the Azores and lasted up to three months. With
this fleet of U-boats the Navy was well equipped to do justice to the
task assigned to it, although England had used the whole of 1916 to
develop her defence. The sinkings of the year 1917 prove this.
They were :

February,	1917	...	...	...	781,500 tons
March,	,,	...	...	...	885,000 ,,
April,	,,	...	...	...	1,091,000 ,,
May,	,,	...	...	...	869,000 ,,
June,	,,	...	...	...	1,016,000 ,,
July,	,,	...	...	...	811,000 ,,
August,	,,	...	...	...	808,000 ,,
September,	,,	...	...	...	672,000 ,,
October,	,,	...	...	...	674,000 ,,
November,	,,	...	...	...	607,000 ,,
December,	,,	...	...	...	702,000 ,,
January,	1918	...	...	...	632,000 ,,
February,	,,	...	...	...	680,000 ,,
March,	,,	...	...	...	689,000 ,,
April,	,,	...	...	...	652,000 ,,
May,	,,	...	...	...	614,000 ,,
June	,,	...	...	...	521,000 ,,
July,	,,	...	...	...	550,000 ,,
August,	,,	...	...	...	420,000 ,,
September	,,	...	...	...	440,000 ,,

The enemy's defence consisted, firstly, in directly combating the
U-boats, and, secondly, in special measures which England adopted

to counterbalance the loss of tonnage. The first impediment our U-boats had to overcome—I am speaking of the activities of the U-boats assigned to the Fleet (the same applies to the Flanders boats), whereas those in the Mediterranean mostly worked under less difficult conditions—lay in the minefields blocking the North Sea. To deal successfully with these the Fleet had had to create a special organisation. In addition to the actual mine-sweepers, whose work it was to keep certain paths through the belt of mines clear, special convoying flotillas had been formed, fitted with mine-sweeping apparatus, which accompanied the U-boats along the routes that had been cleared, till they reached the open sea, and met them at the same spot on their return from their fields of operation to take them safely home again. When attacking steamers the boats had to reckon with their armament, for in spite of the large number of guns required and the crews to man them, nearly the whole of the English Merchant Fleet—at any rate all the more valuable steamers—was armed.

As a further defence, besides the destroyers which were excellently suited to this purpose and were armed with depth charges, a large number of new kinds of boats with shallow draft had been built especially to combat the U-boats. Nets and all sorts of wire entanglements hindered the U-boats in their work near the English coast. The so-called "Q"-boats, intended to serve as traps for submarines, were specially fitted out; they presented the appearance of neutral ships, and on the approach of the U-boat let fall their disguise and attempted to destroy it with guns and explosives. The practice of gathering considerable numbers of British merchantmen together and convoying them added greatly to the difficulties the U-boat encountered in achieving success; these ships were protected according to their size and value either by light craft or by bigger warships.

During the first months of the U-boat campaign I never missed an opportunity of hearing the story of his experiences and adventures direct from the lips of the commander of a returning U-boat; and thus I had opportunity to form an idea of the perseverance, courage and resolution of these young officers who won my highest admiration for the seamanship and the calm intrepidity, which they succeeded in communicating to the crew as well. It is a splendid testimonial to the spirit of the Navy that all who could possibly be considered suitable for the U-boat service, both officers and men, rushed to offer themselves. Even older Staff officers, in spite of

Our U-boats and their Method of Warfare

their many years of service, begged to be taken as commanders of U-boats, even if they had to serve under a flotilla commander younger than themselves.

The three half-flotillas into which the U-boats of the Fleet had been formed at the beginning of the war developed in time into four flotillas. Their commanders were: First U-Flotilla, Commander Pasgnay; Second U-Flotilla, Commander von Rosenberg; Third U-Flotilla, Lieutenant-Commander Forstmann (Walter); Fourth U-Flotilla, Commander Pranse. I should like to mention in connection with these Lieutenant-Commander Bartenbach, who was at the head of the U-flotilla in Flanders, who so often supported the enterprises of the Fleet with his boats. In an exemplary manner, despite all obstacles, he directed the difficult operations of the Flanders boats, against which the British defence was particularly heavy. All who served with him were animated by a spirit of comradeship and readiness for action, which had the most refreshing and grateful effect upon anyone who spent any time with them.

The Chief Director of the U-boats under the command of the Fleet was Captain Bauer; he himself took part in the fighting expeditions of the U-boats in the blockaded area round England, in order to be able to form his own opinion of the circumstances in which the boats under his command had to operate. It is his great merit that he recognised the capacity of the U-boat and brought it to that degree of efficiency to which its later successes are due. When, later on, owing to the increasing activity in construction, the number of U-boats grew to such an extent that their organisation far surpassed that required for a squadron and demanded a corresponding increase in authority, Commodore Michelsen, who had hitherto commanded the torpedo-boats, was placed at the head. His great knowledge and experience of the department of torpedoes designated him as particularly suitable for this post, and he completely fulfilled all expectations in this respect.

The U-boat service was the one which suffered the heaviest losses of the Navy; the number of boats lost on fighting expeditions amounted to 50 per cent. Altogether 360 U-boats and U-boat cruisers were employed in the U-boat campaign, of which 184 were lost in the course of their enterprises. This high percentage of losses was for the most part due to the defence of the enemy, which grew more and more vigorous, as he tried to get the better of the U-boat danger by the use of all sorts of dodges and methods; yet a large proportion is ascribable to the fact that our U-boat commanders

263

could not resist the temptation, when sinking a steamer, to save the lives of those on board as far as possible, even though they so often met with disappointment.

I should like to illustrate the difficulties encountered by our U-boats by a few instances, quoting the official reports concerning them. But it would be impossible to do all the commanders equal justice, for they vied with each other in meeting the dangers which their difficult business involved, and with which the public are already familiar through various popular writings.

The journey to America of the "U 53" is a splendid testimonial to the perseverance of the crew and the high quality of the material. On September 11 this U-boat received orders to lie off the American coast about the time when the U-merchant boat *Bremen* was expected to arrive at New London (North America), in order to search for and attack enemy ships which, in all probability, would be waiting there for the submarine merchantman. After completing this task, the boat was to call at Newport, Rhode Island, but was to leave again after a few hours at most, so as to give the American authorities no excuse or occasion to detain her. There was to be no replenishment of supplies, with the possible exception of fresh victuals. If no enemy warships were met with, she was to carry on commercial war according to Prize Law off the American coast.

On September 17 the boat started on her outward voyage from Heligoland. In the North Sea she had very heavy weather. There was a S.S.W. gale and such high seas that the men on watch on the conning tower of the boat were up to their necks in water all the time.

The supplies of the boat had had to be increased so as to make the voyage possible. Four ballast tanks were altered for use as fuel tanks, so that the oil supply was increased from 90 cb. m. to 150 cb. m.; the supply of lubricating oil of 14½ cb. m. was considered sufficient for the voyage. Added to this, there was the increase in fresh water and food supplies, so that the boat's draught was increased by 40 c.m. So far as her sea-going qualities were concerned, her commander reported that the boat rode very steadily on the whole, but that every sea went over her upper deck, even when the force of the wind was only 4; from almost every direction spray flew over the bridge. Consequently for those on duty on the bridge, the voyage, especially at first, was a tremendous strain. The commander did not think that the officers and petty officers would be able to stand it (the rubber suits that had to be worn almost daily for the first fortnight were not watertight), and he would have turned back if the weather had not improved soon after September 24.

The route for the voyage out had been chosen to run from the most northerly point of the Shetland Islands, which they passed on September 20, straight to the Newfoundland Bank, so as to remain on the northern side of the usual belt of low barometric pressure. Weather conditions were uncertain and changeable. There was often a very high and very steep swell, in which the boat pitched heavily. They, however, experienced following winds nearly all the time, which were favourable for the journey. After reaching the Newfoundland Bank, the boat was carried vigorously to the west by the Labrador current. On the whole the health of the crew was good, until they were nearing the Newfoundland Bank. Then a number were attacked with headache and sickness, which is said to be a common occurrence in these parts.

On October 7 the boat lay before Long Island Sound. No warships were encountered. At 3 P.M. the commander entered the harbour of Newport, Rhode Island, accompanied by an American submarine, which had joined him on the way, and there he paid official visits to Admiral Knight and the commander of torpedoboats, Admiral Gleaves. He wrote in his diary:

"The former received me very coolly, and said that the *Bremen*, as far as he knew, had been sighted about 10 days before between Newfoundland and New York. [That was not correct, as the *Bremen* never reached America.] Admiral Knight obviously thought it most desirable that the ' U 53 ' should leave again the same evening. If I had not announced that such was my intention, I think I should have been given a pretty broad hint on the subject.

"Admiral Gleaves was very friendly and much interested; he inquired about all particulars of the voyage. The adjutants of both admirals returned my visit. At 4.30 P.M. Admiral Gleaves himself came to inspect the boat. I took him over her, as, earlier in the day, I had done several young officers. More than anything else the Diesel engines roused envious admiration. Many officers came on board with their ladies, as did civilians, reporters, and one photographer. The crew received all sorts of little presents. At 5.30 P.M. we weighed anchor. Proceeded to sea at 6.30 P.M. Trial dive. Course, Nantucket Lightship; 270 revolutions—equivalent to 9 knots."

Germany's High Sea Fleet

Nantucket Lightship was reached on October 8 at 5.30 A.M. Very clear, calm weather prevailed. The commander decided to examine the merchant traffic outside territorial waters and to wage cruiser warfare.

At this meeting-point of so many trade routes, the boat was able to stop seven steamers in the course of the day, and after the crews had in every case left the ship, she sank the British steamer *Strathdene* from Glasgow (4,321 tons), the Norwegian steamer *Chr. Knutsen* (3,378 tons) with gasolene destined for London, the British steamer *Westpoint* (3,847 tons), the Dutch steamer *Blommersdyk* (4,850 tons), whose whole cargo consisted of absolute and conditional contraband. According to an American certificate, the *Blommersdyk*, before reaching her destination, was to call at Kirkwall (in the Orkney Islands, a British examining station for merchant steamer traffic). In his log the commander reports as follows:

"Meanwhile, in this narrow space besides the two steamers—there was an English passenger boat as well, the *Stefano*, from Liverpool, 3,449 tons, which had already been stopped and was disembarking her crew—and the 'U 53' sixteen American destroyers had assembled, so that we had to manœuvre with the greatest care. While I was towing back the boat of the *Blommersdyk,* which had brought the officer with her papers, 'U 53' got so near an American destroyer that we had to reverse with both engines to avoid a collision. We cleared one another by about 50 m. When reversing, I cast my tow loose, and her crew did not return to the *Blommersdyk* at all, but went straight on board a destroyer. I had told the officer that the crew would be given twenty-five minutes in which to disembark —till 6.30 P.M. To make sure that no one should be hurt, he was to haul down his flag to show that no one was left on board. Then I approached the passenger steamer to examine her papers, or, in case she had not yet lowered a boat, to dismiss her forthwith out of consideration for the passengers. I had already given orders for the signal, 'You can proceed,' when I realised that the steamer had been abandoned and all on board accommodated on an American destroyer. I then returned to the *Blommersdyk*. By means of a siren and calling through a megaphone I made sure that no one was left on board. A destroyer which lay very near the steamer was asked by Morse signal to move away a little, so that the ship might be sunk. This the destroyer did at once. Hit with torpedo, a depth of 4 m. in hold 4. The steamer was then sunk by a second torpedo."

266

Our U-boats and their Method of Warfare

The passenger steamer *Stefano* was then also sunk. At 10.30 P.M. the boat began her return voyage. Though it would have been very desirable to extend our activities off the American coast as long as possible, yet any further delay would have endangered the whole enterprise because of the fuel supply; for during the short stay at Newport, the boat, in accordance with the general instructions issued to her, had taken in no supplies of any kind. For the return voyage we counted on a consumption of fuel of 60 cb. m., and a certain reserve was allowed in the event of head winds and storms. That this precaution was necessary is proved by the fact that although the weather as far as the Shetlands was very favourable, the boat arrived at Heligoland with only 14.5 cb.m. of fuel. For the return voyage the longer route *via* Fastnet Rock was chosen. In so doing, the unsettled weather conditions that had been encountered in the higher latitudes on the voyage out were avoided; also on this southern side of the belt of low barometric pressure there was less fear of head winds than in the north. After waiting twenty hours at the S.E. corner of the Newfoundland Bank to weather a storm, the boat proceeded with little delay as far as the Hebrides, passing through an area of high pressure (770 mm.) accompanied by a steady west wind. The route then followed was round the Shetland Islands. On October 28, at 3 P.M., the boat entered the harbour at Heligoland. It had covered a distance of 7,550 sea miles and had only stopped once for two and a half hours in Newport. When the boat arrived at Wilhelmshaven next day, I was able to assure myself by personal observation that all her crew were in excellent condition. They might well be proud of their eminent, seaman-like, and technical achievement.

Let us follow this same Lieut.-Commander Rose on his "U 53" on a cruise, during which he waged war according to Prize Law, as still had to be done in January, in pursuance of the instructions issued, before the introduction of the unrestricted U-boat campaign. I will quote his log, omitting what is not of general interest :—

"*January 20th*, 1917.—Left Heligoland. Wind east, force 8, cloudless, clear. Route *via* Terschelling Lightship to Nordhinder Lightship.

"*January 21st.*—Sank to the bottom, 38 m. Conversation (by submarine telephone) with 'U 55.' 6.30 P.M., dark, starless night, wind east, 3-4. Started on normal course.

"*January 22nd.*—11 A.M. Sank French sailing ship *Anna* (150 gr.t.) by thirteen rounds of gun-fire; cargo, road-metalling. 9 P.M., South of Lizard Head. ' U 55 ' reports station. As the presence of U-boats in the Channel is thereby betrayed, tried to report on own station and intentions (valuable for ' U 55 '). Immediately after heard very loud British convoy signals and then the warning, ' German submarine 37 miles south of Lizard.' That could only apply to ' U 53.' 11.40 P.M., south of the Wolf Rock, two ships with many lights, little way, and changing courses at a distance of about 6 sea miles from one another. Apparently guiding ships to show entrance to Channel. After prolonged observation, steered west between the two.

"*January 23*, 12.5 A.M.—A big cargo steamer approaching with a course of 90 degrees. At some distance behind several lights; probably one of the expected convoys. Two officers of mercantile marine who are on board think the ship to be British of about 4,500 gr.t. She is fully laden. Started attack on surface. At first attempt a miss, at second a hit, port amidships. The steamer stops, sinks lower, gets a list, keeps on burning blue lights, then lowers boats. Left soon as further action impossible. Did not observe the sinking of the badly damaged ship. Passed several guardships with different lights. One of them on a course towards the scene of disaster. Let her searchlight play there for a short time. The guiding ships have gone on or put out their lights.

"6.40 A.M.—A steamer with bright lights and funnels lit up steers a zigzag course. She seems to be waiting. Sent Morse message to steamer in English. She is Dutch, with oilcake for Rotterdam. Dismissed steamer before dawn.

"2 P.M.—Avoided a ' Foxglove ' (new type of British U-boat chaser) and the steamer accompanying it.

"11 P.M.—Avoided a guardship. She carried steamer lights on forestay to appear bigger.

"*January 24*.—12 midnight. A smaller steamer, arranged for carrying passengers, steers 200 degrees. Flag illuminated, but not recognisable. Obviously a neutral. Sheered off.

"7 A.M.—A steamer, course 250 degrees, approached, pretending to be a French outpost ship. She is a neutral tank steamer. Sheered off.

"8.30 A.M.—Wind east, but swell; cloudy in parts, visibility good. Dived on account of an airship approaching from the east; it may be a captive balloon broken loose. Voyage under water to the neighbourhood of Ushant (French island at the western end of the Channel).

Our U-boats and their Method of Warfare

"2 P.M.—Wind east, force 7-8. Rose to surface.

"3.15 P.M.—Small sailing ship in sight in southerly direction. Owing to high seas, no opportunity to attack.

"10 P.M.—Wind east, 6-7, swell. Absolutely impossible to fire at night. A lot of water comes over. Dived. Voyage to presumptive meeting place near Ushant.

"*January* 25, 6.30 A.M.—Wind south-east, force 7-8, position Ushant, 50 sea miles to east. Hove to. Waiting off Ushant. A small sailing ship about 30 sea miles west of Ushant. Left her unmolested because of heavy sea. Not possible to fire at night because of high seas. Visibility bad, therefore dived for night journey.

"*January* 26.—Weather unchanged. Dived for night journey.

"*January* 27, 3 A.M.—Wind east, force 8. Visibility bad. Snow from 11 A.M. Boat rolls more and more. Depth 34 m. Position not fixed. Stood out to sea at low speed.

"5 P.M.—North of Ushant. Wind, force 10,* swell. Sighted large steamer of about 200 tons, so far as can be seen, armed fore and aft. Gave way, as impossible to fire at the time and no improvement in weather to be expected for next few hours. Steamer going slow; was painted grey. Apparently one of bigger guardships. Dived for night journey.

"*January* 28, 8 A.M.—Came to surface north of Ushant. Wind E.S.E., force 6.

"6.30 P.M.—Inspected Spanish steamer *Nueva Montana*, of Santander, 2,000 gr.t., from under water, then stopped her with shot. Cargo, iron ore to Newcastle. Crew on board took boats in tow. Set fire to three explosive bombs in engine-room. Steamer sinks slowly, deeper and deeper. As all buoyancy chambers are connected, her sinking only a matter of time. When last seen, the swell was pouring over the after part of the ship. Took crew as far as 12 sea miles west of Ushant; left boats there.

"*January* 29, 7 A.M.—Danish steamer *Copenhagen;* cargo, coal from Newcastle to Huelva. Examined and dismissed.

"6 P.M.—Steamer *Algorta*, 2,100 gr.t., from Segund with iron ore for Stockton. Inspected from under water, then stopped by shot. Took crew in tow. Sank steamer with four explosive bombs.

"10.15 P.M.—Cast off boats in neighbourhood of medium-sized steamer steering about 240 degrees. Called up steamer by star shell.

* Force 10 is a heavy gale. Force of wind indicated according to Beaufort's scale, from 0 = calm to 12 = hurricane.

Germany's High Sea Fleet

"*January* 30.—Course, 340 degrees. Intend activity for next two days in neighbourhood of Scilly Isles. Nothing in sight. At dusk, south of Scilly Isles, steamed on towards Lizard, distance 8 sea miles. Encountered no commercial traffic, only guardships southwest and west of the Scillys.

"*January* 31, 9 A.M.—Stopped Dutch steamer *Boomberg*, about 1,600 gr.t. Coal from Cardiff for Las Palmas; dismissed her.

"10 A.M.—Stopped Spanish steamer *Lorida*, about 1,600 gr.t Cargo, coal from Cardiff to Cadiz. Dismissed her.

"2.30 P.M.—Stopped Norwegian steamer *Hickla*, 524 gr.t. Cargo, pit props for Cardiff. Set steamer on fire. Crew sails for Scilly Islands.

"5.30 P.M.—Stopped a smallish steamer, steering 175 degrees, coming from astern. Steamer returns fire at 80 hectometres from gun of at least 8 cm. calibre. Her shots fall short, but are well aimed.

"6 P.M.—Ceased gunfire after about forty rounds. Distance increased to bounds of visibility, then tried to keep touch at full speed. In dusk steamer gets out of sight and cannot be found again.

"11.50 P.M.—Weather calmer, bright moon. Clear. Stopped Danish motor-boat *Falstria*, about 4,000 gr.t., from Far East *via* Dartmouth. Ship in order; ship dismissed.

"*February* 1.—West of Ushant. Steamed all day over field of search; nothing in sight.

"*February* 2, 5 A.M.—Attacked with bronze torpedo a large fully-laden steamer, about 2,000 t., steering 170 degrees. No marks of neutrality. Hit amidships. Steamer stops; lights go out. No movement or work discernible on deck. After half an hour steamer still afloat. Will probably sink, as she is badly damaged.

"4 P.M.—Near Bishop Rock stopped a French old square-rigged schooner, *Anna Maria*, from St. Malo, about 150 gr.t., by using signal 'Abandon ship.' After a time the mate came on board in a little rowing boat without a keel. The crew try with boots and cups to keep the boat more or less dry. In consideration of the impossibility of rescuing the crew in this boat, the ship was allowed to continue her journey. The mate gave a written promise in the name of the crew not to go to sea any more during this war. The cargo of the ship consisted of salt and wine.

"*February* 3.—West of the Scillys. Wind east, force 2. 8 A.M., stopped Norwegian steamer *Rio de Janeiro*, 2,800 gr.t. Wheat, linseed, oil cakes, tan for Copenhagen and Christiana. Steamer dismissed.

Our U-boats and their Method of Warfare

"11 A.M.—Submarine attack on American steamer *Housatonic*, 2,443 t. Then rose to surface and stopped steamer. Cargo 3,862 tons of wheat from New York for London. Fired bronze torpedo from first tube to sink steamer. The torpedo slips half out of the tube without leaving it. It starts to go, and we can hear the engine running slowly. The boat is stopped. Watertight doors closed. After some time detonation under the boat, without any turmoil of water or column of smoke. The torpedo has left the tube and obviously sunk and exploded at the bottom, at a depth of 110 m. A few rising air bubbles indicate that the airchamber must have separated owing to pressure as the depth of water increased. Steamer sunk by bronze torpedo from 4th tube. Took boats in tow and handed them over to a guardship which was called up by two shots. When retreating from the guardship, which came up at once, we met ' U 60.' ' U 60 ' dives. I intend to draw the guardship past ' U 60.' Guardship sheers off, rescues crew of American boats which ' U 53 ' asked her by wireless to do. ' U 60 ' dives. Exchange of reports with ' U 60.'

"*February* 4, 12.5 A.M.—With gunfire and explosive bombs sank French barque *Aimée Marie*, from St. Servant, 327 gr.t.; cargo, salt and wine for home port. Crew rows to Scilly Isles. Owing to the extraordinary lightness of the night, avoided darkened guardships. Meeting and exchange of reports with ' U 83.'

10 A.M.—Sank with two explosive bombs schooner *Bangpuhtis*, from Windau, 259 gr. t., and ballast from St. Nazaire for Cardiff. Crew sails for Scilly Islands.

"4 P.M.—Examined Norwegian three-master *Manicia*, 1,800 gr.t., from Rosano with linseed for Rotterdam, and dismissed her. Ship at sea since December 1.

"*February* 5, 12.30 A.M.—Wind east, estimated force, 5-6. Surface attack on steamer on which all except navigating lights are out, no lights as distinguishing marks, estimated at 3,000 gr.t. Armament cannot be discerned. On attacking became convinced that size of steamer has been over-estimated. When sheering off recognise Swedish distinguishing marks. Stop steamer by white star-shell and Morse lamp signals. Steamer answers no signal and makes no other sign. After a time steams at full speed out to sea. Stopped anew by two shots. She does not answer Morse signals. Circled round steamer till dawn. By daylight found she was steamer *Bravalla*, 1,519 gr.t. By flag-signals she announced her port of destination as Liverpool, cargo, nuts. If sunk at that

271

spot crew would have been lost. Impossible to tow boats owing to high seas. Therefore gave steamer signal ' Follow.' Further signals giving exact instructions as to behaviour when ship was to be sunk later on, were cut off because as soon as she understood the first words ' I am going to sink you,' she hauled down the answering signal and took no further notice. On the way I had to force the steamer to obedience again, as she tried to sheer off. The sea gradually decreases. Shelter owing to neighbourhood of land perceptible. A guardship is sighted. Signal to *Bravalla*, ' Abandon ship.' She does nothing. Not till four minutes later, when the gun is trained on her, does she hoist the answering signal. A shot before her bows, then one in her forecastle. Steamer lowers boats. Ceased fire. When the boats had hove to, opened fire again. Difficult to aim owing to rolling of boat and target. There is a very heavy hail squall. Steamer hit several times, but does not sink. Although no one is left on board the engines keep going with fewer revolutions. Guardship approaches to a distance of about 40 hectometres, opens fire : dived. Sank steamer by a torpedo, guardship meanwhile rescued Swedish crew."

And so on. These extracts should suffice to show under what difficulties the boats worked so long as they had to consider the neutrality of steamers, and it also shows how many opportunities for sinking ships in the blockaded areas were lost.

To illustrate other kinds of U-boat activities in the restricted U-boat campaign, we will quote from other logs. The first extract is from the log of a U-C-boat that had orders to lay mines along the east coast of England.

"*December* 13, 1916.—Various vessels to be seen ahead, among them one lying with lights out, which I took to be a destroyer. Dived to avoid danger. Broke through guard-line under water.

"9.25 A.M.—Rose to surface. Continued journey on surface. Sighted several steamers which, coming from the south, seemed to be making for the same point as I. It gradually grew very misty, which made it impossible to fix position. Presumed we were near land, as the sea grew calm, the water was dirty yellow in colour, and there was a strong smell of coal dust. After diving quickly several times to avoid steamers, continued under water 270 degrees (course west).

"1 P.M.—Sighted strong surf on starboard bow. A wall was

dimly visible above, and over that a big factory, with several chimneys. At the same time the boat touched the bottom at 10 m. Reversed course, and as I was quite uncertain of ship's position, resolved to rise to surface to get my bearings above water. Hardly opened hatch of conning tower when I see about 600 m. to port at 2.14 P.M. a large destroyer with three funnels and two masts, passing at about 20 knots on a course N.N.W. She seemed to have appeared quite suddenly out of the mist and not to have seen me yet. Dived to 16 m.

"2.20 P.M.—As many steamers were in sight and visibility still bad, gave up intention of finding ship's position. Lay at the bottom, 23 m. water. Boat lay very unsteady; repeatedly heard the noise of screws above me.

"5 P.M.—Dusk. Northerly swell. Rose to 10 m. As it was getting dark and no ships were to be seen, rose to surface to re-charge and pump in air; stood out to sea a bit.

"5.42 P.M.—Several steamers coming from direction of land towards me. Dived.

"6 P.M.—Very dark night. Rose to surface as darkness had fallen completely. The steamers were coming from west by south. So I concluded that the entrance to the harbour must be in the direction from which they came. The course led towards a darkened light, which now and then sent a ray up vertically. On approaching I see the end of the breakwater. The pilot thought he could recognise this as the entrance to the Tyne. As the night was very dark I decided to go close up to the breakwater. First I made for the northern breakwater; just before reaching it I turned to starboard so as to get a bit farther north. In so doing the boat ran aground north of the northern breakwater. Both engines reversed full steam. Boat slipped off.

"6.42 P.M.—Turned hard-a-starboard to 160. Close to the end of the northern breakwater the first mine dropped. Then turned slowly to starboard so as to get as near as possible to the southern breakwater. When this was in sight at a distance of about 80—100 m., turned sharply, let the last mine fall and stood out to open sea, 90 degrees (course east)."

How much more difficult it was for our U-boats to attack when the steamers travelled in convoys, appears from the following extract from the log of "U 82," commanded by Lieutenant-Commander Hans Adam :

"*September* 19, 1917, 3.19 P.M.—I shot past the bows of this steamer towards steamers 4 and 5. Steamer 4 I hit. Steamer 2 had hoisted a red flag, which was probably to announce the presence of the boat; for several torpedo-boats make for the steamer. As there was no chance of firing from the only remaining usable tube (stern tube) I dived. The destroyers dropped about 10 depth charges; one burst pretty near the stern. The attack was rendered very difficult by the bad weather, swell, seaway 5 and rain squalls. The success of the attack was due to the excellent steering under water. Made off noiselessly to S.E. under water.

"4.45 P.M.—Rose to surface. I try to come up with the convoy again, as it is still to be seen. But a destroyer forces me under water again.

"6.37 P.M.—Rose to surface. Two destroyers prevent me from steaming up. Owing to heavy seas from S.E. it is impossible to proceed south so as to get ahead of them. Moreover, sea and swell make it impossible to fire a torpedo. Therefore gave up pursuit."

On July 19 and 20, 1918, two of our U-boats encountered a new and valuable steamer, the *Justitia*, of 32,120 tons, which was very strongly protected on account of its value, and which would accordingly be very difficult to sink. The account of the attacks of the two boats, "U-B 64" and "U 54," is given below. "U-B 64" met the steamer on July 19 and damaged her severely, while "U 54" encountered her the next day when she was being towed into port and finished her off.

As the steamer *Justitia*, being new, was not on the register on board the U-boat, and the number of such large steamers is small, they thought she was the German steamer *Vaterland* which the Americans had rechristened *Leviathan*.

"*July* 19, 1918. 3.50 P.M.—Two destroyers in sight, course 320 degrees (N.W.). Behind the destroyers a convoy. Boat situated straight before them. Attack prepared for double shot at steamer (3 funnels, 2 masts) situated in the middle of the convoy, which numbers about 12 steamers. Protection by destroyers and submarine chasers in large numbers. Convoy zig-zags. Shortly before the shot the steamer turns towards the boat, therefore only stem shot possible. Distance 350 m.; hit behind the bridge port side.

"4.33 P.M.—British steamer *Justitia*, 32,120 tons in ballast. Dived. There follow 35 depth charges, that are well placed.

"5.20 P.M.—Depth 11 m. Steamer has stopped, blows off a lot of steam; apparently hit in boiler or engine. Many destroyers to protect her. Counter course for attack. Destroyers pass over the boat several times.

"6.15 P.M.—Double shot from tubes 1 and 2; distance 2,000 m. Hit midships and astern, port side of steamer, which has stopped. Dived. 23 depth charges which follow immediately on shot.

"7.3 P.M.—Rose to 11 metres depth so as to be able to look through periscope. Steamer has a list to port and is much down by the stern. Started new attack. As destroyers about all the time, cannot show periscope often. In the meantime, the steamer has been towed on a southerly course by large tugs. Steamer towed about 3—4 knots. With course 180 degrees (south) went ahead under water.

"9.48 P.M.—Fired from tube at distance 900 m. Hit on port side. Dived. On a course 0 degrees (north). 11 depth charges. Made off, as battery exhausted.

"10.33 P.M.—Depth 11 m. Steamer being towed. List has increased, also lies lower in the water.

"11.23 P.M.—Came to surface. Charged batteries. Reloaded bow tubes with two torpedoes.

"11.50 P.M.—After the four hits, the steamer must undoubtedly go down. It is only a matter of time until the last watertight doors give way. Towing against the sea must make her engine break away soon.

"*July* 20, 1918.—Before the North Channel (Irish Sea). Kept touch during the night, so as to be sure of observing sinking. As the condition of the steamer grew steadily worse, the course of the tow was altered towards morning to the south for Lough Swilly. Surface attack by night impossible because it was too light.

"4 A.M.—As it was pretty dark and there was a jumble of ships, it was particularly difficult to get in right position for attack. Before 'U-B 64' was ready to attack, steamer was towed along again. Position very far aft. Steamer lay considerably lower. Batteries not in a condition for me to follow under water.

"5.37 A.M.—Depth of 11 m. Steamer lies athwart with considerably greater list.

"8.40 A.M.—Rose to surface. It could now be ascertained that the depth charges had badly damaged oil bunkers, so that the boat left a broad track of oil. Steamer at the moment out of sight. Wireless messages to boats in the neighbourhood.

"11 A.M.—Steamer sighted to port on course 180 degrees. Hardly possible for her to reach the coast. Steamer with heavy list can barely be moved.

Germany's High Sea Fleet

"11.30 A.M.—Observed two high, clear columns of water, closely following one another, behind steamer; must come from two torpedoes. In boat detonation of 35 depth charges was heard.

"2.15 P.M.—Steamer sunk. On looking round ascertain that many protecting vessels, with steamer's lifeboats in tow, are making for land. Other craft have rushed to the floating debris. Made off. Many destroyers in pursuit of me."

"U 54," which fired two torpedoes at the *Justitia* on July 20 at 11.20, reports further:

"11.32 A.M.—In the hail of depth charges that became more intense after the detonation of the first torpedo, of course no further detonation could be heard in the boat. After 122 seconds, the petty officer telegraphist noted the second hit through the submarine receiver. As I had only 2,200 amperes in the battery, I could not possibly make a further attack. I went down for half an hour and found bottom at 59 m. 20 minutes after the shot the British depth charges ceased to explode.

"12.30 P.M.—Rose from bottom till I could use periscope on northerly course. Round about me, near by, many guardships. I immediately dived again. As I assumed they were following me with submarine sound receivers, I remained under water; continued till the large ship was safe. I proceeded north, then altered course to N.W. and then to west.

"3.51 P.M.—Rose to surface. The boat had 50 mm. pressure. As letting off air took too long, I ordered the helmsman to open the conning-tower hatch. The helmsman was blown out, and the central conductor which has a sail attached below, was blown against my arm and crushed it against the top of the conning-tower. The pain was so great that I fainted for a moment. When I heard that the helmsman saw a number of ships, I crept on to the conning-tower and saw that south and astern was full of vessels. I attributed this activity to myself and dived away again, as I could no longer risk being seen.

"6 P.M.—Rose to surface. Far in the south a smoke cloud. I ran farther west, and as soon as my batteries were pretty well charged I sent wireles messages to all U-boats giving course and possibilities of attack on *Vaterland*. There was no object in my following any longer, as I could not have caught her up before the North Channel.

"*July* 21. 10.45 A.M.—U-boat in sight; ascertained to be

Our U-boats and their Method of Warfare

'U-B 64.' Approached within hailing distance. From exchange of experiences I learnt that the day before, at 2 hr. 30 mins. 4½ sec., 'U-B 64' saw the *Vaterland* sunk by my shot, capsising on port side."

In conclusion, here is the description of a fight which "U 84," commanded by Lieutenant-Commander Rohr, had with a steamer which kept her guns hidden and hoped by deceiving the U-boat to be able to surprise and sink her:

"*February* 22, 1917. 1.50.—Tank steamer, about 3,000 tons, with course 250 degrees, in sight. Dived. Torpedo fired from second tube; missed by 700 m.; had underestimated way. Steamer turns upon counter course. Went down. Rose to surface. Stopped her with gunfire. Steamer stops, blows off steam, crew leave the ship in two boats.

"2.30 P.M.—Approached under water. No armament. Boats, about 8—10, are away from steamer.

"2.49 P.M.—Rose to surface near boats which still try to pull away from U-boat.

"2.49 P.M.—Steamer opens fire from four guns. Dive. Conning-tower hit five times: one shot through the bridge, one above the aerials, the third (4.7 cm.) goes through the conning-tower, explodes inside, nearly all apparatus destroyed. Second officer of the watch slightly wounded. Fourth shot smashed circulating water tubes; fifth shot hit a mine deflector. Abandoned conning-tower. Central hatch and speaking tube closed. As the conning-tower abandoned, the boat had to be worked from the central space below the conning-tower. The lifeboats throw depth charges to a depth of 20 m. Switch and main switchboard held in place by hand. Electric lamp over magnetic compass goes out. Boat is top-heavy and oscillates round the transverse axis [because the conning-tower was filled with water]. A number of connections between the conning-tower and hull do not remain watertight. Owing to short circuit the following fail in quick succession: gyro-compass, lamp-circuit [for lighting], main rudder, means of communication, forward horizontal rudder jams. In spite of being 14 degrees down by the stern and engines going full speed, the boat sinks by the bows to 40 m.; compressed air. To get rid of the water, rapid expulsion of air to 20 m. to 16 degrees to load aft. Tank No. 1 gets no compressed air. All hands in the bows to avoid breaking surface.

277

Torpedo coxswain and No. 1 (petty officer) even counter-flood forward. Boat falls 8 degrees by the bow, and sinks to 35 m. depth. Compressed air on forward tanks.

"Meanwhile the spray (from leaks in the conning-tower) is kept off the electric apparatus by sail-cloth, waterproofs, flags, etc. The watertight auxiliary switchboard is the saving of the boat. Boat sinks down by the stern again and threatens to break surface. Steering under water no longer possible.

"3.10 P.M.—Compressed air on all tanks. Starboard electric engine breaks down. To the guns, clear oil motors, full speed ahead ! "

The commander decided, as the boat could not remain under water, to rise to the surface and chance fighting the steamer.

"The steamer is 35 hm. off and opens fire at once. Shots all round the boat. One 7.5 and one 4.7 cm. shell hit the upper deck forward of the boat's 8·8 cm. gun. Second officer of the watch receives other slight wounds. Replied to fire, unfortunately without telescopic sight as the conning-tower is still full of water. Distance quickly increases to 50 hm. Then the steamer follows slowly. To starboard a destroyer which opens fire at 80 hm.; shots fall short. Put on cork jackets. The intention is to continue gunfire till the boat can be sunk in the neighbourhood of a sailing vessel 8 sea miles away, to save the crew from a *Baralong* fate.

"3.17 P.M.—The destroyer is a ' Foxglove,' but cannot steam faster than the boat. At about 75 hm. replied to fire. The ' Foxglove ' soon begins to try and avoid shots; is hit twice, and increases the distance. Her guns only carry about 75 hm.

"3.20 P.M.—Conning-tower can be made watertight; boat cleared; ammunition for gun cleared; except conning-tower, all damage can gradually be repaired. Course 165 degrees. The ' Foxglove ' follows in our wake. Steamer lost to view. At a pinch the boat can dive, but leaves a heavy oil track behind her. If no destroyer comes before night, the boat can be saved.

"6.50 P.M.—The ' Foxglove ' has approached to 70 hm. and opens fire again. Return fire: hit. Enemy sheers off and falls back to over 100 hm.

"8 P.M.—Twilight. Pursuit out of sight. On account of oil track zig-zag course. Run into another oil track, turned to port and gradually on course of 240 degrees."

Our U-boats and their Method of Warfare

The boat then began her return journey and reached home without further incident.

I myself had occasion to inspect "U 84" after her return from this expedition. I realised that it was little short of a miracle that, in spite of such heavy damage, she reached home. It was chiefly due to the assurance with which the commander handled his boat, the perfect co-operation of the whole crew in these trying circumstances, and the excellent practice made by the gunners, in connection with which it must be remembered that the height of the platform of a U-boat, on which the gun is mounted, is only 2 m. above the water-level, and that aiming is thereby rendered far more difficult. Lieutenant-Commander Rohr is, unfortunately, one of the many who have not returned from their voyages.

It would take too much space to quote extracts from other U-boat experiences, or to mention the names of all those who particularly distinguished themselves. Wherever in this war heroism is spoken of, it applies without exception to our U-boat commanders and their crews.

CHAPTER XV

ACTIVITY OF THE FLEET DURING THE U-BOAT CAMPAIGN

BESIDES the direct support of the U-boats which operated from home bases, the Fleet supplied almost the whole personnel required to commission the new boats. It was particularly important for the U-boats to have technical men who were well trained in seamanship. The commanders had to be officers who had sufficient experience to navigate and handle the boats without assistance in the most difficult circumstances. That meant a big demand for officers of the watch, because they, by age and seniority, were best fitted for such service. The Fleet had to train men to take their places, so younger officers were promoted to be officers of the watch, and the training of midshipmen was accelerated. The substitutes for the latter were taken direct on board and received their training as naval cadets with the Fleet. This entailed a very extensive shifting of all ranks which was bound to have a deleterious effect upon the efficiency of the ships.

The project of a raid with the Fleet to the Hoofden in March, 1917, to attack the convoy traffic between England and Holland, never materialised. The weather had been uninterruptedly bad up till March 11. By that time the clear nights were over, which were a necessary preliminary condition for the enterprise. The weather prospects grew worse, so that we could not rely on scouting from the air. A cruiser raid by night had also to be given up, because it was reported from Heligoland that the wind (E.S.E., force 7—9) threatened to become worse.

The second leading ship of the torpedo-boat fleet was sent with Flotillas VII and IX to the Baltic for training in mine-sweeping, as the mine-sweeping divisions did not suffice for the work of escorting the U-boats as well as that of clearing the routes of mines. I considered, too, that Fleet manœuvres were necessary, so that the new commanders might become familiar with handling their ships in co-operation with the rest of the Fleet. I could leave the defence of the North Sea for the time being to the cruisers, as it seemed improbable that the enemy would make an attack on the Bight.

The Fleet during the U-boat Campaign

Meanwhile, the battleship *Baden* had been made ready as Fleet-Flagship, and the Commander-in-Chief of the Fleet had embarked in her. Flotillas III and IV could be spared from the North Sea, while the Fleet was at manœuvres in the Baltic, and were sent to Flanders, where they could be put to better use, by carrying on the war in the Channel from Zeebrugge.

The number of mines laid in the North Sea by the enemy grew steadily greater. Almost daily we suffered losses among the mine-sweeping craft, while among the ships used to escort the U-boats in their passage through the mine-fields there had been so many losses that in March the Fleet had only four such vessels at its disposal. The Secretary of State for the Imperial Navy was asked to raise the number to twelve again, so that they might suffice for the needs of the Fleet by working in four groups of three.

While the battleships were at squadron practice in the Bight of Heligoland on March 5, the *Kronprinz* and the *Grosser Kurfürst* collided and suffered damage which in the case of both ships took several weeks to repair; before these were complete any considerable enterprise for the Fleet was not to be thought of.

While the Fleet was practising in the Baltic, on March 21, the cruiser *Moewe* reported her return in the Kattegat, and entered Kiel on the 22nd. During her cruise of four months she had sunk or captured twenty-seven ships, amounting to 123,444 tons gross registered tonnage. One of the prizes, the *Yarrowdale*, had been brought into Swinemunde on December 31, 1916, and had conveyed news of the success of the *Moewe*, from whom we had heard nothing since she left at the end of November, 1916. The safe return of the successful ship was greeted with great joy.

On March 29 the outpost boat *Bismarck*, leading ship of the special group commanded by Lieutenant Schlieder, ran on a mine and sank; only three of the crew could be saved. Their smart commander also lost his life. He had won great credit by driving submarines out of the Bight.

To illustrate the demands that the U-boat campaign made upon the Fleet, I quote below from the log of the High Sea Fleet, beginning with May 9, 1917:

"*May 9, 1917.*—Wind and weather in the German Bight, E. to N.N.E., force 3—4; weather fine and clear. Seaplane scouting in the Inner German Bight without result. Mine-sweeping according to plan. Scouting Division IV protects the operations in the west. During mine-sweeping operations of the Ems outpost flotilla, the *Mettelkamp* strikes a mine north of Borkum and sinks.

" U-BOAT CAMPAIGN

"Returned from long-distance trips : ' U 82,' ' U-B 22,' ' U-B 21,' and ' U 93.' ' U 93 ' proceeded to sea on April 13, and up to April 30 sank 27,400 tons. On April 30 had a struggle with a U-boat trap (iron-masted schooner), in the course of which the commander, Lieutenant-Commander Vohr von Spiegel, the helmsman, and one petty officer were hurled overboard, and three men were badly wounded. The boat, badly damaged, unable to dive properly, and deprived of its wireless, is brought into List by Lieutenant Ziegler. ' U 46 ' is escorted to the north. ' U 58 ' reports position among mines accomplished; two steamers sunk, three damaged, in 1 degree longitude west; a great deal of convoy traffic.

" *May* 10.—Scouting by our seaplanes without result. No airship observation. Mine-sweeping according to plan. Scouting Division IV protects mine-sweeping operations in the west. H.M.S. *Hindenburg* commissioned.

" U-BOAT CAMPAIGN

"' U-C 76,' while shipping mines in Heligoland harbour badly damaged by mine explosion and sunk. Among the missing is the commander, Lieutenant-Commander Barten. Salvage-boat *Oberelbe* goes from the Ems to Heligoland to give assistance. ' U-C 77 ' back from long-distance expedition; ' U 46 ' has passed the danger zone; ' U 30 ' proceeded to the North *via* Terschelling.

" *May* 11.—Wind E., force 4—5. Seaplane scouting; nothing suspicious. No airship observation owing to easterly wind. Mine-sweeping according to plan. The half-flotilla occupied in sweeping mines from the route to the west, in following up a barrier of mines, has got north of its prescribed route. New mines are observed, and the leading boat of the 5th Half-Flotilla of mine-sweepers strikes a mine and sinks. Four men are missing. Among them the commander of the Half-Flotilla, Lieutenant-Commander Beste. As it has now been ascertained that the English have barred the approach from Horns Reef from N.W. by mines, the officer in command has received orders to lay mines which will bar the approach from northeast and from the west, so as to deprive the English of this meeting-point, which we can do without. A further barrier of mines north

The Fleet during the U-boat Campaign

of Tyl Lightship is to bar the way to mining operations against Nordmannstief. At night a group of barrier-breakers goes along the U-boat route down the Dutch coast to the west, and another group to the north.

" U-BOAT CAMPAIGN

"' U 30 ' passed danger zone; ' U 58 ' back from long-distance trip; ' U 93' enters Wilhelmshaven towed by ' V 163.'

"*May* 12.—Wind E., force 6. Seaplane observation without result. Observation by airships impossible owing to weather conditions. Mine-sweeping only carried out to small extent owing to heavy sea. Scouting Division IV takes over protection of operations in the west. In the course of the morning, the two barrier-breaking groups return from night voyage. No incidents. A boat of the North Sea outpost flotilla reports an enemy submarine; set the torpedo-boat half-flotillas at my disposal to search; submarine ' kite' exploded; result doubtful.

"*May* 13.—Wind N.W., force 2. Mine-sweeping according to plan. Scouting Division IV on patrol in the west. At night the officer commanding the division, with the auxiliary minelayer *Senta*, lays the barrier of mines at Horns Reef and north of the Tyl Lightship according to orders.

" U-BOAT CAMPAIGN

"' U 33 ' leaves for long-distance trip in the west, and ' U-C 41 ' for the Bell Rock.

"*May* 14.—Wind E. to N.N.E., force 3. For protective scouting ' L 22 ' goes up to the west; ' L 23 ' to the north. Mine-sweeping according to plan. Scouting Division II goes for protection of operations in the west to the Osterems. Thunderstorm, 6 P.M. The Staff of the Fleet embarks in H.M.S. *Baden*. High Sea Fleet warships clear. Scouting Division II., with two torpedo-boat leaders, assemble in the course of the evening in Schillig Roads, for the intended manœuvres in the Bight on May 15. No communication from ' L 22 ' since report that she had risen. Thunderstorms in the west. It is possible that she has taken in her wireless mast and can send no message. In the late afternoon thick fog over the whole Bight, consequently not possible to have search made by seaplanes or surface craft. Seaplane No. 859 noted an explosion and a cloud of smoke at 9.50 A.M.

"7.40 P.M.—The leader of the airships reports that according to telephonic information from Borkum this observation is very prob-

283

ably connected with the loss of ' L 22.' A telegram arriving at night from the Admiralty confirms this statement. The probability is that on account of the thunderstorm ' L 22 ' had to remain below the level where the gas would completely fill the cells, and was shot down by British warships.

" U-BOAT CAMPAIGN

"11.40 P.M.—*Orion*, one of the 3rd Mine-Sweeping Flotilla, reports that ' U 59,' which was being convoyed out to sea, and the mine-sweeper *Fulda*, have struck a mine and sunk. Outpost boats of the List Division, the mine-sweepers of the flotilla, and the 17th Torpedo-Boat Half-Flotilla are sent out to meet the *Orion*. The boats receive orders at the same time to pick up ' U-C 51 ' and ' U-C 42,' which are west of Horns Reef on their return journey. ' U-C 44 ' and ' U-C 50 ' put out to sea to the west on a long-distance trip. ' U-C 51,' on return journey, reports position. Operations among mines completed; about 4,000 tons sunk; travels 5 knots only (nature of her damage not made out owing to defective wireless).

"*May* 15.—Wind N.N.W., force 2—3. ' L 16 ' and ' L 37 ' go up for aerial observation. Thick fog forces them to return. Slight visibility at times only. The tactical manœuvres in the Bight are therefore postponed to the 16th; the First Leader of the Torpedo-Boats reports that some of the 1st Torpedo-Boat Flotilla, with auxiliary engines, have broken down, and that the rest are not fit for use outside the Bight. Consequently the 1st Torpedo-Boat Flotilla is instantly dispatched to Kiel for repairs. The officer in command had sent ' U 59 ' to take note of the place of the accident; the List Division of the North Sea Outpost Flotilla, a pump steamer, and a tug, to assist the *Orion;* and a torpedo half-flotilla to control the U-boat route out to sea. The reports of the boats sent to assist do not give a definite idea of the degree of danger to be apprehended from enemy mines in the north. While trying to get into communication with ' U 59 ' by tapping, the outpost boat *Heinrich Rathjen* strikes a mine and sinks; missing, one petty officer and three men. Officer in command receives orders for the time being to stop the work of breaking through the belt of mines in the west, and to clear or test the U-boat route out to sea in the north, with all craft at his disposal. In view of the interruption of the work of breaking through in the west, any considerable enterprise of the Fleet must be postponed. I therefore decide, in order to make use of the time immediately after the evolutions, to send Squadron III,

The Fleet during the U-boat Campaign

and, on the return of Torpedo-Boat Flotilla II, Torpedo-Boat Flotilla V to the Baltic for manœuvres. Both are badly in need of training.

"' U-C 42 ' and ' U-C 51 ' return from long-distance trips. ' U-C 41,' which put to sea on May 11 for the west coast, had to break off expedition owing to engine trouble.

"*May* 16.—Wind N.E. to N.N.W., force 3—6. Seaplane observation without result. Airship observation impossible owing to fresh north-east wind. Tactical manœuvres by the High Sea Fleet in the Bight of Heligoland. On completion, sent Squadron III to the Baltic. In the absence of Squadron III, Squadrons I and IV to take outpost duty in turn. H.M.S. *Kaiser* sent to Kiel harbour for repairs. The officer in command of the Scouting Divisions to discover the cause that led to the loss of ' U 59 ' has ordered mine-sweeping flotilla to test and clear Squares 132, 117, 133, 116. One torpedo-boat half-flotilla is to sweep Squares 134—84. The auxiliary mine-sweeper flotilla is to mark the spot of ' U 59's ' accident and try to get into communication with ' U 59 ' by tapping. In the course of these operations, ' M 14 ' strikes a mine, and in the attempt to save ' M 14,' Torpedo-boat No. 78 does likewise. Both boats sink. The attempts to get into communication with ' U 59 ' must consequently be abandoned.

"In the night, ' S 27 ' of the Ems Outpost Flotilla strikes a mine and sinks while convoying ' U 86 ' on a long-distance trip to the west. ' U 86 ' thereupon returns with the rest of the convoy to Borkum Roads. Wireless reports received: ' U 62 ' left on April 21, position; in April 10,000 tons sunk; in May, 13,000; on April 30 captured commander of U-boat trap ' Q 12.' ' U-C 55 ' left April 28 for west coast, position. ' U 40 ' left May 5. Mine-sweeping operations completed, two explosions heard, nothing sunk. ' U 21 ' left April 19, position; 13,500 tons sunk ' U-C 49 ' left May 2, position; mines laid, 3,365 tons sunk.

"*May* 17.—Wind E. to E.N.E., force 3—6. No airship observation. Seaplane scouting without result. Reports hitherto received about the mines laid by the enemy south of Horns Reef do not yet give a clear idea of the condition of affairs. It seems as if new enemy mines were lying south of the Senta barrier, direction east to west. Consequently, all efforts must be made to clear up the situation and keep this important waterway clear. It is once again apparent how short we are of mine-seekers and sweepers. We shall therefore once again approach the Admiralty and demand that the 'M-boats' (new mine-sweepers) that have been allocated to the Commander-in-Chief of the Baltic, should be handed over to the High Sea Fleet. As a substitute, we will offer boats of the North Sea Outpost Flotilla or trawlers of the Auxiliary Mine-sweeping Flotilla. To make the position of the lightship at List more reliable and easier from the point of view of navigation, a buoy will be placed to mark the position which will be occupied at night by an outpost boat. Further, to help the U-boats on their outward voyage, a number of outpost boats are to cruise continually west of the position of the lightship. During the night two groups of barrier-breakers go out, one to the north and one from the Ems to the west.

" U-BOAT CAMPAIGN

"' U-C 49 ' and ' U-C 41 ' back from long-distance trip; ' U 86 ' leaves under convoy for Flamborough Head; *via* Bruges comes a wireless report that ' U-C 75 ' has sunk 3,500 tons and the English warship *Lavender*. Torpedo-Boat Half-Flotilla XVII puts to sea to meet the damaged ' U-C 40 ' and to bring her home through the subsidiary waterway of Nordmannstief."

And so it went on from day to day. Owing to the pressure of the demands made upon them during the war, the organisation of the mine-sweepers was developed in the following manner:

At the beginning of the war three mine-sweeping divisions existed and were stationed at Cuxhaven. Of these, Divisions I and III took up their activities in the North Sea and Division II in the Baltic. Each division consisted of a leading boat, eight sweepers and two—increased to four later—buoy-boats. (The buoy-boats marked the channels swept by the mine-sweepers for the groups of larger ships that followed.) The boats were without exception small, old torpedo-boats, of the class ' V 30 ' and ' 80.' Their speed was 17—18 knots with a draught of 2.7 m., their armament one 5 cm. gun. In command of the flotillas were Commanders Bobsien and Wolf-soin; later the North Sea Auxiliary Mine-sweeper Flotilla under Commander Walter Krah was added. It consisted of trawlers and small torpedo-boats.

At the end of 1915 and the beginning of 1916 the small old torpedo-boats were gradually replaced by the "A"-boats and "U"-boats which had been built during the war. The "A"-boats had a speed of 23—25 knots, a draught of 1.9 to 2 m., a displacement of 210 to 345 tons and an armament of two 8.8 cm. guns. The "M"-boats had a speed of 16 knots, draught 2 to 2.2 m., displacement 450 to 520 tons, armament three 8.8 cm. or two 10.5 cm. guns.

On September 1, 1916, the Mine-sweeper Divisions I and III were divided into the 1st and 2nd, and 5th and 6th Half-Flotillas respectively. On October 6, 1916, what had hitherto been Mine-sweeper Division II was divided into the 3rd and 4th Mine-sweeper Half-Flotillas. These Half-Flotillas were still grouped under their original Flotillas.

In May, 1917, Mine-sweeper Flotillas I and III were both augmented by a third half-flotilla, consisting of "M"-boats.

In June, 1917, Mine-sweeper Division II with the parent ship *Ammon* and ten motor-boats left the Baltic and joined the North Sea warships. These motor-boats ("F"-boats) have a speed of 11 knots, draught 1 m., length 17.5 m., displacement 19 tons, and an armament of one machine gun. Later on, in January, 1918, the Mine-sweeper Divisions III and IV formed the 3rd Mine-sweeper Flotilla. Mine-sweeper Flotilla II was also augmented by a third Half-Flotilla—No. 9. The Auxiliary Mine-sweeper Flotilla of the North Sea was denominated from the beginning of 1918 onwards Mine-sweeper Flotilla IV, and the trawlers of which it originally consisted were for the most part replaced by new mine-boats. All the mine-seekers and mine-sweeper groups were then placed under the command of one officer, and Captain Nerger, well known as the commander of the auxiliary cruiser *Wolf*, was appointed to this post after his return from his cruise. Further formations were: Mine-sweeper Division II, consisting of the parent ship *Ammon* and twelve boats, and the Mine-sweeper Flotilla VI, one leading boat and two half-flotillas, consisting of an "M"-boat as parent ship, six boats, eleven "U-Z"-boats (small, fast motor-boats) and three large motor-boats. The "F-M"-boats (shallow draught "M"-boats) had a speed of 14 knots, draught 1.3 m., displacement 170 tons, length 40 cm., and an armament of one 8.8 cm. gun. The "U-Z"-boats when towing their apparatus had a speed of 18 knots, draught 1.5 m., displacement 20 tons, length 26—30 m., and an armament of one 5 cm. quick-firing gun.

At the time the Armistice was concluded the following boats were available for the mine-sweeping service in the North Sea:

17 torpedo-boats, 27 "U"-boats, 71 "M"-boats, 4 "F-M"-boats, 23 trawlers, 58 motor-boats, and 22 "U-Z"-boats, 4 parent ships and a repair ship, whereas at the beginning of the war there were only 33 small, old torpedo-boats available.

At the beginning of the war we had three forms of mines, with a charge of from 70 to 150 kilos, capable of use at a depth of from 90 to 115 m. The newest of these existing types of mines could ultimately be used at depths of 345 m. During the war the following types of mine were added: 1. A defence mine against submarines, with a charge of 20 kilos and effective to a depth of 95 m.; 2. A mine in the form of a torpedo that could be shot out of a U-boat travelling under water, with a charge of 95 kilos, effective to a depth of 200 m.; 3. A mine to be laid by U-C-boats with a charge of 120 to 200 kilos, effective to a depth of 365 m. The U-C-boats could carry 12 to 18 mines; 4. A mine for the first big minelayers which, however, was not made after the construction of the U-C-boats.

Mine-sweeping tackle was improved to such an extent that the area swept increased from 45 to 300 m.; the depth to 30 m.; and the speed of sweeping, when the new boats were used, to 15 knots.

As a defence against mines—in the first place for the minesweepers, and later destined for all classes of ships—special apparatus was invented. This was attached to the bows and was intended to cut the mine-cables before the boat struck the mine. It was found of great use.

For defence against submarines a depth charge was made which could be thrown from a boat on to submerged submarines. The charge weighed about 50 kilos; it was detonated under water by an adjustable time fuse.

In addition to this there was a submarine "kite" with an explosive charge of 12 kilos. It was towed by boats on a cable which served at the same time to indicate the direction of the current. It was electrically exploded as soon as the "kite," while being towed, struck a submarine.

To keep off submarines nets of various kinds were made, which were moored to buoys to bar the submarine's path, and lighter nets, provided with gas buoys, which indicated the path of the boat and the spot where it had broken through if a submarine ran against the net.

The convoy service for U-boats demanded large numbers of light craft; about 100 torpedo-boats and smaller steamboats were used for this purpose. They were divided into two convoy flotillas :

1. Commanded by Commander Faulborn, consisted of three half-flotillas, each comprising two groups of five torpedo-boats.

2. Commanded by Lieutenant-Commander Hoppensledt, consisted of six half-flotillas, each of ten to twelve steamboats.

The mine-sweepers and convoying craft deserve great credit for making it possible for the U-boats to carry out their campaign. They suffered many losses which would otherwise have been inflicted on the U-boats. That, however, did not prevent them from performing their dangerous service year in and year out with the greatest trustworthiness in spite of the inclemency of the weather. The officers and men of this service surpassed all others in the Navy in their intrepidity and skill as seamen.

The most successful part of the activities of the Fleet fell to the lot of the U-boats; the battleships, together with the cruisers and torpedo-boats, and especially the mine-sweepers, assisted in overcoming the enemy's defence. Their efforts were primarily directed against the belt of mines which England had laid in the North Sea to prevent our boats from getting out. The accompanying plan shows how the Bight was made to bristle with mines.

It was impossible, in view of its great extent, to clear the whole area. We barely had sufficient ships to ascertain where mines were laid. Our efforts were confined to getting two or more paths through the mines, and keeping them clear; one to the west, following the coast, one in the middle between Terschelling and Horns Reef, one to the north along the Danish coast. This last had the advantage of making it easier for the U-boats to find their way home on their return, as they could feel their way to the coast of Jutland while they were still outside the area sown with English mines, and seek the route that had been cleared, which led along the coast into the Bight. The route off the Dutch coast was the shortest for those boats which chose the Channel passage to get to their station west of the British Isles. It certainly was the shortest route, but also the most dangerous, because of the strong defence in the Channel, and the various obstacles there in the shape of nets and mines. These cleared paths had to be so wide that the boats could find them even in bad weather, when they were unable to determine their position with accuracy; and also they had to be wide enough to allow freedom of movement to the auxiliary boats which accompanied the mine-sweepers; for the mine-sweeping operations were in all probability observed by the English submarines, and if the cruisers in the auxiliary groups had only been able to move slowly up and down in the narrow area they would have presented an easy and welcome target. Consequently we tried to keep a large basin free of mines, situated to the rear of about the middle of the belt, so that it lay in a central position for all the routes. Even this did not guarantee absolute safety, so the boats were always accompanied by an escort capable of clearing away any mines which might after all be encountered.

In July, 1917, the English had extended the area which they had announced to be mined, to the north up to the latitude of Hanstholm (north-west coast of Jutland), in the west as far as four degrees longitude east, in the south to 53 degrees latitude north. By this means the path which had to be kept clear by the boats of the High Sea Fleet was lengthened at the narrowest point by 20 to 25 sea miles.

Up to the end of June, 1917, despite months and months of work, the mine-sweepers at our disposal had not succeeded in breaking through the old danger zone. The demands which, in consequence, were made upon the U-boat convoys—which had to take the U-boats through the mined area into free water—naturally impeded the actual mine-sweeping. In case of need, recourse might be had to the torpedo-boat flotillas, but, after all, they represented material as valuable as that which they were to protect; and in particular the new boats had too much draught to pass through the mine-sown areas except at great risk. (The lighter the draught the less the danger for the mine-sweepers, in the construction of which this point was particularly considered). The new boats constructed in the last months hardly made good the losses, and the number fixed by the commanders of the Fleet as the minimum required had not been attained. These clear routes were not needed for the U-boats alone, but also for communication between Rotterdam and the Elbe and Ems respectively. In the middle of July, 15 to 20 steamers lay at Rotterdam waiting for the word that they might safely cross. It was the Fleet's duty to guarantee that the waterway along the coast was free and to convoy the ships with outpost craft to safety.

In spite of all the difficulties we managed to prevent anything from stopping the U-boats from going out. There were altogether very few days when for safety's sake we had to avoid the direct route into the North Sea and take the roundabout way through the North Baltic Canal and the Kattegat. The small loss of time

The Fleet during the U-boat Campaign

was of no importance compared with the increased safety that was thereby gained. As the boats could replenish their fuel supply in Kiel, they were able to stay in their field of operation for comparatively a long time. But it was not unknown to the English that our boats used this way to get out, especially later when the U-cruiser flotilla had been formed at Kiel; these boats mostly took the route through the Kattegat for their outward and return voyages. That forced the Fleet to extend its mine-sweeping to the Kattegat, and to take counter measures when the English mines were laid from Skagen across to the Swedish coast.

It is obvious on what a large scale English mine-laying was carried on, when it is considered that they set about mining the whole of the North Sea between the Shetland Islands and Norway. As we learnt afterwards it was chiefly American mines that were to be used and American craft to do the work. If they had really succeeded in sowing mines sufficiently thickly in that area the Fleet would have found it an exceedingly difficult task to clear the necessary gaps there. However, the great depth of the water in this part of the North Sea made it possible for U-boats to avoid the barrier by travelling at a sufficient depth below the surface. So far as we could ascertain, we suffered no losses in U-boats from these mines.

The boats, when going out, and before their return, reported their position, so that the commanders of the Fleet and the officers in command of the U-boats knew for certain that the first difficulties had been overcome, and that the boat was making for her actual field of operation; or, as the case might be, was on the homeward voyage after accomplishing her work. Thus it was possible to establish with great accuracy in what period of time boats that were missing must have met with misfortune.

The Fleet considered it its most important task to place all its strength at the disposal of the U-boats on their outgoings and incomings, so as to protect them from the dangers of this part of their voyage. Plenty more awaited them which they would have to cope with alone, once they were in their particular sphere of activity. This was essentially the point of view of the mine-sweepers; they suffered ever greater losses, yet did all that was possible to take upon themselves the main dangers that threatened the U-boats.

In August, 1918, H.M. the Emperor announced his intention to visit the Fleet. Shortly before his visit there had been signs of insubordination among the crews of some of the ships of Squadron

Germany's High Sea Fleet

IV. (*Prinzregent Luitpold* and *Friedrich der Grosse*); this bore the character of mutiny, but thanks to suitable measures taken by the officers it was nipped in the bud before it had assumed considerable dimensions or had injured the efficiency of the ships. Inquiry into the matter, however, revealed that behind these comparatively unimportant outbreaks there lay a movement which must be taken very seriously, and which had as its aim the forcible paralyzing of the Fleet as soon as the political wire-pullers deemed the moment ripe. The judicial inquiry established the fact that there was a connection between the members of the Independent Social Democratic Party and the leaders of the movement in the Fleet. Their first aim was to get a sufficient number of the crews to allow their names to be put on lists which were to prove at the forthcoming Congress at Stockholm that the crews at the front had grown weary of fighting and were ready to join in the political movement. This movement aimed at bringing the war to an end in all countries by overthrowing the existing forms of government. Very cleverly had the leaders sown discontent on certain ships; they had made the most of supposed abuses, especially in the rationing, and had not even shrunk from influencing their comrades by threats of forcible measures. The whole network of the plot was laid bare and those who had stirred up the trouble were punished. In certain cases the court-martial pronounced sentence of death, which was carried out so far as the most guilty parties were concerned. Most of those implicated had not realised the consequences of joining the organisation; to many it had not even been explained. Compared with the total numbers of the crews, those who had joined the movement were very few.

The great danger which lay in this unrest, stirred up in the Fleet by conscienceless agitators, could not be overlooked. Conditions on the big ships in particular unfortunately provided fruitful soil for such activities, as the crews were all the time in close communication with their homes and could, therefore, not be kept immune from the prevailing depression. These men performed the same service on the big ships all the year round, and they lacked the refreshing stimulus of meeting the enemy in battle. On the other hand, they had a daily supply of newspapers and pamphlets which teemed with war weariness and the condemnation of our war leaders. Thus it was unhappily possible to influence their views and make them forgetful of their duty.

The Secretary of State for the Navy arrived in Wilhelmshaven

The Fleet during the U-boat Campaign

on August 17, the day before the Emperor embarked. I made earnest representations to him that it was the duty of the Government to protect the Fleet from this Socialistic organisation, as otherwise the efforts of the officers to shield the men from these disastrous influences would be of no avail. Admiral von Capelle was very doubtful whether it was possible, with the sentiments then prevailing in the Reichstag, to call the leaders of a party to account for their political agitation which, so far as it was subversive of order, was carried on with the greatest circumspection. But he quite admitted the gravity of the situation and promised to see that the necessary protective measures were taken by the Imperial Government. He spoke to His Majesty to that effect the next day, after I had reported these incidents in the Fleet to him. Unfortunately at the discussions in the Reichstag, which followed shortly afterwards, it appeared that the Government was not sufficiently firm to take radical measures and secure the consent of the majority of the people's representatives to them.

The Fleet had to depend entirely on its own efforts to shield the crews from the devastating influences which were brought to bear upon them. The best distraction certainly was active warfare. The crews had never refused to obey the call of this necessity; courage and the joy of battle still prevailed in their old, original form. They were so deeply rooted in the character of the German people that they could not cease at the first onslaught from without.

The influence of enemy propaganda was turned to account by the Independent Social Democratic Party to achieve its own ends. It could be counteracted to some extent but not entirely removed, and its disastrous effects were apparent later. There is a widespread view that the crews had justifiable grounds for complaint in the differentiation in treatment between the officers and men; this is totally unfounded. Service on board makes at least the same demands on the officers, and indeed much greater demands, than on the majority of the members of the crew. On watch and on every other form of service, the proportionate number of officers is employed with every group of men, and they have no alleviations as compared with the crew; on the contrary, they are much more exposed to the inclemency of the weather, and far greater demands are made upon their vigilance when at sea. Even in the unpleasant process of coaling, all the officers co-operate, and there is no difference then between them and the men in respect of their "get-up" and their unavoidable condition of dirt. This practice, introduced in time of peace to attain the greatest possible efficiency in coaling, was of necessity continued during the war, when the unpleasant work of filling the coal-bunkers had to be undertaken much oftener.

His Majesty embarked in the flagship *Baden,* and took his first trip to sea during the war. On this occasion, he visited the Island of Heligoland to inspect the fortifications and harbour works there. He landed again at Cuxhaven, where he spoke to the crews of the mine-sweeper flotillas, and was able to confer decorations upon some of the leaders and crews who had had a brush with enemy destroyers a few days before, in which they so successfully warded off the attack that we did not have to record the loss of a single boat.

CHAPTER XVI

THE CONQUEST OF THE BALTIC ISLANDS AND THE CAPTURE OF HELSINGFORS

IN September, 1917, after the taking of Riga, the Supreme Army Command asked for the co-operation of the Fleet to conquer the Baltic Islands. This offered a welcome diversion from the monotony of the war in the North Sea. The Navy's task was to take a landing corps, consisting of a reinforced infantry division, under the orders of the General Officer of the 23rd Regimental Command, to Oesel, and to land them there.

The right flank of the landing troops had to be protected from the sea by quickly sending ships to the Gulf of Riga; and the attack on the bridgehead of Orissa on the Island of Oesel, to make it possible to cross to the Island of Moon, had to be supported with all the means at our disposal. So long as the Straits of Irben were commanded by the heavy enemy guns at Zerch, the bays of the Island of Oesel in the Gulf of Riga were useless for landing. Consequently, the Bay of Tagga was chosen for the troops' disembarkation. This is the only bay in the north or west of Oesel that can hold a large number of transports and offer them protection from the west winds which prevail there in the autumn.

After the warning example of the landing of the Franco-British army in Gallipoli in the spring of 1915, the attempt to carry the war on land overseas by the help of the Fleet had to be made with the greatest care, and such strong defensive measures were taken that a reverse appeared to be out of the question. We had to prepare ships for the transport of 23,000 men, 5,000 horses, and much material.

The warships had to clear the approaches of mines, so that none of the transports with the troops on board might be lost, also to send flying-men to find out the position of the enemy beforehand, so as to ascertain the most favourable circumstances for the landing, which had to be a surprise. The Russians had recognised the danger which threatened them, and had tried to ward it off by placing batteries on Cape Handsort and Ninnast, at the two entrances to Tagga Bay. Heavy batteries had been strongly built on the peninsula of Sworbe, in the south of Oesel, some time previously.

The warships set aside for this undertaking were placed under the orders of Vice-Admiral Ehrhardt Schmidt, the Commander of Squadron I. He had a special Staff for the occasion, made up of officers of the Fleet and Admiralty Staffs. Captain von Levetzow was nominated Chief of the Staff; the battle-cruiser *Moltke* was the flagship.

Under the orders of Admiral Schmidt were: Squadron III— Vice-Admiral Behncke: Battleships *König, Bayern, Grosser Kurfürst, Kronprinz* and *Markgraf;* Squadron IV—Vice-Admiral Souchon: Battleships *Friedrich der Grosse, König Albert, Kaiserin, Prinzregent Luitpold* and *Kaiser;* Scouting Division II, under Rear-Admiral von Reuter: Second and third class cruisers *Königsburg, Karlsruhe, Nurnberg, Frankfurt, Danzig* and the light cruisers of the Baltic Fleet, *Kolberg, Strassburg* and *Augsberg,* under Vice-Admiral Hopman. Commodore Heinrich was in command of the following torpedo-boats, he himself being on board the second-class cruiser *Emden:* Torpedo-Boat Flotilla II (Commander Heinecke) with 10 boats; Torpedo-Boat Flotilla VI (Commander Tillesen) with a half-flotilla; Torpedo-Boat Flotilla VIII (Commander Nieden), with 11 boats; further, the 7th Torpedo-Boat Half-Flotilla; the 13th Torpedo-Boat Half-Flotilla; Torpedo-Boat Flotilla IX (Commander Hundertmark) the latter with 11 boats; besides these 6 U-boats of the Kurland U Flotilla (Lieut.-Commander Heinrich Schott); Mine-Sweeper Flotilla II (Lieut-Commander Doflein); the Mine-Sweeper Division IV, and a half-flotilla of mine-seekers that numbered a little more than 60 motor-boats. In addition to these there was Captain von Rosenberg's flotilla, who had at his disposal 72 boats—trawlers, and other craft of similar size. Nineteen steamers were requisitioned as troop-transports, the tonnage of which amounted to 153,664 tons.

The enterprise was first mooted on September 12. On October 9 the troops embarked; on October 11 the transport fleet put to sea under the protection of the battleships and small cruisers. The preparatory work of mine-sweeping had been delayed by the bad weather during the end of September and beginning of October, so that those in command waited with impatience for operations to start.

This delay was an advantage for the transports, as it enabled

The Conquest of the Baltic Islands

us to drill the troops in embarkation and disembarkation, which materially facilitated the landing afterwards. The number of steamers was not sufficient to transport the troops and all the baggage in one journey; two echelons had to be formed. This circumstance also made it advisable not to start the expedition until the mine-sweeping operations in the Irben Straits were nearing their close, so that the second echelon might be transported in safety to Arensburg without running risks from submarines.

The manifold preparations for the embarkation of the troops and for carrying out the operations on land in conjunction with the Fleet had been completed, and there had been the most exemplary harmony between the leaders of the Army and those of the Navy. Thanks to them, the conquest of the islands of Oesel, Moon and Dago was carried out according to plan, with the most perfect success.

On October 10 everything was in trim; the fleet of transports lay ready in the naval port of Libau to proceed to sea; the *Moltke,* with Squadrons III and IV, lay in the Bay of Danzig, behind the peninsula of Hela; the small cruisers and torpedo-boats were at Libau.

The battleships were to silence the batteries at the entrance to the Bay of Tagga before the landing was effected, as well as to force the passage of the fortified straits between the islands of Dago and Oesel, and of Soelo Sound, which leads into the Kassar Wick. It was necesssary to command the Kassar Wick, which owing to the depth of the water can only be used by torpedo-boats, so as to secure the passage to Moon from the north, and to prevent Russian warships from leaving the Gulf of Riga and making for the north.

The batteries of the Bay of Tagga were attacked by Squadron III and the *Moltke*; Squadron IV was to destroy the batteries on Sworbe. It was important to land an advance guard in Tagga Bay as soon as possible after the silencing of the batteries in the north, so as to occupy the coast-line and thereby ensure the safe passage of the main transport fleet. The warships and the transports left harbour on the morning of October 11. The night journey through the mine-field passed without incident. The lightships placed by the Rosenberg Flotilla marked the track that the search flotilla had reported clear of mines. It was not until towards midnight that a check in the advance occurred, which threatened to be critical; the leading squadron had approached so closely to the mine-sweeping flotilla that speed had to be slackened. At first the delay was accepted, but finally it was realised that by slackening speed the punctual landing of the advance guard would be jeopardised and at the same time the element of surprise, which underlay the whole undertaking, threatened to be lost. Consequently Admiral Schmidt gave orders to the mine-sweepers to remove their gear and make room for the Fleet; he preferred taking the risk of negotiating the rest of the passage without the security afforded by the mine-sweepers to endangering the success of the whole enterprise. Fortune favoured this decision, for the Fleet succeeded without accident in reaching the positions from which the bombardment was to take place. They passed through a gap in the belt of mines right in front of the Bay of Tagga, the existence of which was only definitely ascertained later on. While taking up their positions to bombard the batteries on the Sound of Soelo, the *Bayern* and the *Grosser Kurfürst* struck mines, which, however, did not hinder them from completing their task.

At 5.30 A.M. the landing was begun. It was a complete surprise to the enemy, and met with little opposition, which was quickly overcome by the fire of the torpedo-boats, supported by the troops. The disembarkation of the advance troops which were on board the ships of Squadron III, was carried out by the motor launches of the ships and three small steamers, one of which was the *Corsica*. The leader of the Torpedo-Boat Flotilla steamed ahead with his boats. The batteries at Hundsfort and Ninnast were quickly silenced and at 8 A.M. in the hands of our troops. The Toffri battery on the southern point of Dago was destroyed by the *Bayern* and the *Emden*, Commodore Heinrich's flagship. At 6.45 A.M. the transports received orders to enter the bay, and the disembarkation was proceeding apace at 10 A.M. On entering the bay, the steamer *Corsica* struck a mine. She was run aground and her crew taken on board by torpedo-boats and landed. This incident showed us that the main part of the Fleet must have passed in safety through a gap in the belt of mines.

The second part of the Fleet's activities consisted in quickly penetrating into the Kassar Wick, and invading the Gulf of Riga. On the very day of the landing, Captain von Rosenberg, with his flotilla, had pushed through the Sound of Soelo, and so proved that it was navigable for torpedo-boats. Under the command of Commodore Heinrich the boats of Flotilla II and of the 12th and 13th Half-Flotillas then drove the enemy back into the Moon Sound. In this they were supported and covered by the fire of the *Kaiser* and

The Conquest of the Baltic Islands

the *Emden*, which lay before Soelo Sound. On this occasion, October 14, the destroyer *Grom* was captured and a gunboat was sunk. We suffered no losses in battle, but three boats were damaged and one was sunk by mines. In many cases boats ran aground in these badly surveyed waters, and in so doing injured the blades of their propellers.

The boats of the Rosenberg Flotilla established communication with the bridgehead at Orissa, and this was maintained until the troops had crossed. The flotilla brought bread and munitions to the pioneers, and later on undertook their transport across to Moon.

It was impossible for our light craft to push on into Moon Sound from the Kassar Wick, on account of the heavy guns of the Russian battleship *Slava*, which bombarded them from the south; consequently Moon Sound had to be taken from the south. For this purpose we had first to destroy the fortifications of Zerel. This task was assigned, on October 14, to the officer commanding Squadron IV, with the battleships *Friedrich der Grosse*, *König Albert*, and *Kaiserin*. The Russian batteries opened fire on them, and our ships returned it until darkness fell. The next morning the Russians had abandoned the position and destroyed the batteries. The landing troops had meanwhile continued their march towards Sworbe and Onesa. It was imperative that our ships should quickly penetrate into the Gulf of Riga, so as to hold the Russians on the Island of Oesel and prevent them from crossing to Moon. The minesweeping operations in the Straits of Irben, conducted by Vice-Admiral Hopman, had made good progress by October 13, although they were under the fire of the batteries at Zerel, and came upon belt after belt of mines. But when there was danger that the Russians might retire too soon to Moon and thence to the mainland, the passage to Arensburg had to be forced. Vice-Admiral Behncke, commanding Squadron III, received orders to support Admiral Hopman's light craft in this undertaking. Thanks to the energy of the officers, he carried these orders out with a celerity that surpassed all expectations. When Sworbe fell on the morning of the 16th, our warships lay before Arensburg, and on the evening of the same day before the southern extremity of Moon Sound. In this way our warships had completely surrounded the Island of Oesel, and made it impossible for the enemy, who had been driven by our troops to the south-east of the island, to escape by water.

On the morning of October 17 Moon Sound was reached; the

batteries there were destroyed, the Russian ships driven off to the north, and the Russian battleship *Slava* sunk. This success deserves the highest appreciation; it was gained under difficult conditions, in waters bristling with mines. The officer commanding Squadron III particularly commended the conduct of the mine-sweepers, who worked admirably under heavy fire.

While the commanding officer of Squadron III forced Moon Sound from the east, Admiral Hopman, with the *Kolberg* and the *Strassburg*, penetrated the Little Sound, ready to render the Army the assistance it required for crossing. In the night of October 17-18 our troops crossed to Moon, and this island, too, was surrounded by our ships on the east, south, and north-west. It was no longer possible for the enemy to escape to the mainland.

As the operations had been so successful, we proceeded to take Dago, which had not been included in the original plan. The Rosenberg Flotilla landed 300 men on the southern point of Dago, and occupied a bridgehead there for the subsequent landing of an infantry regiment. It maintained its position against attack until the troops arrived. For the conquest of Dago 3,700 men, 500 horses, 140 wagons, and a field battery with munitions were landed, and the landing party from the flotilla was withdrawn.

After Tagga Bay had been cleared of mines, an essentially necessary proceeding, the Fleet still had to perform the task of cutting off the Russian retreat from the north part of Moon Sound. Up till then this had been left to the U-boats. They received orders when Squadron III invaded the Gulf of Riga, to assemble before Moon Sound and to attack any Russians who should attempt to get out. "U-C 58" torpedoed the armoured cruiser *Bogatyr*, and "U-C 60" sank a transport steamer. It was not till October 18 that the torpedo-boats could be withdrawn from the Kassar Wick, and the necessary mine-sweepers liberated, both being needful for the protection of big ships.

On October 17 Squadron IV, with Scouting Division II, two torpedo-boat half-flotillas, and the necessary mine-sweepers, was to push forward to the northern exit of Moon Sound. But the weather made mine-sweeping impossible. Consequently the advance through the mine-fields north of Dago could not be carried out. Five boats had simultaneously reported that the enemy was retiring to the north, so it followed that the whole of Moon Sound must be clear of hostile craft; the enterprise was therefore abandoned. The damage that the large ships would probably have sustained from mines was out of proportion to anything that might have been gained by pushing on farther.

This completed the operations of the Fleet. The conquest of the islands attained by this co-operation of the Army and Navy represents a military achievement which was as unique as it was successful. The Navy is especially proud of it, as it gave them an opportunity of lending valuable aid to the Army.

The departure so far east of such a large portion of the Fleet, and its sojourn there for several weeks, was bound to give us a definite idea as to whether the English Fleet would feel called upon to interfere in this enterprise, or to take advantage of the absence of the ships to make a strong advance in the North Sea. In the latter case we should have had to take the risk of our remaining warships in the North Sea being able to ward off an attack that would probably aim at destroying the U-boat base at Wilhelmshaven, or the airship sheds on the coast. On the other hand, if the English Fleet had decided to make a demonstration on a large scale against the Baltic, we should have been forced either to abandon the enterprise in the east, or to oppose the English with very small forces in the west of the Baltic. But the English Fleet did not deem it desirable to pursue either course to divert us from the conquest of the Islands.

The fact that our Main Fleet was thus occupied presented a favourable opportunity for us to make an advance with light craft into the northern waters of the North Sea, since under the circumstances the enemy would least expect it. We, therefore, dispatched the light cruisers *Brummer* and *Bremse* to harry the merchant ships plying between Norway and England, or, should none be met with there, to extend the expedition to the west coast of the British Isles. This enterprise will be described later.

THE CAPTURE OF HELSINGFORS

Once again our Fleet had occasion to support our Army in the east, when, after the Peace of Brest-Litovsk, urgent cries for help against the Russian Red Guards were raised by the Finnish Government. A special division was formed under the command of Rear-Admiral Meurer, which consisted of the battleships *Westfalen* and *Rheinland* (to which the *Posen* was added later), a number of light cruisers, mine-sweepers, as well as barrier-breakers, ice-breakers, and outpost ships. They were to convoy seventeen steamers to the Finnish coast, and to establish a base for them on the Aaland Islands. The chief difficulty of the undertaking lay in the fact that there was so much ice.

On February 28 the voyage to the north was begun, and on March 5 the ships anchored off Eckerö on the Aaland Islands. One ice-breaker was lost owing to a mine explosion.

It turned out that the ice made it impossible for ships to approach the Finnish coast from the Aaland Islands at that time of the year, and the advance had to be made in a direct line from the south.

On April 3 our ships appeared before Russäro, the strongly-fortified island before the harbour of Hangö on the south-west coast of Finland. The Russians declined to oppose us, so that it was unnecessary to fight and demolish the fortifications; the landing of the Baltic Division could proceed without difficulty. From this point they set out on their march to Helsingfors. The warships were to penetrate into the harbour of Helsingfors from the sea; it had been a strongly fortified base of the Russian Fleet.

On April 12 the ships entered Helsingfors, and landed their troops under cover of their guns; there was heavy street fighting with the Red Guards in the town. At the threat of a bombardment, the latter ceased their resistance and capitulated, so that about 2,000 prisoners fell into the hands of the Navy. The taking of the city brought timely relief to an advance guard of troops which had penetrated into the town and was very hard pressed.

When the Baltic Division itself arrived, the Navy had to safeguard the lines of communication between Helsingfors and Reval. According to the treaty stipulations, the Russian Fleet retired into the inner Gulf of Finland to Kronstadt, and thus there was no longer any necessity for the presence of our battleships, as our Baltic light craft seemed to suffice for the assistance of the Baltic Division in their task of liberating Finland.

On April 11 the battleship *Rheinland*, which had remained near the Aaland Islands, ran on the rocks in a fog, when she was going to Danzig to coal; her situation at first seemed very grave, but the bad leak she had sprung was successfully stopped, and the ship was got off and taken into Kiel Harbour. The repairs were so extensive that the *Rheinland* was of no further use for war purposes.

As help to the Finns in their need could only be taken by sea, and as such help must be immediate if it was not to be too late to be of any avail, the liberation of Finland was only possible if we could succeed in overcoming the difficulties presented by the ice, which made operations by sea impossible. A further hindrance was that the battleships had to make a way for themselves, and that no previous sweeping of mines could take place. Admiral Meurer's energy succeeded in overcoming all the hindrances which were due in part to the ice conditions and in part to the difficulties of navigation in these rock-infested waters. The Navy regarded it as a particularly beautiful and elevating task to render timely help to the seafaring nation of the Finns.

CHAPTER XVII

OUR LIGHT CRAFT IN ACTION, AND ADVANCE OF OUR FLEET TO THE NORWEGIAN COAST

CRUISER ACTION ON NOV. 17, 1917

TO obtain information as to British mines and nets outside the belt of mines on the line Horns Reef—Terschelling, so-called test-trips were devised. The object of these test-trips was to ascertain the whereabouts of these barriers, and having done so to find means of circumventing them. Having, as a result of these test-trips, gained a clear idea of the situation of the various barriers (consisting of belts of mines), the next thing was to determine which of them should be cleared away. Every test-trip group comprised mine-seekers and sweepers with their tackle for finding mines, behind them went torpedo-boats with U-boat "kites," with which to locate nets; these were followed by barrier-breakers, and light cruisers with seaplanes for scouting. Heavy warships protected the test-trip groups on routes that were known to be free of mines.

Such a test-trip had been decided upon for November 17, 1917. Led by Rear-Admiral von Reuter, the 6th Mine-Sweeper Half-Flotilla, 2nd and 6th Auxiliary Mine-Sweeper Half-Flotillas, the 12th and 14th Torpedo-boat Half-Flotillas, Barrier-Breaking Division IV and the cruisers of Scouting Division II were to search from about the centre of the line Horns Reef—Terschelling in the direction north by west. Ships of Squadron IV, which was on outpost duty, were to be sent to cover the group. Squadron Commander Vice-Admiral Souchon chose for this task the *Kaiserin* and *Kaiser*, with the commander of the *Kaiserin*, Captain Grasshoff, in charge.

Rear-Admiral von Reuter ordered his group to assemble at 7 A.M. at a pre-arranged meeting point. The commander of the *Kaiserin* reported that at 7 A.M. he would lie west of Heligoland. Airship scouting was impossible, and the cruisers had been unable to take the seaplanes on board in good time because of the thick weather. Of the seaplane stations on land only Borkum was at first able to send out scouts. Towards 8 A.M. the test-trip was assembled at the point of departure, excepting the 2nd and 6th Auxiliary Mine-Sweeping Half-Flotillas.

Our Light Craft in Action

As the latter could only be a few thousand metres behind, the leader of Scouting Division II determined to fetch them up with his flagship *Königsberg*. He had just left his division when it was attacked from the N.W. by guns of large and medium calibre. The western horizon was very misty; the type of attacking ship was very hard to make out at first. In the east it was clearer; probably therefore our own ships showed up distinctly. The wind blew with a force 2—3 from the W.N.W.; the sea was slightly rough. The leader of Scouting Division II on board the *Königsberg* arrived. Scouting Division II, under the command of the senior officer, Captain Hildebrand in the *Nürnberg*, advanced against the enemy on a N.W. course, so as to protect the mine-sweepers. The torpedo-boats struck out N. and N.W. and put a smoke screen between the enemy and the mine-sweepers. "V 45," Lieutenant-Commander Lossman, making use of her favourable position, attacked the enemy at a distance of 40—60 hm. The mine-sweepers let go their tackle and steamed away to the east developing smoke-clouds.

With this the most urgent part of their work achieved, the cruisers and torpedo-boats under heavy enemy fire—range about 130 hm.—started on a south-easterly course, developing smoke and steam-clouds which made the screen between the enemy and the mine-sweepers denser. The enemy, with the exception of a few torpedo-boats, turned aside from the mine-sweepers in their way eastward and followed the more valuable cruisers. Owing to the smoke- and steam-clouds developed by the latter, he was obliged to steer towards the southern wing, that is to the windward, of our cruisers, so as to get a better chance of observation for his guns. These movements which, according to irreproachable observations and bearings, were carried out by hostile cruisers of the "Concord" class with a speed of 33 knots, increased the distance between them and the mine-sweepers. Visibility astern was, of course, very much reduced for our cruisers. The large enemy ships did not go beyond the windward edge of the smoke screen, as owing to the danger from mines they tried to keep within the limits of the waters through which we had passed. They were, therefore, only visible for a few seconds at a time; it was impossible to get absolutely reliable observations of their composition and strength. No doubt light craft were in advance on the windward side of the large enemy ships, apparently also on the lee side.

All took part in the firing. Our cruisers lay in the midst of

U

well-aimed salvos, of medium and heavy calibre. With great skill they avoided being hit by steering a zigzag course without damaging the effect of their own gun-fire. Our batteries replied energetically and with good results.

At 9.24 A.M. explosions resulting from our gun-fire occurred on two of the hostile battle-cruisers. One of them thereupon sheered off. About the same time our light cruiser *Pillau* forced an enemy destroyer that she had hit to retire from the fight. The leader of Scouting Squadron II hoped by going at full speed to separate the enemy light craft from the big ships, and so to get a chance to attack the former, but this hope was not fulfilled; the large ships were able to keep pace.

The U-boats of the Auxiliary Mine-Sweeping Flotilla had meanwhile steamed on in the direction E.S.E. At 8.50 A.M. they had a fight with the northern group of enemy destroyers at a range of 90 hm. After three hits had been observed on the destroyers, the enemy sheered off. Our U-boats again came under fire from 9.5 A.M. to 9.30 A.M., apparently from a leading torpedo-boat; after that they were no longer molested and returned to port. Several U-boats noticed that an English destroyer came to a standstill and that another drew alongside of it. This observation was confirmed later on by a seaplane which reported that it had seen one destroyer being towed by another.

The 6th Mine-Sweeping Half-Flotilla had steamed off to the east. It also came into conflict with the northern group of enemy destroyers at a range of 70 to 75 hm.; an advance of 3 destroyers brought the latter to within 10 hm. The English destroyers scored no hits; ours claimed one for certain. At 9.40 A.M. the enemy destroyers retired. Mine-Sweeping Flotilla VI then returned to port without any further molestation from the enemy. It is not clear why the enemy destroyers did not make better use of their superior armament and speed to destroy our weak mine-sweepers completely.

The fight of the cruisers, in loose echelon formation on a south-easterly course, brought them into the neighbourhood of the trawlers and the 2nd and 6th Mine-Sweeping Half-Flotilla, which at the beginning of the conflict had made off to the south-east at full speed.

The cruisers nearest to them, the *Nürnberg* and the *Pillau*, threw smoke bombs to protect them, and the 14th Torpedo-Boat Half-Flotilla also helped to envelop the mine-sweepers in smoke. The enemy destroyers, which had already come pretty near, sheered off

Our Light Craft in Action

from the smoke. The mine-sweepers steamed off in an E.S.E. direction and were not molested by the enemy. It is possible that the latter suspected poison gas in the smoke.

At 9.50 A.M. destroyers approached Scouting Division II to make a torpedo attack. Judging by bearings and distances, the attack was doomed to failure from the first. The enemy scored no hits. At the same time Admiral von Reuter ordered our torpedo-boats to attack. The boats advanced to the attack in a running fight, scattered as they were. It was not possible to collect for a closed attack owing to the speed at which the fight moved on. Altogether six torpedoes were fired; no hits were recorded with absolute certainty. At any rate the enemy cruisers turned off sharply for the time being, and in so doing unavoidably afforded our light cruisers a welcome alleviation. The *Königsberg* and the *Frankfurt* also fired torpedoes; no result was observed.

At 10.30 A.M. the battleships *Kaiserin* and *Kaiser* hove in sight. Admiral von Reuter tried by holding an easterly course to draw the enemy after him through the belts of English and German mines, so as to get him between our battleships and our cruisers. He would then only have been able to get away to the north and the north-west through the belt of mines. If he chose this route in preference to a retreat to the west he was pretty certain to suffer losses by striking mines. The battleships, which owing to the smoke and steam could not overlook the situation clearly, and did not rightly interpret the signals made by the *Königsberg*, steered on a N.W. course towards the approaching ships in action, unable at first to distinguish friend from foe. Scouting Division II then determined to try to join up with the battleships. The latter meanwhile had opened fire on the light cruisers of the "Concord" class. The *Kaiserin* quickly got the range, and a hit was observed on the leading cruiser. Thereupon the hostile ships sheered off. When Admiral von Reuter went to turn with the *Königsberg* and pursue the enemy along a north-west course, he was still under fire, and a shell hit the *Königsberg*, causing a serious bunker fire.

With this shot the firing suddenly ceased. The action was over. The enemy ran away at full speed to the N.W. In the meantime the *Hindenburg* and the *Moltke*, which on receipt of the news that an engagement was in progress had followed the other two battleships, had reached the scene of action; probably their appearance induced the enemy to break off the engagement. Our boats which started in pursuit did not succeed in getting into touch with the

307

enemy again. An advance with Torpedo-Boat Flotilla VII, undertaken the same night, met with no result either. Torpedo-Boat Flotilla II, which had advanced to the Hoofden the night before and was just returning thence to the Bight, could not be sent in chase owing to lack of fuel.

So far as could be ascertained at such a distance and with the smoke that was developed, the following ships were engaged on the enemy side: 4 battle-cruisers (2 "Lion" and 2 "Courageous"), and 6 to 8 light cruisers of the "Concord," "Caroline" and "Arethusa" classes, as well as 16 to 18 destroyers. According to seaplane observations, confirmed by other reports, behind these cruisers and outside the Horns Reef—Terschelling line there were other heavy fighting ships—at least one battleship squadron which, however, did not dare to enter the belt of mines, while the enemy cruisers kept in a straight line where our ships had passed and thus obtained some security from that danger.

The following hits were observed from our ships: five on the enemy battle-cruisers, six on the light cruisers, and seven on the destroyers. Our cruisers were hit by two heavy shells, one of which was a 38 cm., and by three 15 cm. It was remarkable what little damage the 38 cm. shell caused in the *Königsberg*. It passed through all three funnels of the ship, went through the upper deck into a coal bunker—the inner wall of which it burst; there it exploded and caused a fire. The fragments of this shell were picked up and its calibre determined. This proved to us that the English had built a new class of cruiser armed with a 38 cm. gun. The great speed of the ships was extraordinary. So far as the somewhat doubtful observations of our cruisers went, they had only two turrets, one fore and the other aft. The fact that a battle-cruiser felt obliged to sheer off on being hit by one of our light cruisers seems to indicate that its armour cannot have been very strong; probably weakened to allow of the high speed that was aimed at.

The losses on our side were: 21 killed, 10 seriously wounded, and 30 slightly wounded. The only ship that fell a victim to the enemy was the outpost steamer *Kedingen* which was stationed as a mark-ship at the point of departure of the test-trip. The English directed the fire of their 38 cm. guns on this little boat, so that the crew had to go overboard. She was captured undamaged by the English and carried off.

Our light cruisers amply fulfilled their duty of shielding the mine-sweeping groups and drawing fire upon themselves. Their relative strength, when compared with the enemy, unfortunately made it impossible for them to achieve a greater success, especially as the two battleships came to their support so late. This induced us in subsequent similar undertakings to make the support groups stronger and to send them forward, as far as the mine-fields would permit of such a course. The demands thus made upon the battle-ships of our outpost section increased considerably. The field of operation of the mine-sweepers extended 180 sea miles to the north and 140 miles to the west of the Jade. Work at such distant points was impossible without strong fighting support.

As a rule one-half of these support ships were placed immediately behind the mine-sweepers, and the remainder about 50 sea miles farther back. On days when air-scouting was possible, only half of the outpost-ships were required, but when air-scouting was limited all the outpost forces took part in the operations. In the neighbourhood of the Amrum Bank an anchorage was made secure from submarine attack and surrounded by nets. Here the support ships for the operations in the north could anchor, and thus avoid the long return journey to the Jade or the necessity of cruising about at night and burning unnecessary fuel. But this anchorage was not ready for use until the summer of 1918.

HOLDING UP CONVOYS

While the Fleet was busy with the conquest of the Baltic Islands the light cruisers *Brummer* and *Bremse* received orders to make a raid on the traffic route between Lerwick, in the Shetland Islands, and Bergen, the object being to inflict damage on English trade by surface craft as well as by U-boats. In the event of their encountering nothing there they were to push on at their own discretion to the west of the British Isles into the Atlantic, as far as their fuel supply would allow. These two cruisers had joined the Fleet in 1916 and had originally been constructed in German shipyards as mine-layers for the Russian Government; they were distinguished for high speed. Their engines were adapted for coal or oil fuel. They carried a 15-cm. gun. The mine-laying apparatus, with the exception of the dropping-gear, had been removed so as not to hinder the ships on their cruises. While our other light cruisers could accommodate but 120 mines on deck, when they carried them for a special expedition, the *Brummer* and the *Bremse* were capable of taking thrice that number. The addition of these two cruisers was a very welcome reinforcement, and made it possible to form two scouting groups of light cruisers (II and IV) with modern ships of approximately the same speed, after the alterations of the other light cruisers had been completed and they had received a 15-cm. gun instead of their 10.5-cm. guns, which were too weak.

It was known that neutral merchant vessels assembled in convoys to travel under the protection of English warships, and therefore they might be regarded as enemy vessels, since they openly claimed English protection so as to benefit the enemy and consequently to injure us. Interruption of this traffic was intended to heighten the effect of the U-boat campaign. Apart from depriving the enemy of the supplies he awaited, it would place him under the necessity of affording better protection to the neutral shipping placed at his service, for which more warships would be required; these, again, would have to be taken from among those occupied in the war on U-boats. We might also anticipate that the success of such attacks would have a terrorising influence.

On putting out to sea the cruisers were delayed for a day, because the mine-sweepers who accompanied them found mines in their path, but at dawn on October 17, 1917, they lay in the middle of the fairway Lerwick—Bergen, and before day broke they encountered a convoy of ten steamers under the protection of two or three warships. At the head of the formation, which was in a double row, was the destroyer *Strongbow,* and when she recognised our cruisers as enemy ships she advanced smartly to the attack and was sunk after a few shots had been exchanged.

The steamers had stopped when they realised the position they were in, and began to lower boats in which the crews might find safety. A second British destroyer, the *Mary Rose,* had first made off to the north when the fight began, but changed her mind and returned, after about 20 minutes, to the ships under her protection. She also attacked our cruisers and was sunk after a short fight. The steamers were then sunk as they passed at a short distance, which enabled the shots to be placed on the water line. As two of the steamers had been able to get away in time on noticing the attack, the care of the crews in the boats could be left to them, for our cruisers had to consider their own safety on the long return journey. A further extension of the cruise offered no prospect of success after this incident.

It was to be foreseen that this action would occasion a great outcry among those that had suffered, if only to divert attention from the humiliating fact that German cruisers had appeared in the Northern waters supposed to be completely under English control. If in this starvation war, introduced by the English, the neutrals worked against the German nation and so openly assisted the enemy as to place themselves under the protection of his warships, they must take the consequences of their action. To what an extent they regarded themselves as being on the side of the enemy is shown by the fact that some of these neutral steamers carried guns on the forecastle which they did not hesitate to use.

If England wanted to demand the right to enjoy undisturbed supplies, thanks to the complaisance of the neutrals, or to the pressure brought to bear on them, no one could expect us to look on with folded hands until English sea power had completed its work of destroying our nation by starvation. The counter-measures which this necessitated must recoil upon England as the originator of this form of warfare.

The effect of such action had to be heightened by a speedy repetition of a similar attack. The next time Torpedo-Boat Flotilla II was chosen, which comprised our biggest and fastest torpedo-boats. A half-flotilla was to attack the convoy traffic near the English coast in the so-called "war channel," while at the same time the other half-flotilla was to go to the Bergen—Lerwick route. Flotilla II (Commander Heinecke), accompanied by the light cruiser *Emden* (the ship substituted for the one of the same name that Captain von Müller had commanded), left early in the morning of December 11, at a speed of 19 knots. The weather was clear, sea smooth. At 4 P.M. the half-flotillas parted at the north-east end of the Dogger Bank, and the *Emden* remained behind.

The 3rd Half-Flotilla went north, the 4th steered for a point on the English coast 25 sea miles north of Newcastle. At 6 P.M. a wireless message was received that a convoy with destroyers would leave the Firth of Forth for the south between 8 and 11 P.M. On account of this message the leader determined to go up the "war channel" to the north, about as far as Berwick, so as to meet the enemy on this part of the route between 3 A.M. and 6 A.M. According to other English wireless messages received, there were in the Firth of Forth 8 British cruisers, in the Tyne some destroyers, and in the Humber 2 destroyers with various guard-boats. This, however, did not hinder the leader of the flotilla from pursuing his purpose. Towards 2.30 P.M. on December 12, 1917, before the flotilla had turned into the "war channel," a steamer of about 3,000 tons was sighted coming at a distance of about 25 nautical miles from the coast; it was sunk by a torpedo. The crew of the steamer took to their boats. As the flotilla approached closer to the coast the beacons they expected to see were not visible, so that they could not find their way between the Farne Islands and the land. To have gone out to sea and so round the islands would have meant missing the convoy, so the half-flotilla turned southwards in the direction of the mouth of the Tyne. Although the course ran only 3 to 4 nautical miles from the coast, nothing was to be seen of the land or any towns. It was very misty near the coast. At 4.45 A.M. a steamer with very great draught came into sight on the port bow; her size was estimated at 5,000 tons. This ship was steering on a southerly course down the "war channel," and was sunk by a torpedo; the crew took to the boats. A quarter of an hour later four small steamers came in sight; obviously they were the convoy boats which had already indicated their presence by wireless messages, and were now on the point of entering Tynemouth. Two of them were destroyed by gunfire, the other two escaped because our torpedo-boats were looking around for larger steamers or destroyers that might be in the neighbourhood. As nothing further was found, the boats started on their return journey at 6 A.M. At 5.15 P.M. they rejoined the *Emden,* which had waited at sea for the flotilla.

The half-flotilla under Lieutenant-Commander Hans Holbe had continued on a northerly course after separating from the others on the previous day. The farther north they went the worse the weather became. Towards 10 P.M. there was a heavy swell and a strong freshening wind from the south. The next morning, at 4 o'clock, speed had to be reduced first to 15 and then to 12 knots, because heavy seas came up from the north-west. It was impossible to fire a gun or a torpedo. The leader of the half-flotilla had to give up his plan and steered towards Udsire on the Norwegian coast, so as to be able to fix his position and then to try and catch a convoy announced from Drammen. At 7 A.M. he sighted Udsire. As the barometer fell no lower and the seas seemed to be decreasing, he once more turned upon a northerly course, which, however, had to be abandoned again at 11 A.M., because in the sea then running he could only make a speed of 9 knots. The boats, therefore, once more turned south, intending to stay out of sight of land by day and to approach the coast by night, expecting to meet some merchantmen there. In the course of the morning one boat developed a leakage in the condenser. But the commander of the half-flotilla decided to keep the boat with him and reduce the speed of all his boats to 25 knots, preferring this to sending the boat back home alone from such a great distance.

While he was steaming along on a southerly course, at 12.30 P.M. a convoy of six steamers came into sight which was protected by two destroyers and four trawlers. It was going from Lerwick to Norway on an easterly course.

The destroyer *Partridge*, which was ahead of the formation on the port side, steamed towards our half-flotilla and came under fire at 1 o'clock. The destroyer *Pellew*, which was on the starboard side, had steamed ahead full speed, and the *Partridge* joined her.

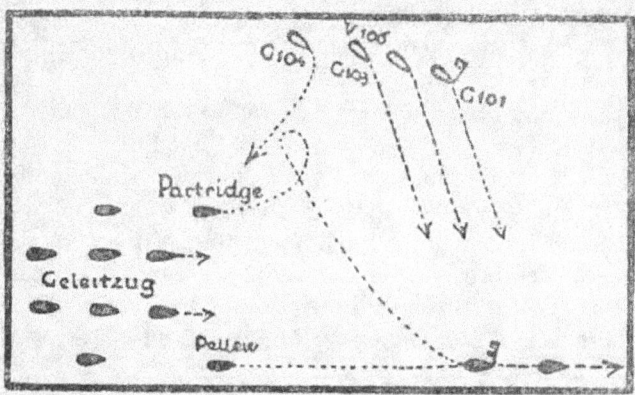

The Attack on the Convoy

The British destroyers left the convoy and the four trawlers to their fate; probably with the idea of drawing our boats away from the latter, and of fighting them. The fire of the British destroyers was not very effective. The fight was carried on at a distance of 50 hm. till the *Partridge*, after a shot in her main steam-pipes, could not continue. She tried to carry on the fight with her torpedoes, but one torpedo stuck in the tube which had been damaged by gunfire; a second torpedo, fired at short range at our boat "V 100," hit, for the shock was distinctly felt in the boat, but it did not explode. While three boats of our half-flotilla took up the fight with the two destroyers, the fourth boat (which could only travel at 25 knots) was sent to destroy the convoy. The destroyer *Pellew*, pursued by the leader of the half-flotilla, succeeded—thanks to her superior speed—in getting out of sight in a squall of rain, and escaped to the land. Four officers and forty-eight men of the *Partridge* and the trawlers, which were all sunk, were taken on board as prisoners of war, as well as 23 civilians. Our casualties were three wounded. The convoy consisted of one English, two Swedish, two Norwegian and one Danish merchant steamer. The shipwrecked men of the latter refused to come on board our boats; of the others some consented to come on board, and then the steamers were all sunk. The whole affair was over in three-quarters of an hour. Owing to the high seas, it cost a lot of trouble to get the English on board when they were floating about in the water, having taken refuge on rafts. The half-flotilla then started on its return journey round Skagen, as a weather report announced stormy weather in the North Sea, and so reached Kiel Harbour. This repeated interference with merchant traffic, which had shown the insufficiency of the protection afforded by English convoys, had the desired result, and compelled the employment of stronger forces.

Information obtained by U-boats was to the effect that American ships were pressed into service for this purpose; they were recognisable by their masts. This confirmed statements received from other quarters that the English Fleet was receiving support from the Americans in the War Zone of the North Sea. Thus there was little further prospect of our light craft being able to destroy any more convoys. Stronger forces would have to be employed for this purpose. This led to an expedition of the Fleet in April, 1918.

AN EXPEDITION OF THE HEINECKE TORPEDO-BOAT FLOTILLA

In February, 1918, Flotilla II was confronted with a new problem, which it solved brilliantly. The Naval Corps in Flanders had sent a request to the commanders of the Fleet begging them to destroy the English light-barrier which had just been instituted between Dover and Calais. In the last months the enemy had, with much expenditure of material, tried to make these Straits impassable for our U-boats. According to the reports of the boats, there were net barriers between Cape Grisnez and Folkestone, and farther south between Boulogne and Dungeness. The nets were guarded by a large number of vessels which, by means of searchlights and magnesium lights, formed a very effective light-barrier all night long. This made it very much more difficult for our U-boats to get through unmolested, and the Straits were actually almost impassable. The forces in Flanders alone were not able to deal a sufficiently effective blow to this Anglo-French barrier to the

The Heinecke Torpedo-boat Flotilla

Channel. For this undertaking the commanders of the Fleet chose the strong boats of the Heinecke Flotilla, which was sent direct from the German Bight without first touching the coast of Flanders, so as to make sure of surprising the enemy.

On the day of the enterprise Flotilla II was to be off Haaks Lightship at 5.30 P.M. and thence proceed in close formation as far as the northern end of the Channel by the Sandettie Bank; there the two half-flotillas were to separate; one led by the Flotilla Commander was to attack the barrier west of Varne Bank, and the other was to attack east of that point. When the attack was over they were to enter Zeebrugge Harbour, take in a fresh fuel supply, and start the return journey to the German Bight the same night.

Owing to bad weather the enterprise, originally planned for February 7, was postponed to the 13th. In the meanwhile the route the boats were to have followed had been made impracticable by new English mine-fields, and they had to go close by the Frisian Islands, thereby running the risk of being seen early in the evening from Dutch territory and their advance being reported. Consequently the misty weather on February 13 was not unwelcome. Flotilla II managed to pass along the mine-swept route at Terschelling with the help of the land, without having been able to see any landmarks; but when it arrived off Haaks Lightship it had to give up the attempt because of the fog; the boats would have had to travel at high speed to reach their goal in time, and this was impossible in the foggy weather. The flotilla anchored during the night north of Norderney. The next day, February 14, it started again in very clear weather. So as not to betray his real goal, the Flotilla Commander set out from Helder on a westerly course; when out of sight of land he steered south, and, after darkness had fallen, down the Dutch coast as far as the Schouven Bank. At the Hook of Holland one of the boats had to be sent back to the German Bight owing to defects in the condenser. In the night of the 15th at 12.30 A.M. the two half-flotillas separated according to plan north-east of the Sandettie Bank. The group led by Captain Heinecke was to circumvent the first and more northerly barrier near the English coast, and begin by attacking the southern barrier presumed to be off Dungeness, and then on his return roll up the northern barrier from the Varne Bank to Folkestone. The latter, being a light-barrier, could be seen from far off. On approaching it, it became clear that it consisted of a large number of craft, anchored or moored to buoys, which were placed right across the fairway, not in one line, but in echelon in a broad band. These boats lit up the fairway all the time with searchlights, and from time to time, about every quarter of an hour, they threw magnesium lights overboard which, floating down the tide for minutes at a time, lit up the vicinity for a distance of two or three miles, so that it was almost as light as day. In and out among these a lot of boats moved without lights, armed trawlers, submarine chasers and motor-boats, to attack any U-boat that might come. At the north-west end of the barrier a searchlight, apparently placed on land between Dover and Folkestone, threw a steady beam of light in the cross Channel direction. In these circumstances it was impossible to get round the barrier, and the Flotilla Commander determined to make a direct attack. He first made for a large boat placed about the middle of the barrier with a specially bright revolving search-light; this he sank from a distance of 300 m. It was an old cruiser, or a special boat of the "Arabis" type. After this the group first swept round to the north-west and then went more slowly along the barrier in a more or less south-easterly direction. In a short time they sank 13 of these guardships, including a U-boat chaser with the number "1113," a small torpedo-boat and two motor-boats, one of which had come up in order to fire a torpedo; these were all sunk at close range by gunfire.

The enemy was taken completely by surprise. Several of the boats sounded their sirens, clearly under the impression that they were being attacked in error by their own ships. No warning was given, and a considerable time elapsed after we had opened fire before all lights were put out. This may have been due to the fact that the big ship that was sunk first had been in command of the whole, or else the ships on guard may have been used to hearing gunfire owing to the frequent fights with U-boats. An attempt to take prisoners had to be abandoned, because owing to the swift tide it proved too dangerous for our ships to go alongside the sinking boats that were in part moored to buoys. The whole affair lasted from about 1.30 A.M. till about 2.30 A.M. Owing to the lateness of the hour it was out of the question to attack the other barrier supposed to lie farther south, from which, however, no lights or searchlights were seen, and so the return journey was begun.

Meanwhile the other half-flotilla had turned towards the southern end of the barrier, and first made for Cape Grisnez. Again one of the boats developed a leakage in the condenser, but the commander of the half-flotilla could not dismiss the boat and had to reduce

The Heinecke Torpedo-boat Flotilla

the speed of the other boats to that of the defective one. Off Calais the group encountered the first guardship, lying close to the coast, a large, armed trawler, and, taking her by surprise, sank her by gunfire. Steering west, they met a number of other boats which were using searchlights and magnesium lights. In several cases the supply of magnesium lights on the guardships caught fire owing to the shots. In this part of the barrier, too, it was some time before the boats realised that the enemy was among them, and retired to the west. Altogether this torpedo-boat half-flotilla sank twelve armed guardships and two motor-boats.

At 2.40 A.M. the half-flotilla started upon the return journey. At 3.30 A.M. the stern lights of six English destroyers were sighted ahead. Owing to his unfavourable position with regard to the enemy and the reduced speed of the one boat, which left him with only two boats that were quite intact, the commander of the half-flotilla was forced to avoid a fight. He turned off and did not reply to the enemy's signal. The latter at first followed in the wake of the half-flotilla, but after altering his course a few times was lost to view. On making for Zeebrugge the torpedo-boat "G 102" struck a mine about 12 nautical miles from the harbour entrance; two compartments filled with water, but the boat was able to reach the harbour without assistance. Three men were killed through this mishap. These were the only casualties of the expedition.

After replenishing their supply of oil fuel in Zeebrugge the flotilla began its return journey in the evening of the same day and reached home without further incident. The damaged boat was temporarily repaired in Flanders, and followed a few days later. The flotilla's success was due to the completeness of the surprise. Besides the direct damage inflicted on the enemy by the sinking of so many boats that were of value to him, we accomplished our aim of breaking the barrier across the Calais—Dover Straits through which our U-boats were again able to pass for the time being. A scouting trip carried out the following day by the torpedo-boats of the Naval Corps showed that the guard had been completely withdrawn.

The demands made on the skill of the officers commanding these boats were very great, as it was difficult to distinguish things clearly because of the gunfire, and particularly because of the smoke on the water from the magnesium lights. The gun-layers did excellent work in shooting down the fast motor-boats which, owing to their speed, could only be discerned at the last moment, but were always

317

knocked out by the first shot. It was a great help to the expedition to be able to break the return journey by running into Zeebrugge, because otherwise the voyage would have had to be made by daylight, and in that case the English would probably have made an attempt to cut our boats off.

When the portion of the Fleet that had been sent east had returned from the conquest of the Baltic Islands, some weeks elapsed before the ships and torpedo-boats had had the damage repaired that they had suffered from mines and from running aground. The winter months brought no change in the activities of the Fleet, which were directed towards supporting the U-boat campaign.

In the spring of 1918, when our army was attacking in the west, English interest was bound to centre in the Channel. Through agents, through the aeroplane service in Flanders, and through following the enemy's wireless messages, we ascertained that he had materially reinforced the warships protecting his transports, and that large ships had been sent to the Channel, and parts of the crews of the Grand Fleet had been sent to reinforce those of the light craft in the Channel. On the other hand, the enemy had carefully improved the convoy traffic between England and Norway since the successful raids of the *Brummer* and the *Bremse,* and of the boats of Torpedo-Boat Flotilla II. Our U-boats had learnt that the steamers were assembled there in large convoys, strongly protected by first-class battleships, cruisers and destroyers. A successful attack on such a convoy would not only result in the sinking of much tonnage, but would be a great military success, and would bring welcome relief to the U-boats operating in the Channel and round England, for it would force the English to send more warships to the northern waters. The convoys could not be touched by light craft. But the battle-cruisers could probably, according to information received, deal with all exigencies likely to arise if they could have the necessary support from the battleship squadrons.

So far as could be made out convoys mostly travelled at the beginning and middle of the week. Consequently Wednesday, April 24, was chosen for the attack. A necessary condition for success was that our intentions should be kept secret. It was enjoined upon the officers in command of the subordinate groups to use their wireless as sparingly as possible during the expedition,

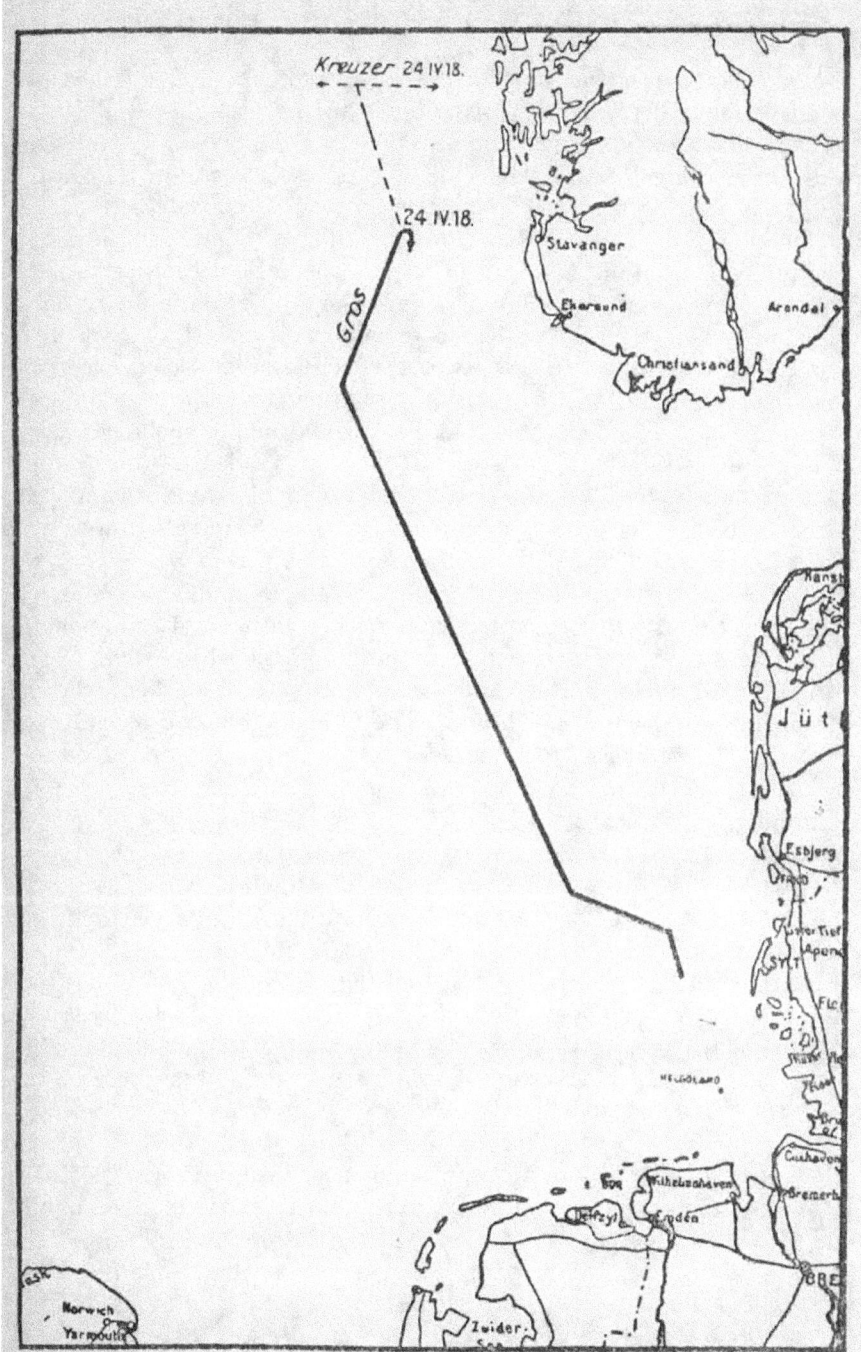

EXPEDITION TO THE NORWEGIAN COAST

which was to extend beyond the Skagerrak up to the Norwegian coast. On the pretext of manœuvres in the Heligoland Bight all warships at our disposal were assembled on the evening of the 22nd in the Schillig Roads. Here the officers in command of the various groups were informed of our intentions and received their orders. The plan was to attack the convoy with the battle-cruisers, the light cruisers of Scouting Division II, and Torpedo-Boat Flotilla II under the leadership of the officer commanding the Scouting Divisions, Admiral von Hipper, while the remainder of the ships took up a position from which, in case of need, effective support could be given to the cruisers. All other flotillas were to remain with the main body of the Fleet. Torpedo-Boat Flotilla V could not be included, as its radius of action was too small. The commander of this flotilla, Commander von Tyszka, was entrusted with the conduct and protection of the convoy service through the mine-fields south-west and west of Horns Reef.

To ensure safety of progress through the mine-fields in preparation for this enterprise, protective barriers had been placed about 70 sea miles west of Horns Reef, running from north to south. The area between Horns Reef and this protective barrier was to be the starting-point of the expedition. The U-boats that had recently put to sea had received orders to seek opportunities for attack off the Firth of Forth and to report all warships and convoys that were sighted.

On the 23rd at 6 A.M. the various groups put to sea, Admiral von Hipper leading with the Scouting Divisions I and II, with the Second Leader of the torpedo-boats and Torpedo-Boat Flotilla II; following him came the main body of the Fleet in the following order: Scouting Division IV, Squadron III, the Flagship of the Fleet, the First Leader of torpedo-boats, Squadron I, Squadron IV, and with the main body Torpedo Flotillas I, VI, VII and IX. Immediately after they left the Jade a heavy fog descended. As far as List the way was clear; from there it led through enemy mine-fields; to get through these it was necessary for the Fleet to be accompanied by mine-sweepers, and therefore a certain amount of visibility was needful—at least two miles. At first we were able to proceed at 14 knots. But when, at 11.30 A.M., we reached the entrance to the mine-field and visibility was only 100 metres, we had to anchor. Half an hour later it cleared up; one could see three to four nautical miles, and the expedition could proceed. The journey through the mine-fields passed off without a hitch. When darkness fell the boundary had been reached, and the mine-sweepers could be dismissed. The poor visibility had so far favoured the enterprise. The enemy line of submarines on guard round the German Bight seems to have been broken through, if indeed it was occupied at all.

During the night it cleared up; daybreak brought fine, clear weather. At 8 A.M. the *Moltke* reported to the High Sea Commander: "Grave damage, speed four knots, position about 40 sea miles W.S.W. of Stavanger." All haste was made to reach the scene of the accident; the *Strassburg*, the foremost ship in the line of advance, was detached to the *Moltke*, and the battleship *Oldenburg* made ready to tow. At 10.40 A.M. the *Moltke* was sighted; soon after von Hipper appeared from the N.W. with his two Scouting Divisions. He had detached the *Moltke* at 6 A.M. to the main body of the Fleet. At that time she could still do 13 knots. He had not received the message that she was reduced to four knots. When towards 9 A.M. he received the news that the *Moltke* could not move and that the Flagship had not made out the signal—which, however, was a mistake—he decided to go to her assistance himself. He sent no report to the main body of the Fleet owing to the orders that the use of wireless messages should be reduced as much as possible. He had the more reason for this course because when he turned he was already in the northern part of the convoy route, and thanks to the clear weather he could see that for the time being nothing was in sight, and that any approaching convoys would not escape him if he made a fresh advance later. As the *Moltke* had now been taken charge of by the main body the Admiral received orders to advance again to the north. On this second occasion he searched the convoy track as far as the 60th degree of latitude but sighted nothing.

At about 11.45 A.M. the *Moltke* was taken in tow by the *Oldenburg*. The manœuvre was carried out without a hitch in the shortest possible time. The main body of the Fleet with these two ships then set out on the return journey; their speed was 10 knots. There were two routes open to us; the one led through the Kattegat, the other straight into the German Bight. By choosing the former the Fleet would presumably have avoided a meeting with the English Fleet which had time to come up and oppose us, as we could only go at a slow speed in order not to leave the *Moltke* in the lurch. But the road through the Kattegat was very roundabout, and in addition the passage through the Belt would have been very difficult for the damaged ship, and in order to protect the tow all our ships would have had to return through the Little Belt. This was undesirable for two reasons, firstly, on account of the Danes, and secondly because it might provoke the English to lay mines in the Kattegat. This latter proceeding might be very unpleasant for our U-boats, and I decided, therefore, to return through the North Sea into the Bight in spite of the possibility of being attacked by superior forces.

Meanwhile the following condition of affairs had been discovered on board the *Moltke*. The inner propeller on the starboard side had been flung off (the ship had four propeller shafts); the turbine had raced, and before the machinery for stopping it could act the training wheel had flown to pieces. Fragments of the wheel had penetrated the discharge pipe of the auxiliary condenser, several steam exhaust pipes, and the deck leading to the main switch-room. The central engine-room and the main switch-room were immediately flooded owing to the damage to the auxiliary condenser, while the wing engine-room made water rapidly. Salt water penetrated into the boilers, and the engines gradually ceased to work. Through a curious chain of circumstances an accident to a propeller, slight enough in itself, had brought the ship completely to a stand, so that it was powerless to move. Two thousand tons of water had flowed into the ship before a diver succeeded at length in closing the valves which controlled the flow of water in and out of the auxiliary condenser. It was not till then that they got the water under control. In the afternoon the port engines were able to run at half speed; but for the time being there was no guarantee that they would continue to run. The ship would have to be towed right into the Bight, and the highest speed attainable by the tow was 11 knots. At this rate of progress we could not reach the belt of mines west of Horns Reef before dawn the next day.

Information received from the Naval Staff at 2 P.M. concerning the times of arrival and departure of convoys indicated that we had not been lucky in our choice of a day to attack them. Apparently the convoys from England to Norway had crossed the North Sea on the 23rd.

At 6.30 P.M. we received a wireless message from a U-boat that eleven enemy cruisers were about 80 miles behind us. But probably the U-boat had mistaken the cruisers that were following us under Admiral von Hipper for those of the enemy.

At 8.50 P.M. the towing cable of the *Oldenburg* broke, which entailed a delay of an hour. For the night the tow was left at the end of the line. At 11 P.M. Admiral von Hipper had approached to within 30 nautical miles of the main Fleet. At dawn all the ships were together. The enemy was nowhere to be seen. The journey through the belt of mines was accomplished according to plan. Mine-sweepers met and convoyed the Fleet back in the same manner as on the outward journey. One mine-sweeper, "M 67," struck a mine and sank; most of the crew were saved.

Off List the *Moltke* was cast loose, and was able to proceed at a speed of 15 knots. About an hour after she had been cast loose, at 7.50 P.M., she was attacked by a submarine 40 nautical miles north of Heligoland and was hit amidships on the port side. She could not avoid the torpedo, but was able to turn towards its course so that it struck at a very acute angle. The injury did not prevent the ship from entering the Jade under her own steam.

Unfortunately the expedition did not meet with the success hoped for. The opportunity of joining issue with our Fleet was not made use of by the enemy, although by the wireless messages which had to be sent owing to the accident to the *Moltke* he must have known of the presence of our ships. The bringing in of the *Moltke* under such unfavourable conditions of sea and weather as arose during the night of the return journey was an eminent military achievement, especially the part played by the *Oldenburg* (Commander, Captain Lohlen) which towed her, and the work done in stopping the leak by the men on board the *Moltke* deserves great praise.

This expedition was unfortunately the last which the Fleet was able to undertake.

CHAPTER XVIII

THE NAVY COMMAND

AT the end of June, 1918, Admiral von Müller, the Chief of the
Emperor's Naval Cabinet, informed me that Admiral von Holtzendorff's state of health made it improbable that he would be able to
hold the post of Chief of the Admiralty Staff much longer. In this
event His Majesty had designated me as his probable successor.

This information released me from the obligation that had
hitherto prevented me from suggesting a change of organisation
in the department which had directed the conduct of the war at sea.
The system was a failure, was not very popular in the Navy, and
our success was less than we had a right to expect. I could not very
well recommend myself as head of this department, all the more so
as the command of the Fleet involved personal danger, and I did not
care to avoid this by getting a position on land. Even the very
frank discussions which had taken place between the Chief of the
Naval Staff and myself had not resulted in the full satisfaction
of the demands of the Fleet. My personal relations with Admiral
von Holtzendorff enabled me to speak to him without reserve. We
had grown pretty intimate by serving together on the same ships at
different times. We were thrown together at sea for the first time in
1884—86 on board the cruiser *Bismarck*, the flagship of Rear-Admiral
von Knorr, when we went to West Africa, East Africa and the South
Seas to visit our colonies there. After that I was navigating officer
in 1895—96 in the cruiser *Prinz Wilhelm*, which Holtzendorff, a
commander at the time, commanded on a cruise to the Far East.
Later on he offered me the post of Chief of the Staff of the High Sea
Fleet, which I held for two years, 1909—11, under his command as
Commander-in-Chief of the Fleet. On all these occasions I had
learnt to appreciate his personality and his capacity as a leader.
For this reason I was grieved at the particular cause for the change,
but as the spell was broken I urged the Chief of the Naval Cabinet
to accomplish it in any case.

I had no occasion to complain of undue influence or limitation
of the Fleet by the Chief of the Naval Staff. But his position
was not clear; he seemed to us to yield too much to political pressure.

324

The Navy Command

The conduct of the U-boat campaign was typical of it. Even at this moment there were serious differences of opinion as to the way it should be carried on. The forces of the Navy were scattered over the various theatres of war, and the Commanders of the Fleet could not see any necessity for this. The Fleet formed a sort of reservoir which was to satisfy all demands for personnel. Naturally there was great opposition to any withdrawal of personnel from the Navy, unless it was clearly conducive to the main aim of the war. But that aim could only be achieved by the Fleet and the U-boats, and the Commander-in-Chief of the Fleet felt himself responsible for this.

There was no post of command superior to his, where full responsibility could be taken for the success of the conduct of war at sea. For the Naval Staff was not a Supreme Command, but an organ of the Emperor as the Supreme War Lord, which could not be bothered with details of the conduct of the war. The relation of the Naval Staff to the Navy was not the same as that of the Supreme Army Command to the Army on land. If, for instance, a plan of campaign in Roumania is carried out successfully, that is essentially to the credit of the Supreme Army Command, for it correctly estimated the strength and capability of troops and leaders, and set them a task proportionate to their abilities. In the war at sea the Naval Staff apportioned the existing ships and boats to the different fields of operation—the Baltic, the North Sea, or Flanders, the Mediterranean or foreign parts, and had to leave the officers in command there to act independently in accordance with their general instructions. On land the Supreme Command permanently controlled the war operations; this was not the case at sea. If the Fleet had been defeated in battle, no one would have dreamt of making the Naval Staff responsible, but only the Commander-in-Chief of the Fleet. But there was need of some body which should regulate the distribution of forces with a view to some definite end, and not leave the success of naval activities to the individual admirals in command in the different theatres of war.

The U-boat campaign had further complicated matters, because all independent officers in command had U-boats assigned to them, and the Chief of the Naval Staff had even placed certain of them, e.g. the U-cruisers, under his own immediate command. There was need of exchange among the different groups. The development ought to be regulated on uniform lines, and the experiences gained by the individual commanders in their boats, including those who specialised on the technical side, ought to be made of benefit to all. Finally, the personnel of all the new boats, at any rate all the officers and petty officers, had to be drawn from the Fleet.

That meant that the Chief of the Naval Staff must be included in the number of those commanders who were directly responsible for the conduct of the war. We felt the lack of a Supreme Command whose orders must be unhesitatingly obeyed. Our organisation in peace time had not foreseen this. In the year 1899 the Supreme Command of the Navy had been done away with, because at that time two powerful authorities, generally pulling in different directions, were detrimental to the development and building up of the Navy. The Secretary of State, von Tirpitz, did not feel able successfully to carry through the policy requisite for the steady development and growth of the Fleet unless it corresponded in every particular with his own convictions. The result was that the Naval Staff, which was all that was left untouched when the Supreme Command of the Navy was abolished, had been thrown into the shade, and the men appointed as Chiefs of the Staff were not for the most part such as would, in case of war, have the authority of chosen leaders, who had proved their ability as commanders of the Fleet.

When the Supreme Command ceased to exist the commanders of the Fleet demanded more and more independence. They did not pay any attention to strategic questions in peace time; tactics and development gave full occupation to their activities. The Fleet commanders' chief responsibility in the war lay in the apportionment of the most important units of the sea forces, for the aim of naval warfare is to deal the enemy Fleet a destructive blow. Success depends mainly on the skill of the leader. He must be thoroughly familiar with the handling and the capabilities of his weapon—the Fleet. How to bring about the encounter with the enemy must be left to him. Neither the place nor the time can be fixed beforehand. For in contradistinction to the war on land, the position and strength of the enemy are unknown.

Consequently it was thought that more or less indefinite general directions would suffice, which the Naval Staff had to suggest and transmit to the Fleet as orders from the Emperor, based on the Staff's strategic considerations. This had been a mistake. The organisation which had appeared useful for the building up of the Fleet in peace time hampered the ability of the Fleet in war. The

The Navy Command

war at sea grew too extensive to be carried on under the personal guidance of the Emperor, as it should have been in view of the relation of the Supreme Naval authorities to one another. Politics, technical matters, and strategy were all closely connected in this. The war against commerce also, which we had to adopt, influenced our relations with the neutrals. Technical considerations were involved in the decision as to whether we were to build submarine or surface ships, but this again was dependent on the course of our Naval strategy.

When it was realised that England did not intend to put matters to the test in a battle, then the time had come to institute a Supreme Command; all the more so when, towards the end of 1914, the views of the Commanders of the Fleet, the Naval Staff and the Secretary of State upon the course we should pursue were all divergent. Energetic measures should then have been taken to provide the Navy with the leadership it required. Grand-Admiral von Tirpitz himself was the most suitable person, for the Fleet would have willingly subordinated itself to him, although he lacked actual experience in handling it. That, however, was not a point of the greatest importance, as the Fleet Commander had that experience. The point was to co-ordinate and make use of all the forces which could contribute to the achievement of the Navy's aim.

The fact that the Grand Admiral was not appointed Supreme Commander of the Navy was no doubt in part due to the differences between him and the Chancellor. These grew more acute owing to our vacillating policy in the U-boat campaign. When Tirpitz was no longer allowed to exercise his influence in all-important questions touching the conduct of the war, and he was not consulted as to the decision with regard to the U-boat campaign in March, 1916, he, who had worked so admirably in organising our Fleet, felt compelled to resign.

As the war was prolonged it became more difficult to provide personnel and material for all the new exigencies; our warfare extended to far distant parts, and there was a danger of diluting the forces collected in the Fleet. The harder our task became, the more the difficulties that were put in our way. There was delay in the repair of ships and U-boats, in the delivery of new vessels, in the fulfilment of urgent demands and improvements. The Auxiliary Service Law was not calculated to increase the power of production of the workpeople, and this also suffered from the deterioration in food. It cost endless trouble to obtain from the

Army Command technical workmen who were badly needed. Naturally urgent Army needs had the preference. But convincing representation of the needs of the Navy might have met with success in many cases, for goodwill and understanding were certainly not lacking in the sister service.

Between the Admiral in command of the Naval Corps, the officer in Supreme Command of the Baltic, and the Fleet, absolute understanding and the most friendly spirit prevailed whenever help was asked for. But that was a lengthy way of arranging matters, and was an insufficient substitute for a Supreme Command that could overlook the whole situation and give orders accordingly.

The gradual decline in the monthly sinkings accomplished by the U-boats filled one with anxiety. Many a U-boat with a splendid and experienced commander did not return. The new commanders had to gain experience under considerably less favourable conditions.

Day by day the commanders of the Fleet noted down the positions of every single U-boat; its departure and return were followed with care and suspense. All our thoughts centred on finding ways and means to keep up the standard of their achievements and to increase them. There was never a day when we were at sea that the commanders of the Fleet did not discuss this with the officer in command of the U-boats and his Staff of picked professional men. We felt that we were responsible for the attainment of such an end to the war as had been promised to the German people, and that we could achieve it by this means alone. The Fleet was animated by one sole idea—we must and will succeed. Every single vessel, battleship, torpedo-boat, minesweeper, cruiser and airship, with their crews—all were permeated with the gravity and importance of this task which I impressed on officers and men on every occasion. New forces must be found which would undertake to complete the work, which threatened to be a failure when handled as hitherto by the Naval Staff and the Naval Cabinet.

A change of Secretary of State (Admiral von Capelle) seemed also very desirable. It was not to be expected that a man who was convinced that he had done all that was humanly possible would pledge himself, without reserve, to carry out new proposals which would bring him into opposition with his previous conduct of affairs.

It had taken six months of urging, from July, 1917, to December of the same year, before a central organisation for U-boats—the U-boat Office, demanded by the Fleet Command—had been instituted. Such was the delay in carrying out demands or suggestions

The Navy Command

as the case might be—whether they referred to personnel, armament, or technical matters pertaining to ship-building and so on; the working of the different departments was inadequate for the needs of the times.

Though I was bound to the Fleet by such close ties, yet I was ready to take over the post of Chief of the Naval Staff, provided that in that capacity I should have definite powers of command. The Chief of the Naval Cabinet objected that the Emperor would never consent to give up the Supreme Command—a point on which I never insisted—but his doubts were not justified. For the Emperor consented to the request without hesitation. It was of course understood that the Supreme War Lord should be informed of the general trend of matters and of important projects, and that his consent should be obtained thereto. The practice hitherto followed of giving orders in the Emperor's name on matters outside His Majesty's sphere of interest was rather derogatory to the dignity of the Imperial Supreme Command. This incident proves how little foundation there was for some reports as to the Emperor's attitude, reports which emanated from those in his immediate entourage, and easily led to unpleasant decisions being kept from him.

I had such an experience when commanding the Fleet in January, 1917. The matter in question was the design of a new first-class battleship. I happened to be in Berlin for a consultation at the Admiralty, and the Emperor had commanded my presence when the Secretary of State made his report, the Chief of the Naval Staff also attending. The Secretary of State brought two designs for the projected ship, a so-called battleship-cruiser; that is to say, a ship which should combine the qualities of both kinds of ship—gun-power, power of resistance and speed.

Unless such a ship were of gigantic dimensions none of these qualities could be fully developed. This was the reason why up till then two distinct types had been built; the cruiser, with powerful guns, and high speed attained at the expense of its power of resistance, and the battleship, with the most powerful guns, and great powers of resistance at the expense of its speed.

The Emperor had repeatedly stated that he considered it necessary to merge these two types in one; hence these designs. The principle of the ship uniting all these qualities was to be accepted, but a choice was to be made between the two designs for carrying this into effect. The Chief of the Naval Staff and the Secretary of State were of opinion that the Emperor would not budge from what they supposed to be his attitude towards the matter, and they urged me to submit to it. But I had expressed myself to the contrary beforehand, and I repeated my arguments. The Emperor was soon convinced by our war experiences that we must continue to have two types of fighting ships with different speeds, and the Secretary of State thereupon received orders to have new designs made on the old lines that had proved successful—much to the gratification of his Chief Constructor.

Nor was it ever my experience that the Emperor rejected unpleasant information. In the two months of September and October, 1918, the unfavourable reports far outnumbered all others; His Majesty always received them with the greatest calm and common sense.

If I had foreseen the rapid development of events I would have preferred remaining with the Fleet rather than organising the conduct of the war at sea, for my plans never reached fulfilment. Nor do I think it impossible that I might have succeeded in making the Fleet obey my orders and exert its full powers at the eleventh hour. My only excuse for this lack of foresight lies in the fact that my observation of the spirit of the crews was based on the undiminished readiness to undertake any warlike enterprise which they had always shown up to then, and further that no hint of the widespread disintegration in our domestic conditions ever reached the leaders of the Fleet from any reliable political source, just as little as it reached the Admiralty later on.

On July 28 I was summoned to General Headquarters at Spa; Admiral von Holtzendorff had again, on the advice of his doctor, asked His Majesty to relieve him of his post, and his request had been granted. At the same time a decision was to be reached as to whether the U-boat campaign should be extended to America. The Naval Staff had urgently recommended the declaration of a blockade of the American coast, as this was a necessary preliminary to carrying out a successful U-boat campaign. As the chief ports concerned all lay on a strip of 300 nautical miles in length, it was thought that it would be easier to get at the traffic there. The troopships—the immense supplies that went from America to the Western theatre of war, and the large amount of coast traffic from South to North America were to be attacked there.

The Secretary of State for Foreign Affairs made strong objection to the declaration of this blockade. If Chile and the Argentine were thereby also induced to join the Entente, Spain would follow, and that was the only country that still protected German interests abroad. Quite apart from the political reasons urged by the Foreign Secretary, I did not think it probable that any good could result from extending the war to the American coast; for the declaration of a blockade entailed vigorous and decisive warfare. We could not count on stationing more than three boats there until the end of the year. No great success could be expected from that, especially as there was the added risk of the long voyage out. Moreover, to carry the war over to America would open prospects of an extension of the war that were out of proportion to our strength. Ours was a war of defence in Europe. America's interference in this quarrel was contrary to all her best traditions. No doubt there were a great number of intelligent Americans who did not approve of America's taking part, when they calmly and impartially considered the circumstances that had led to the world-war. Perhaps they remembered their own subjection—the manner in which England, their Mother Country, had deprived them of freedom, and how they had fought for their independence and successfully gained it with German assistance. If American troops were injured off the French and English coasts that was but the inevitable result of American interference in European quarrels.

But the feeling against us in that country would be very different if we began an enterprise which we had not the power to carry through successfully, and which must have an unnecessarily irritating effect. Three U-boats off the American coast could effect no essential amelioration in the results of our U-boat campaign. The decision in that campaign would be reached simply and solely by reduction of tonnage, and it must be sought in the main blockade area round England.

The officer commanding the U-boats was quite of my opinion that every possibility of adding to the achievements in this area must be made use of. From all the seas ships crowded to the British Isles. It was easier to deal an effective blow there than to follow the far-reaching trade routes, and try to attack them at their points of departure. And if an American transport now and then fell a victim to a U-boat on setting out from the American coast, that would not ward off the danger which threatened us from that source. It would be very easy for the transports to get through the dangerous strip near the coast by night, or to gain the open sea at any time under protection. We had learnt in the Franco-British blockaded area, where all the trade routes of the oceans meet, how difficult it was to pick out the transports from among all the other shipping for attack. If, for that purpose, we directed our main attention to the southern ports of France, as had been tried several times, the traffic was simply diverted as soon as the U-boat danger became manifest, and our U-boats were stationed there in vain, and achieved no results in the war against commerce. It was only by concentrating our activities on the main area round England that we could ensure success by sufficiently intensifying the ever-growing lack of means of transport, the effects of which were evident in so many directions. The French ports were not, however, left unmolested, and the mine-layers in particular were busy there. The increase in the number of seaworthy and efficient U-cruisers that could stay at sea for months ought to bring more success in our activities against English convoy traffic. They were able to seek the convoys far out at sea, to keep in touch with them, and call up a considerable number of U-boats, as soon as their sphere of activity was reached. Hitherto the attempts at co-operation between the smaller U-boats without U-cruisers had been a failure because of the lack of suitable boats to lead them.

We had long desired to apply to the U-boat war the principles of scouting and keeping in touch, which were applied by the surface warships. Now we had the opportunity to do so, and we must not let it slip by diverting the boats suitable for this purpose to a far distant field of operations. That was the decisive factor which induced me to oppose the declaration of a blockade of the American coast, and the scheme was accordingly abandoned. The Supreme Army Command did not care what means the Navy used, so long as it achieved success. Their desire to have more transports sunk could only be realised by raising the total number of sinkings. The U-boat must attack whatever happened to come within range of her tubes. Naturally the enemy protected the transports especially well, and took them through the danger zone at times which were most awkward for the U-boats.

The greater the number of steamers sunk the more likelihood there was that a transport would be sunk. We should more quickly attain our end with the U-boat campaign by keeping the blockaded area round England and the coast of France under the greatest possible pressure than by extending the blockaded area to include the American coast.

August 11 was the date fixed for me to take over the affairs of the Naval Staff. Before that I had to take leave of the Fleet. I had to hand over the command to my successor and make all preparations for the organisation of the conduct of the war at sea. Admiral von Hipper had been chosen as commander of the Fleet. His great experience in matters appertaining to the Fleet, his efficiency in all the tactical situations in which he had found himself with his cruisers, seemed to point to him as the most suitable person to whom I could confidently hand over the weapon from which I never thought to be separated in this life. The signs of faithful affection shown to me by the Fleet made my parting no easier, but I hoped to be able to continue to serve it in my new position. I felt the parting from my Staff especially keenly. The Chief of the Staff, Rear-Admiral von Trotha, on this occasion again showed his unselfishness by giving up some very important colleagues to assist me in the Navy command. The former Chief of the Department of Operations in the Fleet, Commodore von Levetzow, who had meanwhile been promoted to the command of Scouting Division II, had placed himself at my disposal as Chief of my Staff.

I took from the Naval Staff the necessary personnel for my command under the supervision of a Special Chief of Staff, and transferred them to General Headquarters where it was possible to keep in constant touch with the Emperor and the Supreme Army Command, as this seemed most desirable to me in view of situations which demanded prompt decisions. As a substitute for the Chief of the Naval Staff, Rear-Admiral Friedrich von Bülow was appointed to supervise matters in Berlin, which dealt mostly with reports, the supply of personnel and material, and political affairs. That did not entail any real change in the organisation, but only a regrouping of the Naval Staff for the purpose of the war. The fundamental improvement lay in the powers of command that the Chief of the Naval Staff was allowed to exercise in the "conduct of the war."

On August 12 I went to the General Headquarters of the General-Field-Marshal to introduce myself to him in my new capacity and to consult with him and General Ludendorff upon the situation and further plans for the conduct of the war. Both officers were much impressed with the gravity of the events which had occurred on August 8, and had placed our war on land definitely on the defensive. They both admitted that the main hope of a favourable end to the war lay in a successful offensive of the U-boats, and General Ludendorff promised, in spite of the great lack of personnel in the Army, to do his utmost to help to develop it further.

Until the necessary accommodation for my Staff had been found at Spa, and they could move there, the business of the Naval Staff was concerned with the new regrouping, and initial preparations were made for the extension of the U-boat campaign that had been planned. The results of the last months had shown that the successes of individual U-boats had steadily decreased. This reduction in successes was due mainly to the stronger and more perfect measures of defence taken by the enemy, and also to the loss of some of the older and more experienced commanders. Taking into consideration the then rate of U-boat construction, we had to expect, in spite of the steady increase in the number of U-boats, that the figures of the monthly sinkings, which had already diminished to 500,000 tons, would be still further reduced. Judging by the reports as to building, it was to be feared that within a short time the newly-constructed tonnage would be greater than the amount sunk. The success of the U-boat campaign might thereby be greatly diminished. A mere defensive could not help us to tolerable peace. It was, therefore, absolutely necessary for us to develop our only means of an offensive with all the strength at Germany's disposal, so as to attain our goal—a tolerable peace. In view of the Peace Conference, it seemed also advisable for us to have a strong weapon in the shape of U-boats with which we could bring pressure to bear on our enemies.

But if we wanted to achieve great things with the U-boat campaign then the whole industrial power of Germany must be at our disposal for the accomplishment of our task. I had got into communication with the principal controllers of industry, and at a conference with them and the Imperial Ministry of Marine had drawn up the following figures as the indispensable minimum for the increase in U-boats :

In the last quarter of 1918 ...	... per month	16
,, ,, first ,, ,, 1919 ...	... ,, ,,	20
,, ,, second,, ,, 1919 ...	... ,, ,,	25
,, ,, third ,, ,, 1919 ...	... ,, ,,	30

When I asked the U-boat Office why in January, 1917, when the unlimited U-boat campaign was decided on, more boats were not ordered to be built than was actually the case I received the following answer :

"As a result of the decision in favour of an intensified U-boat campaign no orders for boats on a large scale were placed. In

The Navy Command

February, 1917, only the following were ordered: 6 U-boats of the normal type, 45 U-B-boats, and 3 commercial boats. The large order for 95 U-boats was not given till June, 1917."

No definite information as to the reason for this building policy was forthcoming, but it was certainly strongly influenced by the opinion of the Chief of the Naval Staff that the boats would achieve their effect within a definite period of time, and that the existing U-boats would suffice. Moreover, in the Imperial Ministry of Marine the opinion prevailed that the capacity of the workmen for production was no longer to be depended upon.

After the U-boat Office had been instituted on December 5, 1917, 120 boats were placed on order the same month, and in January, 1918, a further 220 boats. During 1918 the monthly return of the boats supplied was still influenced by the earlier building policy :

January	...	...	...	...	...	... 3	boats
February	...	...	...	...	...	... 6	,,
March	...	...	...	...	...	... 8	,,
April	...	...	...	...	...	... 8	,,
May	...	...	...	...	...	... 10	,,
June	...	...	...	...	...	... 12	,,
July	...	...	...	...	...	... 9	,,
August	...	...	...	...	...	... 8	,,
September ...	...	...	...	...	... 10	,,

With these numbers the losses were covered, but no noticeable increase in the actual total of boats was achieved. We needed a greater number of new boats than an average of eight a month in order to raise the monthly amount of tonnage sunk to more than 500,000 tons. To a further question as to whether it would have been possible for the U-boat Office to get a larger number of boats, and if so from what quarter, I received the following reply:

"The U-boat Office exerted itself unceasingly to obtain a larger number of boats and had only been able, with the number of workmen assigned to it, to provide for a monthly supply of 23 boats up to the end of 1919. The hindrance lay in the supply of workmen Although the War Office did everything possible, and the U-boat Office never ceased to urge its needs, it was not possible to obtain a sufficient supply of workmen from the Supreme Army Command, either in regard to numbers or quality."

Germany's High Sea Fleet

A telegram from the Supreme Army Command in June, 1918, gave the following reason for refusal:

"I learn from the War Office that the Imperial Ministry of Marine has demanded the immediate supply of 2,200 skilled workmen for the Imperial shipyards at Danzig, Wilhelmshaven, and their Reiherstieg shipyard in Hamburg, and a further supply of nearly 900 skilled workmen for October 1. The Army cannot afford to be deprived of any more workmen; the people at home must supply the Army with more and more men, but cannot by a long way cover the demand caused by losses. The most urgent need of the hour is the supply of more men for the Army. Consequently it is improbable that the country will be able to spare skilled workmen from among those employed at home. Therefore I earnestly beg that you will carefully examine the supply of workmen now at your disposal, and that you endeavour to manage as far as possible with them. I also beg you to consider the possibility of employing skilled labour from neutral countries and the occupied territories (Reval, Libau, etc.)."

As Fleet Commander, when on several occasions I tried to effect improvements in the position of the Imperial shipyards by a better supply of workmen, I met with a refusal on the ground that the necessary workmen could not be produced; and I had the impression that there was not a sufficiently close understanding between the higher Navy Commands in Berlin and the Supreme Army Command, and that in consequence the needs of both could not be so adjusted as to assure the attainment of the great end for which both were working. That was the decisive reason why I established myself at General Headquarters, so that by my constantly keeping in touch with the Supreme Army Command all the resources of the country, both in men and material, might be applied to such work as would be of the greatest possible benefit.

When the centre of gravity of the war moved to the west as a result of the events of 1918, there was no reason why those in charge of the conduct of the war at sea should remain in Berlin, and thereby give up all possibility of close co-operation. The plans of the Supreme Army Command must be made to include all possible advantages to be derived from the war at sea, and full use must be made of them. If they admitted that the U-boat offensive could gain a decisive success, then the Army could very well spare some

The Navy Command

workmen for the needs of the Navy. This was the stage we had reached owing to the force of circumstances. Of course, the Navy must first of all give up every man that could be spared for the construction and commissioning of U-boats. That could only be done if the Navy Command took ruthless action.

Despite the menacing situation on our Western Front, the First Quartermaster-General drew the necessary conclusions, as soon as it had been proved to him that it was within the range of possibility to carry out the new U-boat programme if we could depend on obtaining 40,000 to 60,000 workmen. For the next few months a considerably smaller number would suffice to ensure the more rapid delivery of the boats now under construction.

To supply the men for the new U-boats we had to draw to an even greater degree than before upon the existing personnel of the Fleet, and had to take the necessary steps at once for training the commanders and officers of the watch for U-boat service, for it took several months for them to become familiar with the technical apparatus of the boats and acquire the necessary skill in marksmanship.

All U-boats at the time in home waters, as well as the U-cruisers, were placed under the control of the Fleet. In this way the officer commanding the U-boats gained the necessary influence over the training of the whole U-boat personnel, and he had the support of the Fleet in picking out suitable men. For the Fleet now bore the chief responsibility for the carrying out, and consequently for the success of, the U-boat campaign.

In place of the Secretary of State Admiral von Capelle, who had retired, Vice-Admiral Ritter von Mann-Tiechler, hitherto head of the U-boat Office, was appointed Secretary of State of the Imperial Ministry of Marine, in view of the fact that the chief task of this office now lay in furthering the construction of U-boats; and the building of reinforcements for the surface warships, which could no longer exercise any influence on the success of the war, was either given up or postponed, so that our entire capacity in shipbuilding was devoted to this one task.

Immediately after September 10, when I and my Staff had moved to General Headquarters, an opportunity occurred of proving the advantages of the close personal exchange of ideas, when a decision had to be made as to the handling of the Spanish question. Influenced by the Entente, the Spanish Ministry seemed inclined to abandon the correct attitude of strict neutrality which the Spanish

Germany's High Sea Fleet

Government had hitherto maintained; for they made demands arising from the U-boat campaign which were plainly in the interests of the Entente. The Spanish Government made a claim to seize an equivalent amount of tonnage from among the German merchant vessels in their harbours for every Spanish ship sunk in the blockaded area. We were ready to give compensation in kind for ships sunk outside the blockaded area, and so as not to distress Spain by a lack of supplies, we were willing to compensate at once and not make the payment dependent on the inquiry as to whether the ships had been justly or unjustly sunk, leaving that to the Arbitration Court. But we had to repudiate most energetically the demand for compensation for ships sunk in the blockaded area, because otherwise the other neutrals would quite justifiably have made the same claim, and the successes of the U-boats would have been illusory. It would have been nonsensical for us to have given up our own ships in compensation for those which we had justifiably sunk. It was a question altogether of 875,000 tons of German shipping which in this way would have passed automatically into neutral trade that went to England, and we should have derived not the slightest advantage from it. On the contrary, there was the danger that these ships would also be victims of our U-boats.

Another proposal of the Foreign Office was equally impracticable; this was to the effect that Spain should send her ships under convoy through the blockaded area, and that these convoys should be free from attack of our U-boats. Verbal discussions between the Supreme Army Command, the Foreign Office, and the Navy Command soon produced unanimity as to the attitude to be adopted towards Spain, by which, without departing from our fundamental principle that the blockaded area must be maintained, we should not incur the risk of making Spain side with our enemies. The discussions carried on for weeks among the authorities in Berlin had led to no result. But the claims that were then raised were but the regrettable consequence of the conciliatory attitude formerly adopted in this direction, an attitude which could only lend such encouragement to just claims that it was difficult to refuse them without a danger of serious conflicts.

On September 16 and 17 I visited the sea-front in Flanders and I was much impressed with the excellence of the measures taken by the commander, Admiral von Schröder, for defence against a landing of the enemy. I also became acquainted with the arrangements for the U-boat base at Bruges. The value of this position consisted

The Navy Command

in the flank protection it afforded, which had become necessary since the fighting front extended to the coast. So that this position could not be taken in the rear by the landing of troops, the whole stretch of coast from Nieuport, at the mouth of the Yser, up to the Dutch frontier had been strongly fortified. Zeebrugge, which was connected with Bruges by a deep sea canal, was a U-boat base. As any attack on this strong position would in all probability be made mainly from the sea the occupation of these coastal fortifications had been assigned to the Navy. The land defence on the extreme right wing of the front, closely connected as it was with the coastal defences, had also been undertaken by the Navy, which had formed regiments of able seamen. The English recognised the great strength of the position, and had not dared hitherto to risk battleships in the bombardment of the harbour and locks. They had built special craft for this purpose, monitors with shallow draught, and armed with a gun of heavy calibre, but these had not once succeeded in inflicting serious damage, though they had made many attempts. The naval base which had in time developed at Bruges became such a thorn in the side of the English that they did not grudge the enormous sacrifices they made in the various attempts in Flanders to break through our front in this sector.

They only attempted once, on April 22 and 23, 1918, to block the sea canal at Zeebrugge and the harbour of Ostend, and so make it impossible for our U-boats to get out. But this attempt was a failure. The attack, which was made with great pluck under the protection of artificial fog, found our guards at their posts. Two old light cruisers, that had penetrated as far as the mouth of the canal, were sunk before they reached their actual goal—the lock gates, which were uninjured. It was found possible for the U-boats to get round the obstruction, so that connection between the harbour at Zeebrugge and the shipyard at Bruges was never interrupted even for a day.

Another cruiser, the *Vindictive*, whose commander had succeeded with great smartness and seamanlike skill in laying alongside the Mole, landed a detachment of 400 marines, who were ready on the deck with scaling ladders; but this enterprise also met with no success. After suffering heavy loss, he was obliged to withdraw; only 40 men had been able to get on the Mole, where all, with the exception of one captain and 12 men, were killed in a fierce fight.

The *Brilliant* and the *Sirius*, which were dispatched against Ostend at the same time, did not attain their object, but stranded in flames east of the Mole. An English submarine succeeded in reaching the bridge of the Mole at Zeebrugge, and in blowing up the framework, so that for a time the outer end of the Mole was disconnected from the land. The object was to cut off the garrison on the Mole from all assistance from land, but this, too, was a failure, thanks to the courage of those in command of the guard.

Complete safety from such surprises is impossible of attainment, for it is difficult for those in the coastal fortifications lying farther back to be in time to overcome ships which come at night through the mist. But we had to count on a repetition of such attempts. The defences of the Mole therefore were strengthened so that a fresh attempt would probably have met with as little success. We did not lay mines farther out to sea to stop vessels from approaching, for this would have endangered our U-boats.

Although at the time there were no signs that a land attack was imminent, in view of the general situation, we had to reckon with the possibility that the defences which our land front afforded to the U-boat base might be broken through, because we had very slight reserves at our disposal. All the more so when the enemy realised that for the next period of the war we intended to concentrate mostly on our U-boat offensive. The loss of Bruges would have been a very disagreeable blow to the U-boat campaign, especially as the assistance we received from the shipyard there, employing 7,000 workmen, would no longer have been at our disposal. The U-boats, however, could have set out from the North Sea, so that at a pinch we could have got over the loss.

Hitherto the U-boats in Flanders had been responsible for 23 per cent. of the total results. They had sunk 3,342,000 tons in all, which does not include sinkings due to mines.

On my return on September 18 I had a conference with General Ludendorff on the subject of the danger with which the position in Flanders was threatened. I was informed that the situation at the front would probably make an abandonment of the position in Flanders necessary. Under the arrangements made by the Naval Corps such a withdrawal would take eight to fourteen days if the valuable supplies of war material and shipyard fittings were to be saved. The Supreme Army Command could not undertake to give warning in good time, and as the danger did not appear imminent, the Navy Command decided to take the risk of losing this material (in case a hurried retreat were necessary) so as to carry on the U-boat campaign from Flanders as long as possible. The Supreme Army

The Navy Command

Command undertook to inform us in good time of any indications which might point to the necessity of abandoning the position. We took care not to increase the stocks there, but only to keep them up to the level that was absolutely needful.

In the course of September the discussions with employers of industry and the shipyards were continued, to ascertain whether it would be possible to carry out the extended programme of U-boat construction. On September 24 the Ministry of Marine informed the Naval Command that the possibility of carrying it out had, on the whole, been established.

In view of the greater importance that now attached to the U-boats, seeing that they were to give a favourable turn to the end of the war, I suggested to His Majesty that he should visit the U-boat School at Kiel. His Majesty accordingly left General Headquarters on September 23 for Kiel, and on the 24th he inspected first the torpedo workshop, and then the establishment of the Imperial shipyards, which had been very considerably enlarged for the purposes of the U-boat war.

At the beginning of the war the torpedo factory at Friedrichsort had been the only place where our torpedoes were manufactured; but during the war the engineering works (formerly L. Schwartzkopff) in Berlin, which in earlier years had also manufactured torpedoes, was converted into a torpedo factory, as were other works as well. Under the direction of the Chief of the Torpedo Factories, Rear-Admiral Hering, the enormously increased demand for the manufacture of torpedoes was fully satisfied, so that the supplies of the Fleet and of the torpedo-boats were kept at the requisite level. Moreover, they succeeded in making considerable improvements. The ships of Squadron II ("Deutschland" class) and the older torpedo-boats built at the same time, were still armed with torpedoes which had a charge of 120 kilos of gun-cotton, a range of 2,200 m. and a speed of 24 knots. But most of the ships now carried a torpedo of 50 cm. calibre with a range of 10,300 m., and a speed of 28 knots. In the newest ships, like the *Baden* and *Bayern*, the range was still greater, as much as 16,500 m. with a speed of 25½ knots and the calibre was increased to 60 cm. The explosive charge of these newest torpedoes was 250 kilos of a material that had three times the explosive power of gun-cotton.

The U-boat School was established at Eckernförde. Its object was to familiarise the new U-boat crews with the handling of their boats and their armament, and especially to train them in marksmanship. The crews of all newly-commissioned boats were first sent to the U-boat School for a time to obtain practice in the military tasks with which they would be confronted. The Bay of Eckernförde was particularly suitable for diving practice, because of its uniformly great depth. Moreover, in consequence of its remote situation, there was little traffic there, and it was not, like Kiel Harbour, shut in by barriers, the passage of which always entailed a certain loss of time. In the large basin of Kiel Bay, which lay before the school, practice attacks on a large scale and in war conditions could be carried out. Special convoys were formed which were surrounded by guardships on the English model; the ships were painted in the manner which in course of time had been adopted by English ships to deceive the marksmen at the periscope. The ships were painted in all sorts of extraordinary colours, so as to deceive the observer, both as to the size and the course of the steamer, and hence to lead to bad aim in shooting.

At the time there were over 200 officers in training there, who were to find employment on U-boats as commanders and officers of the watch. The school was conducted by Commander Eschenburg, who succeeded remarkably well in imparting an excellent training to the men of the U-boats which were so precious to us, and ensuring the greatest possible number of hits with the torpedoes they carried. He achieved this in spite of the small number of boats at his disposal, because, of course, everyone was eager to make use of really well-equipped boats at the front.

The impression which His Majesty received of his visit of inspection was reflected in his address, before he departed, to the assembled commanders on board the school ship. He was clearly deeply conscious of the gravity of the task he had to impose on this band of courageous and self-sacrificing men, when he expressed his conviction that the Fatherland would not be disappointed in the hopes that must be put in the U-boat commanders. Involuntarily one thought: *Morituri te salutant*. None of us had the vaguest notion that the situation in the war on land was such that the cessation of all hostilities would soon be urged, and that in a few weeks the U-boat campaign would be abandoned.

On the return journey to General Headquarters, *via* Berlin, we received news, the day after starting, that the Bulgarian Front had broken down. This roused the gravest fears as to the steadfastness of our other Allies, and meant that our southern front was endangered. This news induced His Majesty to proceed to Spa on

The Navy Command

the morning of September 29 after staying for a short time in Cassel.

On the journey to Spa I met the Secretary of State for Foreign Affairs, and he informed me that the situation had become extremely serious, and that a decisive conference with His Majesty as to what further measures should be taken was to be held that very morning. Although I expressed to him my desire to take part in this Conference, I was not present, and only learned in the afternoon what had taken place. General Ludendorff informed me that the Supreme Army Command had announced to His Majesty that the situation demanded the immediate initiation of negotiations for an armistice and peace. The Chancellor would deal with the consequences arising therefrom which would affect home politics.

We had no detailed conversation on the subject. I knew what the feelings of the General must be. After years of glorious battles to be confronted with this result as the end of those activities that he had pursued with such an iron will! I therefore contented myself with such information as immediately concerned the Navy.

The points to be considered were the withdrawal from Flanders and the carrying out of the big U-boat programme. To retain Bruges for the U-boat campaign was not to be thought of. On the other hand, General Ludendorff was in favour of keeping to the plan of strengthening the U-boat weapon. The threat it contained might be useful for securing the armistice desired by the Army, as it would be useful in case of a refusal, when all our powers would be strained to the utmost.

Thereupon I at once went to His Majesty to secure his consent to our withdrawing from the U-boat base in Flanders, and our adhering to the design of building more boats. His Majesty gave his consent.

Considering the very grave decisions which the Emperor had had to face this day, His Majesty's bearing was admirably calm and steady. When the questions relating to the Navy had been disposed of, he spoke to me somewhat as follows:

"We had lost the war. He had hoped that God would ordain otherwise, and he hoped that the German nation would stand by him loyally. The Army and the people had behaved splendidly, but unfortunately the politicians had not. The Imperial Chancellor had informed him that he must go. His Majesty, therefore, had requested Count Roeden and the Chief of the Cabinet, von Berg, to suggest a new Chancellor. It would be difficult to find the right man. The new Ministry would have to be formed on a broader basis, and representatives of the parties of the Left would have to be admitted to it."

The same evening orders were sent to the Naval Corps to abandon Flanders as a U-boat base, and to carry out the evacuation according to plan; the Supreme Army Command would reckon on the evacuation of Flanders step by step; for the present there was no intention of giving up Antwerp.

A meeting had been arranged for October 1 in Cologne, with representatives of industry and of the shipyards. The Secretary of State of the Imperial Ministry of Marine, Ritter von Mann, and Colonel Bauer, representing the Supreme Army Command, also attended. Everyone agreed that it would be possible to carry out the extended U-boat programme, so long as the requisite number of workmen, amounting to 69,000 altogether, was forthcoming; these men were chiefly wanted in the shipyards. For the year 1918 only 15 to 20,000 men were asked for. There was no lack of the raw materials required, but such materials as had hitherto been used for other purposes, e.g. bridge-building in Roumania, would henceforth not be available for work of that character.

The representative of the Supreme Army Command declared that the Army was ready to further the undertaking with all the means at its disposal.

I did not feel myself called upon to make any statement on the changed situation on the Army front, but I pointed out that all those in charge of the conduct of the war were unanimous in their desire for us to adhere to this plan, whatever events might occur on the Army front, for the collapse in the South-East might well have serious consequences for us.

On October 5 the new Imperial Chancellor, Prince Max of Baden, sent a Peace Note to the President of the United States, and a Commission for Armistice negotiations was set up by the Navy Command, which was to deliberate with the Commission appointed by the Supreme Army Command under the chairmanship of General von Gündell. Rear-Admiral Meurer was appointed chairman of the Naval Commission, and Captains Vanselow and Raeder and Lieutenant-Commander Kiep were added to it.

In an interview that I had with General Ludendorff on October 6 to determine the general lines on which our common deliberations should be conducted, I asked him what concessions the Supreme Army Command was prepared to make in order to obtain the Armistice, saying that I presumed that these would not go so far as to make it impossible for us to resume our arms in case of need. General Ludendorff fully confirmed this. The Supreme Army Command would consent to an evacuation of the occupied territory in the West by stages, and would accept as the first stage the line Bruges—Valenciennes, and as the second a line from Antwerp— to the Meuse west of Namur. The Supreme Army Command could not accede to a demand to give up Metz to the enemy. The Imperial Chancellor had wished to make further concessions, but had agreed to the Supreme Army Command's proposal in view of the technical difficulties involved.

The important question for the Navy was whether the U-boat campaign was to cease during the Armistice. As the Foreign Office declared that without this concession no armistice could possibly be concluded, I declared my readiness to stop the U-boat campaign during the Armistice, but emphasised the point that in return for this we must obtain other concessions in the shape of the return of valuable shipping lying in neutral ports and supplies of raw material and food. The continuance of the blockade would be unfair if we stopped the U-boat campaign.

As regards the disposal of the Naval Corps, the following plan was arranged: those sections which could be employed in the field, viz. the regiments of able seamen and marines, as well as the transportable batteries of the marine artillery, were to be placed at the disposal of the Army; all the other men were to return to the Navy.

Thus the Naval Corps in Flanders ceased to exist. It had been instituted under the leadership of Admiral von Schröder on September 3, 1914, and played an honourable part in the taking of Antwerp on October 10, 1914. The General Command had its headquarters at Bruges. The infantry of the Naval Corps consisted of three regiments of able seamen and the marines. The latter in particular had played a distinguished part in the great battles in Flanders in 1916 and 1917. The sea-front was guarded by regiments of marine artillery. Thirty guns of the heaviest calibre had been set up there, among them five of 38 cm., four of 30.5 cm., and besides them a large number of quick-firing guns of from 10.5 to 21 cm. calibre. Hitherto they had repelled every attack from the sea.

The U-boat flotilla in Flanders was first established on March 15, 1915. As many as 37 U-boats had belonged to it at one and the same time. The great results attained by this flotilla were achieved at the expense of heavy losses; no other flotilla suffered such losses, and against it the enemy's most vigorous defence was directed.

In addition to this, two flotillas of large torpedo-boats and many mine-sweepers had been active off Flanders. They had made their mark in numerous night raids on the coast of the English Channel, and the bombardment of fortified places like Margate, Dover and Dunkirk; they had also been continually occupied in clearing away the barriers laid by the English to prevent our U-boats from coming out. Among the torpedo-boats the losses due to mines and bombs dropped by flying men were appreciably higher than those in the other theatres of war.

The evacuation of the shipyard at Bruges and the establishments at Zeebrugge had been carried out according to plan and without interference. The ships had returned through the North Sea to Wilhelmshaven; eleven large and thirteen small torpedo-boats; all U-boats excepting four had already been dispatched to the North Sea and had arrived there without incident. Four other torpedo-boats, which required some repairs before being ready for sea, were to follow within the next few days. Four U-boats and two large torpedo-boats had to be destroyed as they were not in a condition to be transported. In the shipyard at Ghent there were three large torpedo-boats whose condition made it impossible to take them into the North Sea. These were to be taken to Antwerp and either blown up or interned in Holland. The fast torpedo motor-boats which had distinguished themselves as lately as August by a successful raid on Dunkirk, had gone to Antwerp and were sent on from there by rail to Kiel. The sea-planes of the Naval Corps had made their way by air back to the North Sea. The aeroplanes and the rest of the Naval material that was capable of employment in the field went over to Army Command IV. Of the heavy guns on the sea-front only ten 29 cm. guns running on rails could be transported; all the others had to be blown up when the batteries were evacuated.

Just as the retirement on our West Front resulted in the abandonment of the base in Flanders, so events in the Balkans led to a withdrawal of our forces there as soon as the Turks concluded a separate peace, and we could no longer dispose of the U-boat bases in the Adriatic.

The battle-cruiser *Goeben* was the last reserve in the defence of the Dardanelles. Turkey had our promise that the ship should be handed over to her after the war. Therefore there could be no question of withdrawing the ship until there was danger of her falling into British hands. The Imperial Chancellor had admitted that this must be avoided for the sake of our military reputation. Consequently the officer in command in the Mediterranean, Vice-Admiral von Rebeur-Paschwitz, had received orders to send the *Goeben* to Sebastopol if her further stay in Constantinople would be of no use.

Some of our naval mechanics stationed at Sebastopol had tried to make seaworthy the warships which the Russians had handed over to us in accordance with the terms of the treaty, but they met with great difficulties owing to the neglected condition of the ships. Among these were the battleship *Volya* and several torpedo-boats and mine-sweepers which we wanted put in order to assist in the transport of large numbers of troops that were to be taken across the Black Sea from the Caucasus and Southern Russia to Roumanian ports. But the development of events in Turkey was such that the idea of keeping the *Goeben* was abandoned. In order to secure better armistice conditions for the transport of our troops fighting in Syria, our Government decided to hand the *Goeben* over to Turkey. The English had made this one of the main conditions, so as to be able to get possession of the ship.

In the Mediterranean our U-boats were busy until well on into October; at the same time all preparations were made to evacuate Pola and Cattaro in good time. The officer in command there, Captain Püllen, was left to decide as to this on his own responsibility. On October 28 the boats that were ready for sea began their journey home to Germany.

Altogether there were 26 of them there, of which 10 had to be blown up because they could not be made ready in time.

The further continuation of the U-boat campaign, if it should appear desirable, was thus dependent on the home bases—in the North Sea and the Baltic—and from these points it could have been directed against the shipping off the French coast and round the British Isles. In this case the whole strength of the U-boats could have been concentrated on this one main object.

The new Government formed at the beginning of October, under Prince Max of Baden as Imperial Chancellor, had approached President Wilson with a request for the conclusion of peace; at the same time they had undertaken to secure the cessation of hostilities as quickly as possible, and to obtain acceptable conditions of peace. But the manner in which they addressed themselves to this task, and their attitude during the negotiations, did not lead to the desired goal. The ever-increasing desire of our enemies to reduce our power of resistance till we were helpless was manifest in these negotiations. If the Government had determined to put a stop to the unduly exorbitant demands in good time, they might have secured important turning points in our fate, as the Imperial Chancellor had promised in his opening speech on October 5. On that occasion he concluded as follows:

"I know that the result of the peace proposals will find Germany determined and united, ready to accept not only an honest peace which repudiates any violation of the rights of others, *but also for a struggle to the death which would be forced on our people through no fault of their own, if the answer which the Powers at war with us make to our offer should be dictated by the desire to annihilate us.*"

The decisions which the Government reached, and the information and advice supplied by the proper military quarters, may be summarised as follows:

To our first request for mediation with a view to peace, sent on October 5; on October 8 we received the answer:

"No armistice negotiations so long as the German armies remain upon enemy soil."

On October 12 the reply from our Government:

"We are prepared to accept the enemy's suggestions for evacuation, in order to bring about an armistice."

Wilson's next Note of October 14 contained the demand:

"Cessation of U-boat hostilities against passenger ships and change of the form of Government in Germany."

The German Government's reply of October 21:

"U-boats have received orders which exclude the torpedoing of passenger ships, and with regard to the form of Government: The responsibility of the Imperial Chancellor to the representatives of the people is being legally developed and made secure."

Thereupon the answer from Wilson on October 23:

"Only such an armistice can justifiably be taken into consideration as will place the United States and the Powers allied to them

The Navy Command

in a position which will make it possible for them to enforce the fulfilment of dispositions that shall be made, and make it impossible for Germany to renew hostilities. Further, the demand that the Monarchy shall be abolished is plainly expressed, otherwise peace negotiations cannot be contemplated, but complete surrender will be demanded."

The attitude of the Supreme Army Command was responsible for the acceptance of the first demand for the evacuation of occupied territory, and it had signified its agreement with the text of our reply in our Note of October 12. No decisive influence could be exerted by the fears of the Navy regarding the danger which would threaten our industrial relations and also our U-boat base in Emden with the withdrawal from the Western Front; for the Army was unable to give any guarantee that it would be able to hold the Western Front in its then advanced position. That was the immediate reason why an armistice was needed. In order to satisfy this need, the Navy had agreed to stop the U-boat war during the Armistice, although the enemy would derive the most advantage from that, if at the same time the English blockade were not raised or considerably loosened.

But Wilson's new claim on October 14 went much further, for the demand that passenger boats should be spared must result in practice in the cessation of the U-boat campaign. Wilson, however, did not offer in return to cease hostilities, but had declared that he would not enter upon negotiations if this preliminary condition were not fulfilled by us. In so doing we should lay aside our chief weapon, while the enemy could continue hostilities and drag out the negotiations as long as he pleased.

It was to be expected that the Government would agree to sparing the passenger steamers, for this concession seemed insignificant. But its consequences might be very serious, for, according to former experience, if the U-boats were again reduced to cruiser warfare, their effectiveness was lost, and so far as one could see, it would be impossible, if hostilities continued, for us to resume the unrestricted U-boat campaign. The following, therefore, was the attitude adopted by the Navy to the new Note: "Sacrifice the U-boat campaign if—in return—our Army obtains an armistice; otherwise, we strongly disadvise compliance."

On October 16 I had occasion to visit the new Imperial Chancellor and to communicate my views to him, which he seemed to understand and share. He invited me to the conference of the War Cabinet which was to take place the next morning in the Imperial Chancellor's palace, when General Ludendorff would report on the military situation upon which the Government had decided to make their attitude to Wilson's Note depend.

The statements made on this occasion as to our powers of resistance were calculated to weaken the unfavourable impression of those made on September 29. The answer to be sent was discussed on broad lines. We were unanimous on the point that the accusations of inhumanity, etc., must be repudiated. The devastation of districts that were to be evacuated was a consequence of the war, so was the killing of non-combatants who went on ships into the blockaded areas. It should be suggested to the President that he should put an end to the horrors of war on land and sea by effecting an immediate armistice, and that he should clearly state his conditions. Germany was not prepared to accept conditions which would dishonour her. The fact was also emphasised that the tone of our answer would have a great influence upon the *moral* of the people and the Army.

It would now become manifest whether the President intended to negotiate honestly on the basis of his Fourteen Points, or whether he wanted to make our military situation worse than was permissible by prolonging the negotiations unduly and by constantly increasing his demands. In the latter case, the German people must be ready to take up the fight for national defence and continue it to the death. Such was the lofty mood of the members of the Government and their military advisers at the end of the session.

The next day I had occasion to report to His Majesty at Potsdam; he had already been informed by General Ludendorff of the outcome of the conference. Confident that the Government would not alter the decision arrived at on October 17, General Ludendorff had returned to General Headquarters. I considered it necessary to obtain the Emperor's approval for the further actions of the Fleet in case, for any reason, we should after all be forced to abandon the U-boat campaign, either temporarily or permanently. In these circumstances the obligations imposed on the Fleet by the necessity for protecting the U-boats would disappear. If hostilities at the Front continued, it would be neither possible nor permissible for the Fleet to look on idly; it would have to try and relieve the Army to the best of its abilities. His Majesty agreed that in this case the Fleet would have freedom of action.

The Navy Command

At the conclusion of my interview, a remark made by the representative of the Foreign Office, Counsellor of the Legation von Grünau, had struck me as odd. He had asked my Chief of Staff, Commodore von Levetzow, who accompanied me, whether there could not be a statement in the Note to the effect that the U-boat campaign would in future be conducted on the lines of cruiser warfare. According to that, the Foreign Office had not adopted the view that the cessation of the U-boat campaign was to be offered in exchange for the Armistice. I therefore determined to stay in Berlin so as to make sure that the text of the reply Note was in accordance with the decisions made on October 17.

On October 19 the War Cabinet deliberated upon this Note prepared by the Secretary of State for Foreign Affairs, Dr. Solf. Contrary to what had been agreed upon on October 17 it contained the sentence:

"The U-boat campaign will now be carried on upon the principles of cruiser warfare, and the safety of the lives of non-combatants will be assured."

The Vice-Chancellor, von Payer, opposed this draft most vigorously, as it was equivalent to an admission that our actions hitherto were contrary to law. "The U-boat campaign," he said, "must not be abandoned; the Navy must not stop fighting before the Army. Moreover, the whole tone of the Note misrepresented the feeling in the country." The Secretaries of State, Groeber and Erzberger, spoke to the same effect.

I made a counter-proposal based on the principle that the U-boat campaign must only be sacrificed in return for the Armistice. It ran as follows:

"The German Government has declared its readiness to evacuate the occupied territories. It further declares its willingness to stop the U-boat campaign. In so doing it assumes that the details of these proceedings and the conditions of the Armistice must be judged and discussed by military experts."

The majority of the representatives of the Government were in favour of the point of view defended by von Payer and myself, and Dr. Solf received instructions to draft a new Note to this effect to be laid before the Cabinet at its afternoon session.

Before this took place, the Ambassadors, Count Wolff Metternich, Count von Brockdorff-Rantzau, and Dr. Rozen, were invited

351

to express their views; the representatives of the Navy were not present. Their statements very soon brought about a complete change in the views of the Cabinet. They now urged that the U-boat campaign should be sacrificed without any return being demanded. The new draft Note unconditionally consented to spare passenger ships.

I again emphatically expressed my grave fears with regard to this dangerous concession, pointing out that by omitting to fix any time-limit they made it possible for Wilson to prolong the negotiations, while the U-boat campaign must, as a fact, cease, and the pressure upon the Army would continue. By conceding this we should admit that we had hitherto acted wrongfully, and would set free hundreds of thousands of people in England who had so far been bound by the U-boat campaign. But I did not succeed in getting my view accepted; even the telegram sent to the Imperial Chancellor by the Supreme Army Command that they could not in any circumstances dispense with the U-boat campaign as a means of obtaining an armistice could not alter the decision of the Cabinet. They were all firmly convinced that they could not justify themselves before the German people if negotiations with Wilson were broken off, and that this would be inevitable if we did not unconditionally concede what was demanded of us.

The form of the Note determined at an evening session contained the sentence :

"In order to avoid anything that might make the attainment of peace more difficult, at the instigation of the German Government all U-boat commanders have been strictly forbidden to torpedo passenger ships."

I declared to the War Cabinet that if we were loyally to carry out this concession, all U-boats sent out to make war upon commerce must immediately be recalled.

I required the consent of the Emperor to issue this order. As His Majesty was convinced of the serious military consequences, he used his personal influence to try and induce the Imperial Chancellor to alter the decision of the Cabinet. But the Emperor did not succeed in making the Chancellor change his opinions, so that His Majesty then informed me through the Deputy Chief of the Ministry of Marine that the Imperial Chancellor had represented the situation as such that the U-boat campaign must be abandoned.

The Navy Command

An attempt on my part to make the Imperial Chancellor at least put a time limit for the concession in the Note in the same manner became fruitless. He declared that we were not in a position to make conditions, and the Navy must bow to the inevitable and at all costs avoid provocative incidents. I assured the Chancellor that we should do our best and that all U-boats should be recalled from the campaign against commerce. This decision as to the limitation of the U-boat campaign was very important because the further operative measures of the Navy Command depended upon it; the High Sea Fleet must again now obtain complete freedom of action.

So long as hostilities continued at the front, and there was for the present no indication of their ceasing, the Navy must not remain entirely inactive, while the attacks of the enemy on our Western Front grew ever fiercer, unhindered by any fear of U-boats. A success at sea must have a favourable influence upon the terms of peace, and would help to encourage the people; for the demands of the enemy would depend on the powers of resistance that we were prepared to oppose to them, and upon the consideration whether their own power was sufficiently great to enforce their demands. Anything that would impair their power must be to our advantage.

The U-boats liberated from the commercial war materially increased the Fleet's power of attack, and by choosing the point of attack wisely it was highly probable an expedition of the Fleet might achieve a favourable result. If the Fleet suffered losses, it was to be assumed that the enemy's injuries would be in proportion, and that we should still have sufficient forces to protect the U-boat campaign in the North Sea, which would have to be resumed if the negotiations should make imperative a continuation of the struggle with all the means at our disposal.

On October 21, when the Note had been dispatched to President Wilson, the U-boats received orders of recall, and my Chief of Staff, Commodore von Levetzow, was commissioned to inform the Fleet Command in Wilhelmshaven of the course of the negotiations, and to take to them the order of the Navy Command: "The forces of the High Sea Fleet are to be made ready for attack and battle with the English Fleet." The Commander-in-Chief of the Fleet, Admiral von Hipper, had already drawn up plans for such a proceeding, as its necessity was foreseen. A plan directed against the English Channel received the preference and my assent;

it was to be carried out as soon as possible. The execution, however, had to be delayed for a few days owing to necessary preparations; the U-boats had to be sent to their stations, and the cruisers fitted out with mines to be laid along the enemy's probable line of approach. The Fleet was finally assembled for this enterprise in the outer roads of Wilhelmshaven on October 28.

Meanwhile, at noon on October 24, President Wilson's reply had been made known, and this quite clearly demanded complete capitulation. Animated by the same views as the Supreme Army Command I went with my Chief of Staff together with the General Field-Marshal and General Ludendorff (on the former's invitation) to Berlin, in order to be on the spot in case we were wanted for the deliberations arising from the new situation. We could not imagine that the Government could do otherwise than reply to this new demand of Wilson's by a direct refusal, consonant with the honour of the nation and its power of resistance.

Immediately on their arrival in Berlin in the afternoon of the 25th, the General Field-Marshal and General Ludendorff had been sent for by the Emperor. At this interview General Ludendorff received the impression that the Emperor would adhere to the suggestions of the Government, so that all that was left to us was to discover from the Vice-Chancellor, von Payer (the Imperial Chancellor himself had fallen ill), what decisions the Government would take.

This interview took place in the evening of the 25th, but its results were entirely negative. In spite of the most urgent arguments on the part of General Ludendorff, which the General Field-Marshal and I endorsed, it was impossible to convince von Payer that our national honour and our honour as soldiers made it imperative that we should refuse Wilson's exorbitant conditions. The Field-Marshal and General Ludendorff declared they would hold the Western Front through the winter. It was in vain. Herr von Payer would not believe Ludendorff's assertions; he wanted to hear the opinion of other generals at the front. But, above all, he had lost all faith in the powers of resistance of the people and the Army.

The discussion had to be broken off without result, as the Vice-Chancellor could not be moved to make any concessions. Even when asked if, when the full conditions—in so far as they were tantamount to capitulation—came into force, the people would not be

The Navy Command

called upon to make a last struggle, Herr Payer only answered: "We must first see what the situation would then be."

At an interview the next morning, granted by His Majesty to the Field-Marshal and General Ludendorff, the latter tendered his resignation, which the Emperor accepted.

The Government's answer to Wilson's latest demand was as follows:

"The German Government has duly noted the reply of the President of the United States. The President is aware of the fundamental changes that have taken place and are still taking place in the German Constitution. The peace negotiations will be carried on by a Government of the people, in whose hands the decisive power actually and constitutionally lies. The military forces are also subject to it. The German Government, therefore, looks forward to the proposals for an armistice, which shall lead to a peace of justice, such as the President has defined in his utterances."

The expectation that the negotiations would take a favourable course, as the Government seemed to imagine, was doomed to disappointment. General Ludendorff's prophecy was amply fulfilled; he predicted that if we continued to yield, the end must be disastrous, *because the Government had neglected to steel the will of the people for a supreme effort.*

But we suffered the bitterest disappointment at the hands of the crews of the Fleet. Thanks to an unscrupulous agitation which had been fermenting for a long time, the idea had taken root in their minds that they were to be uselessly sacrificed. *They were encouraged in this mistaken belief, because they could see no indication of a will to decisive action in the bearing of the Government.* Insubordination broke out when, on October 29, the Commander-in-Chief of the Fleet was making preparations to weigh anchor for the planned attack. As always, the intentions and aim of the expedition had been kept secret from the crews, until they were at sea. The mutiny was at first confined to a few battleships and first class cruisers, but it assumed such dimensions on these ships that the Commander-in-Chief of the Fleet thought it incumbent upon him to desist from his project. By seizing the agitators and imprisoning them in the meantime in Wilhelmshaven, he hoped that the ships could be calmed down. The crews of the torpedo-boats and the U-boats had remained thoroughly loyal.

Germany's High Sea Fleet

The Commander-in-Chief of the Fleet reported these events to the Navy Command on November 2, saying that they were due to a Bolshevist movement, directed by members of the Independent Social Democratic Party, on board the ships. As a means of agitation, they had made use of the statement that the Government wanted peace and the officers did not. Every provocation of the enemy by attacks of the Fleet would hinder the peace; that was why the officers wanted to continue the offensive. The officers wanted to take the Fleet out and allow it to be annihilated, or even annihilate it themselves.

Since October 29, when the first signs of dissatisfaction had become manifest, the movement had continued to spread, so that he did not think it possible to undertake an offensive with the Fleet. The Commander-in-Chief of the Fleet, therefore, detached the individual groups, sending Squadron III to the Jade to place them in the keeping of the commanding officers there.

After that, quiet seemed to prevail again in Wilhelmshaven; but when Squadron III reached Kiel, disturbances broke out there on the evening of November 1. The Governor, Admiral Souchon, succeeded in preserving order for a little while, but on November 3 the disturbances grew, because they met with no vigorous opposition. Even the deputies sent by the Government to Kiel could not achieve any permanent improvement in the situation; just as little effect was produced by the proclamation of His Majesty the Emperor which the Imperial Chancellor now published, and which announced his complete agreement with the Government. Energetic measures against the agitators, which might at the beginning have met with success, were only possible under the protection of strong bodies of troops which the Ministry of War dispatched. But the troops proved untrustworthy. Nor did they arrive in sufficient numbers to produce the desired effect.

I have no official reports of the details of the Revolution which soon blazed up at all the Naval stations, for the military authorities were deprived of their power of command. The instructions issued by the Navy Command to the commanding officers to sink ships hoisting the red flag were not forwarded. They would, at least, have been of guidance to such officers who were in doubt as to what they should do, if they still possessed the power to do anything. Nothing but energetic action on the part of the superior officers who were on the spot could have saved the situation. Whether they failed in their duty, or whether the extent of the movement was underestimated, is an open question. Only when the history of the Revolution is written shall we get full information on the point. The evil-doers who picked out the Fleet as the means by which to attain their ends committed a terrible crime against the German nation. They deprived it of the weapon which at the decisive hour might have saved us from the fate which now weighs upon us so intolerably.

CONCLUSION

"I HAVE no longer a Navy."

With these words the Emperor repudiated my objections when on the afternoon of November 9 I urged that if he resigned the Navy would be without a leader. Deep disappointment sounded in these words, the last that I heard from His Majesty.

In the evening of the same day the Armistice conditions were published, among which was the demand for the surrender of the German Fleet and of all the U-boats. No opposition could be expected from the Revolutionary Government. It consented to everything in order to get rid of the hated "Militarism," and delivered the defenceless German people into the hands of its enemies. A curse lies on the Navy because out of its ranks Revolution first sprang and spread over the land; and many who regarded its deeds with pride are to this day at a loss to know how such a change can have been possible.

The conditions of life on the large ships, the close quarters in which the men lived, favoured the propagation of this agitation, which was spread by any and every means. Further, the crews were most easily exposed to temptation because of their close connection with the Homeland. But the most important and the decisive cause was this: the war-weariness of the whole nation, increased by hunger and all sorts of privations, had become so widespread that even the fighting forces had lost faith in a happy end to the war.

On the day when the German National Assembly accepted that fatal peace which perpetuates hatred the deed accomplished at Scapa Flow once more gave evidence of the spirit which inspired the Navy, as it did the Army, in the days when they rejoiced in battle. However much we are bowed down, we can still do justice to all the great things that were achieved. That is the only comfort that we can take in regarding the dark future that awaits us; it is the foundation-stone upon which to build up new hopes. The strength which the German people developed enabled us to withstand the onslaught of overwhelmingly superior forces for four and a half years, to keep the enemy out of our own country, to fell the giant Russia, and even to bring England, who thought herself unassailable, to the

358

Conclusion

brink of destruction; this strength of ours was so mighty that our downfall could only be accomplished by extraordinary means: we had to inflict defeat upon ourselves.

The credit of inventing this expedient belongs to England, and the surrender of our Fleet appears as the great triumph which her sea power has won. History will not find much that is worthy of praise in the way England waged the war at sea; it may laud her ultimate success, but not the means by which it was achieved. The very surrender of our ships is the best proof that we were not defeated until in the Homeland the will to continue the struggle had been so sapped by hunger and privation that the people were susceptible to the poisonous ideas spread by enemy propaganda, of which an unscrupulous Revolutionary party made use to attain its selfish ends. It was England's privilege to extend the war to the economic sphere in an unheard-of manner. The fight for sea commerce was to lead to the strangling of the whole German people. For that purpose violence had to be done to the rights of the neutrals, whose power, compared with that of the ring of our enemies, was of no avail. England's policy of alliance placed her in a position to carry out her plan of starvation, without any fear of a protest from civilised society. She cleverly diverted attention from the enormity of her proceedings by simultaneously opening a campaign of lies about Germany's atrocities and Hun-like behaviour. Widespread financial operations, moreover, united American with English interests.

It was the task of our Fleet to defeat the English blockade, or to neutralise the effects of it by the damage it inflicted on the enemy. The latter method was chosen. The U-boat proved to be a suitable means to this end. We must be grateful that the technical development of the U-boat had reached such perfection, just in the nick of time, that these craft could be sent out to such distances and for such length of time as the war against commerce demanded. Fault-finding is an objectionable quality of the German. Many a time he has scorned and belittled the great work of twenty years of building a Navy which should be able to meet the English Fleet in battle. The accusations made are false and prove nothing but the ignorance or the ill-will of those who make them. No doubt our ships had faults—no naval authorities can make a claim to infallibility—but they were of absolutely no account compared with the fact that the material, as well as the spirit and training of the crews, were so good that our Fleet was able to hold its own against the English.

Germany's High Sea Fleet

Only a ship-building industry like that of Germany, which, as the German Fleet developed, produced such super-excellent ships, could have helped to supplement our Fleet during the war by the construction of a new U-boat fleet. The reliability of the material, and the manner in which the boats were built, increased the courage of the crews who, with full trust in their weapon, could dare all.

In view of England's plan of campaign, there was no alternative but to inflict direct injury upon English commerce. We could not build a sufficiently great number of additional large ships to compensate for the inevitable losses which we were bound to suffer in the long run in a conflict with the numerically superior English Fleet. In carrying out their blockade, that Fleet had the advantage of choosing its field of action in the Northern waters, far removed from our bases. After their experiences in action, the English left the southern part of the North Sea for us to deploy in, and contented themselves with warding off the U-boat danger. Throughout they were forced to be on the defensive. We ought to have tried earlier what the result of a victory by our Fleet would be. It was a mistake on the part of the naval leaders not to do so. It was only after we had been proved in battle that we gained sufficient confidence to send the U-boats permanently into the North Sea to wage war on commerce against England, and in the teeth of the resistance of her Fleet.

The earlier the U-boat campaign was started in full earnest, the greater was the prospect of being able to go through with it; it was wrong to wait until the endurance of our people had been tried to the utmost by the effects of the blockade. The number of boats at the beginning of 1916 would have been amply sufficient for the purpose. The success of a U-boat campaign does not depend solely on the number of the boats, but rather upon their quality and the skill of their navigators. U-boats of great speed and unlimited powers of remaining at sea, which could not be caught, would soon paralyse the sea traffic of an Island State like England. As such an ideal was not capable of full attainment, the greater number of boats had to make up for the lack of perfection. The results achieved fulfilled, and even surpassed, expectation, even though a criminally long time was allowed the enemy to organise his defence. That we did not reach the limit of England's endurance in time was due, not to the ill-success of the U-boats, but to the encouragement which the enemy found in his hour of need in our political attitude and that

Conclusion

of our Allies. Why should he lower his colours when in July, 1917, we cried to him : " We want peace,"—which in his ears sounded like "We need peace "—and when we let Austria and our enemies know that the country could not continue the war longer than the autumn of that year? The worse the enemy fared, the more boldly he bore himself. We, unfortunately, adopted the opposite attitude.

From the very first a large proportion of the people had been nervous as to the disadvantageous effects of the U-boat campaign. This had become a party question, owing to its treatment in Parliament and the Press. The leading statesman's dislike of it was openly acknowledged everywhere; he left the decision to the Supreme Army Command, who were to fix the date in accordance with the general military situation, and he put the responsibility on their shoulders. True, the nation had absolute confidence in the Supreme Army Command, because the generals in command had earned this confidence. In this question of life and death, too, they formed an opinion in common with the Naval Staff, and decided upon action when no other means of breaking the enemy's resistance was to be found. But to succeed we had need of the confidence and co-operation of the whole nation, so that they might hold out until success was ensured. The Reichstag resolution of July, 1917, must have been viewed by the enemy as a proof that this confidence did not exist.

From then onwards there was no question of the enemy's yielding. Now, a year after the conflict has ceased, we get indications from England every day of how hopeless the situation seemed there. But realising their weakness, they were able to weather the critical period in the autumn of 1917 by seizing enemy shipping for their own ends, and they strove zealously to intensify the disintegrating forces which were at work amongst us. This war has taught us to what an extent a nation can limit its economic needs. For more than a year after the conclusion of the Armistice we bore the burden of the blockade although huge quantities of supplies had to be left in enemy hands, or were idly squandered, when our troops retreated.

Our situation would not have been worse had the war continued, while the enemy would have kept on losing an amount of tonnage that could not be replaced.

But his will to endure was stronger than ours, for he recognised the weakness of our Government, whose leaders, unlike those of the enemy Cabinets, did not have the whole-hearted support of representatives of the majority of the people.

The World-War was to be a test for the German nation, whether it could hold its own as a factor of civilisation overseas. The British tried with might and main to oust it from its position, when the might of the German Empire was behind it. They felt the danger that lay in our superior diligence, the excellence of German work, and the sterling qualities of German education and culture as compared with the shallow civilisation of the Anglo-Saxons meant for nothing but effect. Our peaceful penetration was met with violence. How great they thought the danger is shown by the mighty efforts of our enemies to crush us.

They have attained their object, because our leading statesmen at the outbreak of war did not recognise the magnitude of our task, or—which is worse—looked upon it as beyond our strength. If the great aim had been rightly realised then, if it had been pursued with all the forces and strength at our disposal, and if the nation's will to victory had been continually directed towards it, we might have been sure of success.

The enormity and baseness of the methods with which our downfall had been planned, inflamed the sense of antagonism in our people to a degree which it could not otherwise have attained. The nation, however, could not fail to grow weary of its efforts when the only aim that was left to it after long years of fighting and starvation was that of self-preservation; it was deluded by enemy craft and wiles into thinking that this could be secured by other means.

Thus dissension arose at home, and our strength was exhausted in internecine strife for a phantom of national freedom; and the only palpable result of all this, brought about by the Revolution, is the helplessness of that freedom, deprived as it is of the power to defend itself from foreign aggression.

Toil and labour must start afresh to raise the honour of the German Navy. In this task the Fatherland will feel the lack of many capable men, who cannot live in the straitened circumstances that have been forced upon us, and who will migrate elsewhere. But our hopes are centred on these, that they will not deny their love of home, but will preserve their loyalty to their enslaved Fatherland and will cherish it in their descendants until the vitality of Germany, oppressed and overwhelmed as it now is, has won through to a new development.

Conclusion

The Englishman may now think himself entitled to look down upon us with scorn and contempt, yet in his feelings of superiority there will always be the sting that he was not victorious in battle, and that his method of waging war is one that must recoil upon his own head.

Other World Powers will appear upon the scene who will only concede a prerogative at sea to him who, as in Nelson's days, can assert his pre-eminence in open conflict.

INDEX

Index

Index

Index

Index

Index

Index

Index

Index

Index

374

Index

Index